I0478963

Food Science and Engineering

Food Science and Engineering

Editor: Dorothy Green

www.callistoreference.com

Callisto Reference,
118-35 Queens Blvd., Suite 400,
Forest Hills, NY 11375, USA

Visit us on the World Wide Web at:
www.callistoreference.com

ISBN: 978-1-63239-969-4 (Hardback)

Cataloging-in-Publication Data

Food science and engineering / edited by Dorothy Green.
    p. cm.
Includes bibliographical references and index.
ISBN 978-1-63239-969-4
1. Food. 2. Food industry and trade--Technological innovations. I. Green, Dorothy.
TX357 .F66 2018
641.3--dc23

Table of Contents

Preface

Food science studies the nature of food in terms of its processing, engineering and packaging. It includes the processing involved in developing new food items and various new techniques of packaging. Food science and food engineering can be sub-divided into branches like food chemistry, food physical chemistry, food microbiology, food preservation, etc. This book elucidates the concepts and innovative models around prospective developments with respect to food science and engineering. It will help new researchers by foregrounding their knowledge in this branch. This book attempts to assist those with a goal of delving into this field.

This book has been the outcome of endless efforts put in by authors and researchers on various issues and topics within the field. The book is a comprehensive collection of significant researches that are addressed in a variety of chapters. It will surely enhance the knowledge of the field among readers across the globe.

It gives us an immense pleasure to thank our researchers and authors for their efforts to submit their piece of writing before the deadlines. Finally in the end, I would like to thank my family and colleagues who have been a great source of inspiration and support.

Editor

Improving Food Quality through Institutional Innovations: Using a Free-Rider Approach for Collective Action

Ernst-August Nuppenau[1]

[1] Institut für Agrarpolitik und Marktforschung, Justus-Liebig-University, Germany

Correspondence: Ernst-August Nuppena, Institut für Agrarpolitik und Marktforschung, Justus-Liebig-University, Germany. E-mail: Ernst-August.Nuppenau@agrar.uni-giessen.de

Abstract

This paper applies an innovation from institution economics to agribusiness: profit (cost) sharing as arrangements to promote public good provision. It outlines how a team work approach can be employed to promote food quality. In institutional economics it was recently suggested to use sharing arrangements to overcome the problem of externalities, in particular by using the above mentioned team work approach. Cost sharing as "team work" is analyzed in this paper as a novel institution to improve food quality, then by indirectly creating incentives to overcome the public good character of "quality". It is assumed that consumers recognize industry quality; not individual by firm in case of asymmetric information. We translate a general approach of negative externality to a positive one. (1) We make a reference to the current state of art on how food quality depends on joint efforts of an industry. The aim is to get a better image and discuss the extent to which group efforts are needed to improve quality. (2) We present a mathematical approach on team work for quality. Quality is seen as positive externality and (3) team building is modelled as a likely a process of forming groups sharing efforts in food industries. Finally some remarks are made on how to actively stimulate the needed process of team formation. Also the role of the government is addressed. At the core of the paper we see the argument that *free riding* on quality images can be avoided if collective action prevails. A team is modeled as partnership of producers in which costs for image raising are shared. A prerequisite is that *economies of scale* and jointness in image exist.

Keywords: food quality, team

1. Introduction

There is a huge discussion on means to promote food quality and images (Henson & Reardon, 2005). A central theme is the role of governments and public interventions (e.g. on standards, monitoring, control, etc.: Codron et al, 2005). From a classical viewpoint, the discussion implies using a rigorous approach to show that (1) quality is a common or public good (phenomena: Martinez et al., 2007), that (2) the market does not provide the social optimum and (3) governments have a role to play (Martinez et al., 2007; Narrod et al., 2009). (4) A mechanism should be outlined, which sets incentives for the private sector (Holleran, 1999) to overcome the *free rider problem* in food quality assurance. In this paper we argue that a framing of quality as a common good problem is an eye-opener, yet, on improving institutions governing quality and to throw light on possible new institutional arrangements.

Frequently, the *belief* is that quality is a self-evident criterion of food that can be easily appreciated by consumers (graded) and that market segmentation develops at minimal intervention (just fixing standards for grades). Then, on the basis of *objective* grading that shall work effortlessly (for example see meat in Germany: Sönnichsen et al. 2000), consumers make easily choices. Perhaps that is fine for minimum standards. But do consumers know the industry's potentials to go beyond current standards? Or what happens if the industry does not meet even professed standards, taste is not linked to visible characteristics, and perhaps manipulations can prevail? In the latter case consumers loose trust. For instance the image of the meat industry in the EU has declined over years (especially when BSE was strong in public awareness: Rausser et al., 2011, Chapter. 17). Though we will not go into detail on free riding, further reasons such as the role of media, ignorance, public request for standards, etc. (see on the influence of media: Verbeke et al., 2000), it seems that some food industries (mainly the meat industry) have an image problem that corresponds to low quality image (as compared to thought standards).

Under this condition, rational producers may employ cost minimization strategies (such as the use of low quality feed, minimal maturing times for animals, low hygiene, etc.). This results in even lower quality. There seems to be no mechanism to change behavior; though consumers often express concerns and might pay more if quality improves.

However, vice versa, it is the hypothesis of this paper that the image (quality as public standard) of an industry can be built up if all producers would jointly increase quality, resulting in consumer confidence. Building up quality also means that costs increase, for instance by buying better feed or forage, and this has a price effect. The individual producer is faced with the dilemma that his colleagues may prefer *free riding* and do not contribute to quality, notably as a rent seeking strategy (Nizan, 1991). The costs of individuals without collection action are too high to improve images; and because they fear *free riding*, calculations are biased. So the question is how can one reduce the costs of individuals and assure *team* work in for quality. A suggestion is to share the additional costs that are to be potentially incurred in realizing higher quality (we follow Platteau and Seki, 2000 by reversing their argument of production, output sharing, to cost sharing), and use economies of scale. To frame a cost sharing arrangement, we suggest a re-invention of sharing arrangements such as in resource management (negative externalities: Heintzelman et al., 2008. There the idea is to reduce efforts to reach an optimum). Here we reverse the argument and will work out the economic logic for sharing as mechanism to get a social optimum. This shall be above current quality.

The suggested approach requires that we have to set references for quality as well as to show that additional cost bearing improves quality and revenues. Reference for quality choices in food markets are normally based on two criteria sets: (1) individual benchmarks associated with taste (subjective) and (2) generic criterions linked to image (objective); image can be seen as pre-knowledge to quality beyond preferences. Producers face a situation where prices and quality, which are achievable in a market, are negatively correlated with images of industries. Then efforts of all participants count in image building and up-keeping of standards. On real markets (based on assumptions of imperfect knowledge) instruments of coordination are needed to promote quality as price and quality may not fit (Hanf & Wersebe, 1994).

It is the objective of the paper to give a theoretical outline of a team work to improve quality in a food industry which fails to get better prices (for instance, beef or pork). The sector is perhaps regionally focused and has a *bad* image, to which farmers collectively contribute. We start with a problem statement, then give a model outline and provide derivations of individual and team behavior. Finally some hints for policy and implementation will be given.

2. Problem Statement and Approach

We start with the following ideas to envision the problem: (1) consumers may not be able to reasonably distinguish different produces by simple quality criteria (mostly visual one, if products are bought in markets; consumers even may not see better quality in the market at all) and they have to place trust on food that they purchase; (2) characteristics don't contain full information at visible scales, but images play a role; (3) some characteristics are even hidden; (4) quality and (5) prices remain below expected value and potential levels (Hanf & Wersebe, 1994); but (6) consumers buy food as generic product with low involvement. In this setup, image has a pivotal role; it can be improved as a result of collective action of farmers.

Against this background we introduce the notion of *quality* as a collective standard (image) that is achieved by cost sharing and which can be achieved with economies of scale (campaigns, etc.). Costs beyond minimal costs are associated with others and can be pooled (efforts to improve quality: better fodder, less adds to food, perhaps better husbandry, etc. effect at industry level as image). The sector image matters for prices. In contrast, if an individual producer wants to improve quality he is a marginal contributor. We assume a cost function that includes individual and joint efforts to increase quality. The basic idea is that costs for quality and image advances can be made sharable. Sharing costs means that individual firms can externalize costs (efforts): I.e. "though bearing some costs for others, it pays off". A prerequisite is that *economies of scale* exist in image creating and the image is of joint for the industry.

Yet, to make the approach realistic, standard production costs and cost of quality enhancement must be (made) distinguishable. We refer to costs of market development (image) and quality as those to be shared. If quality is a matter of collective action, it means that revenues of individual firms are determined by joint efforts of all market participants. I.e. price and quality determine revenues on individual and collective levels. Prices are indirectly changeable by collective action (image advances). High quality adds up from individual contributions towards quality advances and finally changes producer revenues. As mentioned above, as assumption we state that one can distinguish costs for quality and ordinary production. The focus of the paper is on min! (quality) costs. Yet, there is a link to be realized between quality and production costs. For us, to produce in bulk is neg. non-linear in quality (in a cost model).

Measures to improve quality are manifold. For instance, in special cases (meat) we can easily presume that better ingredients (feeds) are relevant and they are more expensive as well as we may take the difference in costs as quality effort based. In physical terms, the amount of better fodder in the diet of an animal is an effort "e" and this effort can be measured as a change in recognized ingredients. For instance for beef, a higher percentage of grass in the diet of beef cattle, instead of maize, increases the quality of the meat and we can take the grass percentage in feed as *effort*. However, most case can be generalized and we take a cost function approach that is dual of a production function (Sheppard, 1990: see below). Then the quality, as related to joint efforts, follows a non-linear expression (efforts as regressed on quality see below).

Further assumptions are: (1) the marketing for quality (image) advance is collectively organized (for instance labelled as "better" beef from the association, team of farms having a brand name), (2) there is a willingness to pay, and (3) a premium is paid to producers. The problem is that only cost sharing lifts quality if numerous producers participate. Group (team) formation is negotiated. (5) A further assumption is that a team can be formed by making claims to partners to contribute to image. Then (6) producers are capable to reclaim additional efforts (costs) and pool them, i.e. *costs* (in team) depend with whom one share costs for quality improvement (notably only additional costs to production cost matter). Sharing is done according to group composition and membership in teams. The size of a team (group) is not predetermined, rather it must be simulated. Hence a three-layer-decision problem occurs for individuals and groups (see Heintzelmann et al., 2008). The layers are: (1) the level of effort is decided assuming each farmer is a team member; (2) the group or team formation is depicted as voluntary choice to participate and the number of members in each team is *decided* in terms of options to move between team; (3) then the number of teams is derived endogenously. In the subsequent analysis we follow the analytical outline of the above authors, but reverse their argument from output to cost sharing. Further, we add the production (quantity) decision of farmers on operational size, since we have heterogeneous producers and link it to quality

3. Model Outline and Social Optimum

The formal outline starts with a sector approach on defining production, quality, relationships, objective functions and the "social optimum". Based on that (modified: see Heintzelmann et al., 2008) we work with an average cost function which is: unit cost multiplied by effort. (Notify Heintzelman et al. worked with a production function and looked at revenue.) We reverse the argument and state that costs can be shared. In contrast to Heintzelman et al. (2008) who pursues the idea that sharing will reduce effort in favor of the environment, our issue is to increase costs (quality) based on cost sharing (externalizing). Because an option is to share, costs will result in larger cost for groups and this is hoped to get images needed; the team provides scope for higher grading and better prices as well as incentivizes producers at group level to improve on quality. The lure is that participation in the team through economies of scale (costs) will attract firms. In this respect we follow the assessment that a rigorous institution economic problem analysis would deliver the assertion that quality is a *public good* and too "few efforts" might be currently provided, (i.e. to work with positive externality, yet than negative).

A complication (if one works on quality) is the fact that *quantities* and *qualities* are linked and simultaneous decision variables. About cost vs. quality leadership Porter (1985) put this as strategic options. Here, price levels become disposable if quality improves (prices increase with quality: hopefully). A generic problem is that food prices are considered to be low. (The question arises why farmers should invest in quality. Also, quality is mostly obscured by standards that give minor price signals to consumers.) Thus, the modeling should include quantity quality, and price effects guaranteeing that farmers' decisions are paying off later. Hitherto, the complexity of the problem, in order to become policy relevant, has to be increased. On the other hand it is necessary to maintain the core description of the quality enhancing argument.

3.1 Quality Measure and Objectives

Also we need some remarks concerning measurement of quality: quality can be considered an index to which farmers contribute. As part of production, quality can be achieved, for example, by shares of better fodder in the diet of animals (cut of concentrates), outside grazing, longer feeding periods, mobility and health support, good micro-climates in stables, etc. As part of animal health, reduced additives are also relevant. Cuts of many of modern inputs are called efforts. The cuts of individual farmers are summed up to give a total quality index. Individual actions correlate with efforts at individual and group level.

Subsequently, for simplicity, firstly revenues are described as the product of price multiplied by quantity and quality: P·Q·Y. Quality is given as an index which lifts up average prices and which is associated with a corresponding willingness to pay by consumer for higher quality. Hereby, we assume that there is an anonymous, average market price P; i.e. the average market price level of the sector is not influence-able by the team (rather for any average kind of product the anonymous sector price prevails and the team can only strive for a niche of quality products which gives them a higher price as mark-up and based on image and branding; apparently the practical part is beyond this analysis). Received prices (above average) can be only raised along consumers' willingness to pay for a quality Q which is truly communicated, assured and established: then P·Q is assigned to quantities Y; i.e. those consumers who get good quality accept a higher price P·Q (Rausser et al., 2011: striving for higher quality is limited to segmented markets for quality products and all prices cannot be indirectly lifted).

In a second version we will investigate the situation of a directly (though only partly) influenced market price at a broader range of raising quality of a sector (see below). Though the effect on average price might be relatively small, with spreading of teams it might be possible to improve the image of a whole sector. In that case free-riding problems become more pertinent, teams should serve internalizing them, and the effect (size of price increase) is an empirical problem. However, as long as quality is determined in a competitive and a to-be-assured market, producers have rarely strong power on increasing price. Eventually it can be supposed that several teams form the sector. Only then, if all producers contest, prices increase.

About production costs, as dependent on quantities produced, we refer to the usual curvature of cost functions: being convex. On quality (image, costs) we see economies of scale (concave) and benefits from joint effort. Since we have to infer sizes of teams (shown later) and numbers of teams, we likewise assume that no single firm can establish "quality" for getting better prices (yet, only for a basis image, in a special game, a single firm can form a team). Due to size limitations firms lack the overall capacity to establish *industry quality*. But shared action can make it. So, we model a typical medium sized sector in a competitive food market.

Further assumptions are: (1) we look at a single layer value added chain product. For the moment we do not distinguish producers, processors and retailers. Yet, we see rules and team building on raw material producing farmers and set contracts with processors and retailers. (2) Since it is our intention to discuss an alternative to external standard setting, we abstract from outside treaties, hierarchies. etc.; though it might be still plausible to integrate processing and production as well as other contracts and to focus on decision units: Y and Q, only. (3) The next step is that we want to make the model operational for numerical simulations. For this sake we suggest introducing a linearized version of an objective function which is mathematically treatable. Thence a needed step is to clarify on *quality* for a mathematical outline of efforts and measurements. For this, it is assumed that quality (as index) is linked to effort: $q=\gamma \cdot e$ or at totaled level: $Q=\gamma \cdot E$; next, assuming that weighting of individual quality contributions "q" can occur, it delivers an overall index $Q=\Sigma w_i q_i$, where we assume that $0 < Q < 1$.

3.2 Social Level (Industry) Specification as Well as Quality and Effort Specification

For a further drafting it has to be mentioned that the approach starts at *social* (aggregated) level. The group (team) problem is usually: maximizing the value added (surplus) of a sector composed of producers which form associations (i.e. producers have a common goal). In our case the goal is exploiting quality increases of food (incl. image for a segment of a market which is characterized by produce offering better returns). From putting efforts in quality improvement producers can expect higher revenues and by cost sharing total costs are lower than in individual cases. In fact we have to clarify on quality as a variable entering the objective of a potential group of producers. In equation (1) we specify the revenue as price multiplied by the quality index as value P·Q and Y. Quantity Y and quality Q vary. In this simple version consumers appreciate Q (are willing to pay for quality) and down grade lower quality. Actually in the simple version P·Q is a superficial new pricing. The condition is expressed as:

$$L^s = P \cdot Q \cdot Y - c_a Q \cdot F(Q(E), Y) - C(Y) \tag{1}$$

where: P: price

Y: quantity produced

Q: Quality

E: Effort

C_a: average costs for quality

C: usual production costs

In this specification sector surplus (inverse of production function, i.e. cost function) is defined as the production coefficient function F(…). It is flexible with respect to quality (see (1')).

$$C = \gamma \; E \frac{F(E,Y)}{E} = E \; \frac{[\alpha_{11} E^2 + \alpha_{12} Y \; E \;]}{E} = E \; [.5\alpha_{11} E \; + \alpha_{12} Y] \tag{1a}$$

where: E: effort

F: inverse of production function

γ: technology with measurement $\gamma=1$ to simplify

Furthermore we use a Taylor approximation for the remaining unit costs and get a simplified version (2) as an expression that helps us to apply the analysis of Heintzelman et al. (2008).

$$L^s = P \; [Y_0 \gamma \; E + Q_0 Y] - c_a \gamma \; E \; [.5\alpha_{11} E \; + \alpha_{12} Y] - \alpha_{21} E \; Y \; - .5\alpha_{22} Y^2 \tag{2}$$

In equation (2) we used a specific *Taylor Approximation* for a concave cost function, i.e. made the goal quadratic and resumed that current, minimal quality 'Q$_0$' (for instance: Q$_0$=0.1) is already given as well as current output 'Y$_0$' (serving as benchmarks). These are benchmarks as references: They serve to get an approximation which fits the actual sector cost function (been estimated). Applying the approximation has several justifications: (1) Unlike the usual cost function formulation, we assume a joint effect in increasing quality and quantity on costs; (2) we split the total effect into two branches: ordinary effects (like in any formulation cost functions are separable) and (3) plus α_{21}EY is the joint effect. (Note, for sequential individual optimizations, individual quantities and qualities are of linked character with the team levels.)

Additionally we can supplement the approach with a dependency of sector price P on quality. This expresses that the overall market follows an image that is translated into prices. It is:

$$P = \varepsilon_{11} Q - \varepsilon_{11} Y \tag{1b}$$

As said the reliance of the sector price P on Q and Y might be marginal and it is empirical to get ε's. In vector description (1b') individual efforts are weighted. Efforts give a price increase.

$$P = \varepsilon_{11} \gamma_w' e - \varepsilon_{11} Y \tag{1b'}$$

In Equation (1b) a linear approximation is assumed which is reflecting industry "knowledge".

Notice: in case of making a product price dependent on Y and Q the welfare of the industry (producer surplus) is in the focus. Dropping the dependency on Y and assuming Q is entering welfare of consumers, we even could speak of social welfare as market equilibrium, i.e. as if quality is demanded and the demand corresponds to consumer welfare (Rausser et al., 2011).

4. Social (Industry) Optimization

Given the discussion on an objective (goal) function we can start with a derivation of a social optimum (in the sense of getting a collective one: group or team optimum, i.e. for a number of firms cooperating). It is an "as if" optimization and the firm number is endogenous. Still the optimum serves as reference. Taking the derivative to Q or E, respectively, this would provide an optimum of quality that is analog to the optimum given in Heintzelmann et al. (2008):

$$P \; [Y_0 \gamma \; E + Q_0 Y] \; - c_a Y_0 \; F(Q^*, Y_0) - Q^* F'(Q, Y_0) = 0 \tag{3}$$

Our problem of a social (group) optimum, here also depends on Y: the production level. Hence we must do a double optimization: on Y and Q. We take an explicit instead of implicit function (used by Heintzelman et al., ibid). For function F(Q(E),Y) we assumed a depiction such as F(Q(E),Y) = 0.5α_{11} E^2+ α_{12}Y E (see above) and then, for the first derivatives towards E and Y (we inserted E as cause of Q), we receive:

$$\frac{\partial L}{\partial E} = P \; Y_0 \gamma \; -_a \gamma \; \alpha_{11} E \; - c_a \gamma \; \alpha_{12} Y \; - \gamma \; c_a \alpha_{11} E \; - \gamma \; \alpha_{21} Y \; = 0 \tag{3a}$$

$$\frac{\partial L}{\partial Y} = P \; \gamma \; E_0 - c_a \gamma \; \alpha_{12} E - \alpha_{21} E - \gamma \; \alpha_{22} Y \; = 0 \tag{3b}$$

Solving the second equation for Y

$$Y = \alpha_{22}^{-1} [P \; E_0 - [c_a \alpha_{12} + \alpha_{21}] E \;] \tag{4b}$$

and inserting in the first equation the optimal effort for *social* (team) quality is a solution as:

$$P\ Y_0 - c_a \gamma\ \alpha_{11} E\ + [c_a \alpha_{12} + \alpha_{21}] \alpha_{22}^{-1} [P\ E_0 - [c_a \alpha_{12} + \alpha_{21}] E\ = 0 \tag{4a}$$

Hereby we resumed E_0 and Y_0 already exist as current ones and it is about improving quality. One can build on the reference, as given in a market situation, i.e. without the suggested institutional innovation of a team work. Notably we receive an optimal effort E* from (4a). In this version quality is a sales argument (WTP) and it does not necessarily change usual prices.

The eventual second step (given what has been said already about influence on the whole sector) is to couple the *co-ordination* problem of team work with sector pricing. If, in particular we like to address the total image issue, better quality and reduced quantity may enable higher prices at industry level. The combination of sales promotion by increased quality which is Q (already tackled above) with a handy knowledge on image and price lifting (5) helps here. For this we presented: price dependent on quality Q (above). For an inverse quality-price equation

$$P\ = \varepsilon_{11} Q - \varepsilon_{11} Y \tag{5}$$

it enables linking (with the above specification of the optimal quantity). It delineates a further link amid pricing and quality (as well as effort for quality) in equation (6). In (6) we insert E

$$P\ = \varepsilon_{11} \gamma_1 E\ - \varepsilon_{11} \alpha_{22}^{-1} [P\ E_0 - [c_a \alpha_{12} + \alpha_{21}] E \tag{6}$$

$$P = [\varepsilon_{11} \alpha_{22}^{-1} E_0 + 1]_0^{-1} \varepsilon_{11} \gamma_1 [E\ - \varepsilon_{11} \alpha_{22}^{-1} [c_a \alpha_{12} + \alpha_{21}] E\ ] \tag{6'}$$

and this formulation can be inserted into the result for E (using equation 4a), here to get a price effect. Then equation (7) provides us with a residual depiction of effort at team level:

$$Y_0 [\varepsilon_{11} \alpha_{22}^{-1} E_0 + 1]_0^{-1} \varepsilon_{11} \gamma_1 [E\ - \varepsilon_{11} \alpha_{22}^{-1} [c_a \alpha_{12} + \alpha_{21}] E\ ] - c_a \gamma\ \alpha_{11} E\ + [c_a \alpha_{12} + \alpha_{21}] \alpha_{22}^{-1} [[\varepsilon_{11} \alpha_{22}^{-1} E_0 + 1]_0^{-1}$$
$$\varepsilon_{11} \gamma_1 [E\ - \varepsilon_{11} \alpha_{22}^{-1} [c_a \alpha_{12} + \alpha_{21}] E\ ] E_0 - [c_a \alpha_{12} + \alpha_{21}] E\ = 0 \tag{7}$$

The *Optimal effort* in (7) gives quality endogenously by reinserting. And (7) can serve as a reference for the team building process which will be discussed next as a game approach.

5. Cost Sharing and Individual Optimum

As been discussed by Heintzelman et al. (ibid) sharing arrangements work as game approaches. Hence, we have to formally express the institution of cost sharing as game. The next step serves to implement the rule of cost sharing mathematically and shows its implications. The idea is to have two layers of sharing: (1) at upper level teams form themselves and share costs; (2) at lower level, in teams, members share costs on equity basis. And (3) shares in costs are determined by the size of the teams. The overall efforts are taken as reference for (i) the group-wise and (ii) the in-group sharing. Then the team members make decision on joining teams. As a result individual objective functions are to be specified such as:

$$L_{ik}^s = p_{ik} [y_{0,ik} \gamma\ e_{ik} + q_{0,1} y_{ik}] - \{\frac{1}{m_i} [\frac{e_{ik} + E^{-k}}{E}]\} \{.5 c_a \gamma\ \alpha_{11} E^2 - c_a \gamma\ \alpha_{12} E\ Y\ \} - \alpha_{21.ik} e_{ik} y_{ik}. - 5 \cdot \alpha_{22,ik} y_{ik}^2 \tag{8}$$

and

$$L_{ik}^s = p_{ik} [y_{0,ik} \gamma\ e_{ik} + q_{0,1} y_{ik}] - \{\frac{1}{m_i} [e_{ik} + E^{-k}]\} \{.5 c_a \gamma\ \alpha_{11} E\ - c_a \gamma\ \alpha_{12} Y\ \} - \alpha_{21,ik} e_{ik} y_{ik}. - 5 \cdot \alpha_{22,ik} y_{ik}^2 \tag{9}$$

The argument for team formation is backward: Though objective function (9) is optimized at a second step after group formation, it serves as an "as-if" condition for the team formation. The outline is reciprocal: first, we optimize function (9) provided a behavioral function exists for each individual team, and second we show how the team formation works. Optimization towards individual rationality is summed up later and number and sizes of groups are derived. For the individual firm, however, activities of the groups (teams) are given (presumably). Firms choose team membership. How it works will be argued later. For the moment we optimize (9), above function, where we see y_i as part of Y, i.e. $Y = \Sigma y_i$, and $Q = \Sigma w_i q_i$ and $E = \Sigma w_i e_i$. The inclusion of a weighting for the quality index can be done by industry knowledge, given to all firms (else, we can work assigning equal weights; weights reflect brand characteristics).

$$\frac{\partial L}{\partial e_{ik}} = p_{ki}y_{0,ik}\gamma - \frac{1}{m_i}\{5c_a\gamma\,\alpha_{1\,1}[w_{ik}e_{ik}+w^{-ik}E^{-ik}]+c_a\gamma\,\alpha_{12}[y_{ik}+Y^{-i,k}]\} - \frac{1}{m_i}[e_{ik}+E^{-ik}]5\{c_a\gamma\,\alpha_{1\,1}w_{ik}\}-\alpha_{2\,1ik}y_{ik}. = 0 \quad (9a)$$

$$\frac{\partial L}{\partial y_{ik}} = p_{ik}q_{0,ik} - \{\frac{1}{m_i}[e_{ki}+E^{-ik}]\}\{c_a\gamma\,\alpha_{12}\}-\alpha_{21,k}e_{ik}-\alpha_{22,ik}y_{ik} = 0 \quad (9b)$$

The formulations in equations (8) and (9) are comparable to a participation game. In the optimization (9a and b) we only recognize elements of e_{ik} and y_{ik} being controllable by the individual contingent on the group decisions. It means that a producer has only scope about his optimization and can influence his quality related efforts quantities; others (sum of group activities) are given to him; notably, vice versa, as related to joint quality check he brings in his activities recognized by the others. A problem is how to work with different efficacy of efforts. Information and optimization are limited to the knowledge on own technologies.

In the given case calculations can be reduced to the behavioral description towards efforts for quality. For this purpose we solve the second equation for y_{ik} and insert it in the first equation.

$$y_{ik} = \alpha_{22,ik}^{-1}[p_i\,q_{0,i,k}-\{\frac{1}{m_i}[e_{ik}+E^{-k}]\}\{c_a\gamma\,\alpha_{12}\}-\alpha_{21,ik}e_{ik}] \quad (10b)$$

Subsequently we use

$$p_{ki}y_{0,ik}\gamma - \frac{1}{m_i}\{5c_a\gamma\,\alpha_1[w_{ik}e_{ik}+w^{-ik}E^{-ik}]\}+\frac{1}{m_i}[e_{ik}+e_i^{-k}]\{5c_a\gamma\,\alpha_kw_{ik}\}+\frac{1}{m_i}c_a\gamma\,\alpha_{12}Y^{-i,k}-[\frac{1}{m_i}c_a\gamma\,\alpha_{12}+\alpha_{2\,1ik}]y_{ik}. = 0 \quad (10a)$$

in order to get

$$p_{ki}y_{0,ik}\gamma - \frac{1}{m_i}\{c_a\gamma\,\alpha_1[e_{ik}+E^{-k}]\}+\frac{1}{m_i}c_a\gamma\,\alpha_{12}w_{ik}Y^{-i,k}+\alpha_{22,ik}^{-1}[p_i\,q_{0,i,k}-\{\frac{1}{m_i}[e_{ik}+E^{-k}]\}\{c_a\gamma\,\alpha_{12}\}-\alpha_{21ik}e_{ik}] = 0 \quad (10)$$

In equation (10) "e_{ik}", the effort of the producer "k" in team "I", is a function of E^{-ik} and Y^{-ik}, which are the sum of efforts and quantities of the other group members. These efforts are contributions of other team members and reliance is mutual. "m_i" is the size of team members; it prevails as sub-team. In a simple case the weight "w_{ik}" is given by head count, i.e. 1/n, with n farms. Yet, in concentrated industries weights might be different; for instance, given by previous revenue shares. Note the optimization is contingent on the behavior of the any firm (firms). In total it leads to the necessity to individually optimize and do joint optimization.

6 Team Formation and Team Optimization

The crucial thing for a simulation of any team formation (through modeling which shall be based on rational behavior of individuals joining teams sub-groups), is the determination of the number of teams "n". By this we seek a stable equilibrium. Following this, the sizes "m_i" of the sub-teams can be derived firstly and then secondly the numbers have to fit into the total number N giving n. In Heintzelmann et al. (2008), for instance, the argument is that a sector is presented by the sum of the individual costs/revenues (benefits in our case). For us it is:

$$\sum p_i\,y_{0,i}^a\gamma_i = p_0y_0^a N \Leftrightarrow p_0y_0 = \frac{1}{N}\sum p_i\,y_{0,i}^a\gamma_i \quad (11)$$

Note, in equation (11), y_o^a stands for the average production without an incentive scheme promoting quality. As consumer trade-off indirectly quality and quantity, y might be reduced. For clarifying the size of operation a social optimum can be used. We state that the social optimum is given by n groups and N participants. Team optimization follows sequentially. For a group optimization, a differentiation of "e" and "y" in a *team's objective function* is needed:

$$m_i\,L_{ik}^s = p_i[y_{0,}\gamma\,e_{ik}+q_0y_i]m_i - [e_i+E^{-i}]\{.5c_a\gamma\,\alpha_{11}E - c_a\gamma\,\alpha_{12}Y\}-\alpha_{21,i}e_ky_i-5\cdot\alpha_{22,i}y_i^2 \quad (11)$$

In equation (11) the objective is those of a group (team). We need this objective to specify the number of teams (n) and members (m_i: consecutively) in a team. The next step is to optimize the *behavior* of the team. Note results are similar to the individual optimization; though now it is assumed: the *team as a whole* is optimized. Joint behavior is perceivable as a team formation: *team* is a special subject. For the moment the optimization of the team's objective yields:

$$\frac{\partial m_i L_i^s}{\partial e_i} = m_i \, p_i^a y_{0,i}^a \gamma - c_a \gamma \, \alpha_{11}[e_i + E^{-i}] - c_a \gamma \, \alpha_{11}[y_i + Y^{-i}] - \alpha_{21,i} y_i = 0 \qquad (12a)$$

$$\frac{\partial m_i L_i^s}{\partial y_i} = p_i \, q_{0,1} m_i - e_i \, c_a \gamma \, \alpha_{11}[e_i + E^{-i}] - \alpha_{21,i} e_i - \alpha_{22} y_i = 0 \qquad (12b)$$

For the result (12a and b) of the optimization of objective (11) it is once more assumed that the group has knowledge on quantity and quality elements (in the cost function for the generation of quality). This is a limitation to the team. (We can perceive it as subject). Yet, solving of (12b) for y; here the quantity produced by the team gives (12b'). It results in:

$$y_i = \alpha_{22}^{-1}[p_i \, q_{0,1} m_i - e_i \, c_a \gamma \, \alpha_{11}[e_i + E^{-i}] - \alpha_{21,i} e_i \,] \qquad (12b')$$

Then the equation (12b') enables us to specify the quality e_i that should prevail at team level:

$$m_i \, p_i^a y_{0,i}^a \gamma - c_a \gamma \, \alpha_{11}[e_i + E^{-i}] - c_a \gamma \, \alpha_{11} Y^{-i} - [\alpha_{21,i} + c_a \gamma \, \alpha_{11}][\alpha_{22}^{-1}[p_i \, q_{0,1} m_i - c_a \gamma \, \alpha_{11}[e_i + E^{-i}] - \alpha_{21,i} e_i \,] = 0 \quad (12a')$$

In principle one can establish the effort of a whole team "e_i" for quality. However, on the one hand the determination is based on the size of the group m_i which is not established yet. On the other hand, nevertheless, we know the socially optimal effort of the whole sector and see the rest of the teams' efforts. The next steps will allow us to get the team numbers.

7. Team Numbers and Size

In this section we show how the team size can be modeled. We again follow the argument of Heintzelmann et al. (2008), who have shown that larger groups have smaller do less. Also we indicate how one can derive the total number of teams. As been further argued (Heintzelmann et al., 2008) membership in groups may not differ between teams. Then, there are reference equilibria along the concept of: "no incentives to switch between teams". These are theoretical and empirical arguments and several sub-teams will prevail because of transaction costs.

The theoretical arguments go along the following line: Defined as quality contribution by individual efforts, we have different contributions of producers that, apparently, are distinguishable and producers have a different stake in sector's quality performance. Performance was introduced by a weighted quality index. It means that a team of producers can be of equal size contributing to quality. As we defined Q_i as equal $Q=\gamma E$ and $Q=\Sigma w_i \gamma e_i$, individuals in the group may differ with respect to the efforts offering; yet they control each other. This respectively implies, we cannot maintain equal quality contributions by teams. Then, knowing the socially optimal quality and dividing it by the number of team efforts is core to the analysis.

For the following analysis a theoretical outline is sketched. In this outline the complex mathematical presentation is reduced to equations for which coefficients in front of variables are to be calculated. For instance, if we take the sorted variables m_i, e_i, E_{-1}, and Y_{-1} in equation:

$$[p_i^a y_{0,i}^a \gamma - [\alpha_{21,i} + c_a \gamma \, \alpha_{11}][\alpha_{22}^{-1} p_i q_{0,1}]]m_i - [[1 + \alpha_{21,i} c_a \gamma \, \alpha_{11}] c_a \gamma \, \alpha_{11} - \alpha_{22}^{-1} \alpha_{21,i}] e_i - c_a \gamma \, \alpha_{11} Y^{-i} + [c_a \gamma \, \alpha_{11}] E^{-i} = 0 \quad (13')$$

it is a detailed account of optimality and correspond to a reduced version of joint coefficients:

$$\theta_{10,i} + \theta_{11,i} m_i + \theta_{11,i} e_i + \theta_{13,i} E + \theta_{14,i} Y = 0 \qquad (13'')$$

The coefficients θ_{ii} in equation (13'') can be recalculated from version (13'). We assume that there are only marginal contributions of individuals to E and Y; i.e. it does not make sense to count Y and E as with or without the contribution of a producer under study. Subsequently we can rewrite equation (13'') in a generalized way if we sum up over the number of teams:

$$\sum \theta_{10,i} + \sum \theta_{11,i} m_i + \sum \theta_{11,i} e_i + \sum \theta_{13,i} E + \sum \theta_{14,i} Y = 0 \qquad (14)$$

In this general representation of a sector, as a sum of teams, the number of teams is implicitly contained (Heintzelman et al., 2008). For instance, if we postulate that the averages of the coefficients (i.e. gains of team work are given as higher marginal revenues) are obtainable by

$$\theta^a = \frac{1}{n}\sum \theta_{10,i} \Leftrightarrow n\theta^a = \sum \theta_{10,i} \qquad (15')$$

the sum (in 15') translates into an average coefficient multiplied by the number of teams. This enables us to postulate a slightly modified version of equation (15) which contains n and it is:

$$n\theta_{10}^a + N\theta_{11}^a + E\theta_{11}^a + n\theta_{13}E + n\theta_{14}Y = 0 \qquad (15'')$$

From this equation the number of teams is "n" retrievable. Mathematically the equation (15'') can be solved for "n" given coefficients θ, N, E and Y. As a result one gets the number of teams dependent on the number of potential producers "N" and we refer to a calculated social optimal "E" and "Y" (see above). In this theoretical derivation different variants can be discussed (Heintzelman et al., 2008): A special variant is the case of merely solo-producers: in other words only single players (team) work. This implies (if the above equation is solved for "n" which is the number of teams) a recursive calculation of team sizes enables simulations.

8. Determination of Team Size

Having advanced to a determination of number of groups, the final step is to determine (simulate) the number of members in each team (as remaining problem in team formation), which has to be done separately for any sub-team. We work along the given concept using an objective function and generalize the detailed findings for the individual, team and sector level. Remind that the technical outline of determining variables went along the principle that sub-games and optimization of individuals resulted in teams (summing up) and teams' characteristics help to specify the team number. Knowing numbers of teams (15'') the analysis can come back to member behavior in teams and from the membership number we can re-derive the individual behavior. Finally individual behavior (as sum of effort for quality of each producer) delivers a sector's quality. Notify the approach is embedded in methodological individualism and no mysterious "deus-ex-machina" (coercion) is needed for team formation.

According to Heintzelmann et al. (2008) one can state that teams will form along a concept of achievable equilibria of joining and leaving teams. Incentives are: cost sharing opportunities in various teams due to economies of scale for image which is a common and collective benefit. For practical reasons we can assume that the quality (image) is shared and contributed by each team based on own calculi. However the member number in any team m_i is not yet determined. But deliberation (16) helps us to establish the sizes of (representative) teams (given quality index Q based on E). As described above, the crucial matter is that having "n" teams "n" equations of similar type as equation (13") are given. Then, as definition an average effort in team i is e^a_i, a system of "n" equation (16) for determining the m_i's in a system prevail. Here we make a reference to the fact that the sum of all members is N. We add one equation completing the system. In turn it would mean one team member equation has to be canceled.

$$\theta_{11,i}m_1 \qquad = \theta_{10,1} + \theta_{11,i}e^a + \theta_{13,1}E + \theta_{14,1}Y$$

$$\theta_{11,2}m_2 \qquad = \theta_{10,2} + \theta_{11,i}e^a + \theta_{13,2}E + \theta_{14,2}Y \qquad (16)$$

$$\dots$$

$$m_1 + m_2 + \dots m_n = N$$

A system like (16) would provide us all sizes of sub-teams. Note for decisions to join a team the size is eventually of bigger importance than the costs: Big teams enable producers to externalize; small to control each other. Technically a problem is that the system is over-determined. Over-determination can be avoided if we include a correction measure c_p. This measure can be considered a public intervention toll (incentive) that guarantees that the individual teams will form.

$$\theta_{11,i}m_1 \qquad + \theta_{15,1}c_p = \theta_{10,1} + \theta_{11,i}e^a + \theta_{13,1}E + \theta_{14,1}Y$$

$$\theta_{11,2}m_2 \qquad + \theta_{15,2}c_p = \theta_{10,2} + \theta_{11,i}e^a + \theta_{13,2}E + \theta_{14,2}Y \qquad (16')$$

$$\dots$$

$$m_1 + m_2 + \dots m_n \qquad = n$$

The measure can be a special (subsidy) taken by government. Once "c_p" is then endogenous.

So far we assumed E and Y are known from social optimum. Another way of solving the problem is to reason for stratifying the problem. For instance we start with two teams and get

$$\theta_{11,1}m_1 \qquad +\theta_{11,1}e_1^a \qquad = \theta_{10,1} + \theta_{13,1}E^* + \theta_{14,1}Y^*$$

$$\theta_{11,2}m_2 \qquad +\theta_{11,2}e^a = \theta_{10,2} + \theta_{13,2}E^* + \theta_{14,2}Y^* \qquad (16")$$

$$e_1^a \quad +e_2^a = E^*$$

$$m_1 \qquad\qquad +m_2 = 2$$

which is a system of four equations and four dependent variables (solvable). It implies that a special quality comes out of two teams as simulated. In this case we have assumed that two teams are established first and members, who join them, are the important formation category.

However our previous result was given by "n" groups. So note that further approximations are needed. A way to go ahead is to assume that the number of teams can be approximated by 2^x. Then from the first upper layer of two teams we can descend to the next (lower) layer assuming 4 teams and check for scope of forming. Forming becomes hierarchical. The constraint for a new team is e_i^a or e_2^a. Hence we assume that new teams form from given teams (splitting). It means that for the first sub-teams $m_{11}+m_{12}=m_1^*$ and $e_{11}+e_1=e_1^*$ holds. Similar things can happen to team "2" and we can proceed. With a third layers eight teams are established and with four layers, 16 teams are formed, etc. The closest number in layer creation will give an approximation to number of teams (perceivable form the social optimization equation 15).

A special problem, involved in this procedure, is that a pre-selection of membership in a team and its size determination are given. Since we need to establish the averages for coefficients, this might be arbitrary (dependent on subjective team choices). Again we have to approximate and in reality teams of producers may form according to similarity between themselves. The likelihood for grouping in teams (itself) is size. Starting with a fist layer the discrimination of joint productivity is small and large teams form easily. Then we break down to less large and *bigger numbers* in the group of large ones. Note we do not know the behavior of any individual contributing as producer in advance. Though, the idea is to simulate the outcome.

9. Policy Involvement

So far the discussion has been on the institution of sharing as a rule (instrument without direct intervention). Sharing shall be *invented* by a community itself and tested at the level of individuals; and then, hopefully, members voluntarily join the scheme. Even it could include participation constraints. This probably is the ideal establishment of an economist who does not like government interventions; though it may be possible to influence promotion of the rule and establishment of the community by active involvement. Still no standards are needed and the teams monitor themselves. Yet, the role of government or governance as involvement becomes another issue: are sharing schemes independent from governance? What are the measures to promote and can one use subsidies being helpful as stimuli? For the moment we may assume that governments could also take a share in cost. This share or contribution must be specified according to the delivery of Q and its valuation by consumers. Then we receive

$$L^s = P \ [Y_0 \gamma E + Q_0 Y] - c_a \gamma \ [E \ [.5\alpha_{11}E \ + \alpha_{12}Y] - r_a^g Q_a] - \alpha_{21}E \ Y \ -.5\alpha_{22}Y^2 \qquad (17)$$

Equation (17) is a new objective. Here r_g is a payment that is provided as a cost refund from government based on achieved quality and also, the government can make a cost-benefit-analysis. r_g can be considered as an incentive scheme to be added to prices (freely). Yet involvements of incentives raise the question of a cost benefit analysis. But this is beyond the current analysis.

10. Scope for Application

As been indicated in the introduction, there are many cases in which the quality of food is considered lower or has declined in the last decade. In particular, consumer concerns have been expressed on minimal standards and they try to raise awareness in campaigns sponsored by consumer councils, etc. As well many industries are currently struggling with bad images (for example for beef see: Fearne et al., 2001; for pork see: Gilg & Batterhill, 1998; and also potatoes and vegetables may allow applications: Willersinn et al., 2015). Additionally we foresee applications in case of scandals (Wales et al., 2006) where trust has to be rebuilt. Especially issues like *quality, trust, consumer confidence, industry responsibility, etc.* (Hartmann et al, 2015) have been high on the agenda, recently. In these cases the institutional framework of sharing efforts might have many requests. Especially since a collective responsibility can be created within teams which might be even include risk bearing, internal control, etc. Uses of rules for sharing efforts and responsibility, perhaps, will be even spontaneously emerging and the presented theory should help to better understand the background and behavioral aspects. This particularly relates to the game theory aspect tackled.

11. Summary

This article was discussing cost sharing as an alternative institution to public intervention for quality improvement in food industries. Cost sharing, as a rule, is introduced as an analogy to benefit sharing in environmental economics. There, benefit sharing is a mean to reduce efforts while cost sharing shall stimulate efforts. In our case we presumed that current efforts are too low in food industries to get socially optimal quality. The paper develops a scheme of team work in order to get higher levels of quality. A mathematical outline was provided on how to specify objective functions for collective and individual actions. Also, it was shown how the number of teams and the membership in each team can be simulated on the basis of behavior.

References

Codron, J. M., Giraud-Héraud, E., & Soler, L. G. (2005). Minimum quality standards, premium private labels, and European meat and fresh produce retailing. *Food Policy, 30,* 270-283. http://dx.doi.org/10.1016/j.foodpol.2005.05.004

Fearne, F., Hornibrook, S., & Dedman, S. (2001). The management of perceived risk in the food supply chain: a comparative study of retailer-led beef quality assurance schemes in Germany and Italy. *The International Food and Agribusiness Management Review,* 4(1), 19-36. http://dx.doi.org/10.1016/S1096-7508(01)00068-4

Gilg, A. W., & Battershill, M. (1998). Quality farm food in Europe: a possible alternative to the industrialised food market and to current agri-environmental policies: lessons from France. *Food Policy, 23*(1), 25-40. http://dx.doi.org/10.1016/S0306-9192(98)00020-7

Hanf, C. H., & Wersebe, B. (1994). Price, Quality and Consumer' Behavior. *Journal of Consumer Policy, 17,* 335-348. http://dx.doi.org/10.1007/BF01018967

Hartmann, M., Klink, J., & Simons, J. (2015). Cause related marketing in the German retail sector: Exploring the role of consumers' trust Cause related marketing in the German retail sector: Exploring the role of consumers' trust. *Food Policy, 52,* 108-115. http://dx.doi.org/10.1016/j.foodpol.2014.06.012

Heintzelman, M. D., Salant, S. W., & Schott, S. (2008). Putting Free-Riding to Work: A Partnership Solution to the Common-Property Problem. *Journal of Environmental Economics and Management, 57,* 309-320. http://dx.doi.org/10.1016/j.jeem.2008.07.004

Henson, S., & Reardon, T. (2005). Private agri-food standards: Implications for food policy and the agri-food System. *Food Policy, 30,* 241-253. http://dx.doi.org/10.1016/j.foodpol.2005.05.002

Holleran, E., Bredahl, M. E., & Zaibet, L. (1999). Private incentives for adopting food safety and quality assurance. *Food Policy, 24,* 669-683. http://dx.doi.org/10.1016/S0306-9192(99)00071-8

Martinez, M. G., Fearne, A., Caswell, J. A., & Henson, S. (2007). Co-regulation as a possible model for food safety governance: Opportunities for public–private partnerships. *Food Policy, 32*(3), 299-314. http://dx.doi.org/10.1016/j.foodpol.2006.07.005

Nizan, S. (1991). Collective rent dissipation. *The Economic Journal, 101*(409), 1522-1534. http://dx.doi.org/10.2307/2234901

Platteau, J. P., & Seki, E. (2000). Community arrangements to overcome market failures: Pooling groups in Japanese fisheries. In M. Aoki & Y. Hayami (Eds.), *Markets, Communities and Economic Development.* Oxford.

Narrod, c., Roy, d., Okello, J., Avendaño, B., Rich, K., & Thorat, A. (2009). Public–private partnerships and collective action in high value fruit and vegetable supply chains. *Food Policy, 34,* 8-15. http://dx.doi.org/10.1016/j.foodpol.2008.10.005

Porter, M. E. (1985). competitive advantages: creating and sustaining superior performance. New York.

Rausser, G., Swinnen, J., & Zusman, P. (2011), Political Power and Economic policy. Chapter 17: The Political Economy Lens on quality and Public Standards. Cambridge. http://dx.doi.org/10.1017/CBO9780511978661

Sheppard, R. W. (1970). Theory of Cost and Production Functions. Princeton.

Sönnichsen, M. Augustini, C. Dünckel, R., & Spindler, M. (2000), Handelsklassen für Rindfleisch. Bonn AID.

Verbeke, W., Ward, R. W., & Vianne, J. (2000). Probit analysis of Fresh Meat Consumption in Belgium: Exploring BSE and Television Communication Impact. *Agribusiness, 16*(2), 215-234. http://dx.doi.org/10.1002/(SICI)1520-6297(200021)16:2<215::AID-AGR6>3.0.CO;2-S

Wales, C., Harvey, M., & Warde, A., (2006). Recuperating from BSE: The shifting UK institutional basis for trust in food. *Appetite, 47*(2), 187-195. http://dx.doi.org/10.1016/j.appet.2006.05.007

Willersinn, C., Mack, G. Mouron, P., Keiser, A., & Siegrist, M. (2015). Quantity and quality of food losses along the Swiss potato supply chain: Stepwise investigation and the influence of quality standards on losses. *Waste Management, 46*, 120-132. http://dx.doi.org/10.1016/j.wasman.2015.08.033

Stability of Haskap Berry (*Lonicera Caerulea* L.) Anthocyanins at Different Storage and Processing Conditions

Rabie Khattab[1,2], Amyl Ghanem[2] & Marianne Su-Ling Brooks[2]

[1]Food Science Department, Faculty of Agriculture (Saba Basha), Alexandria University, Alexandria, Egypt

[2]Department of Process Engineering & Applied Science, Dalhousie University, B3H 4R2 Halifax, NS, Canada

Correspondence: Marianne Su-Ling Brooks. Department of Process Engineering & Applied Science, Dalhousie University, B3H 4R2 Halifax, NS, Canada. E-mail: Su-Ling.Brooks@dal.ca

Abstract

The effect of freezing, frozen storage (–18 °C for 6 months), thawing, juice extraction, and hot-air drying on the anthocyanin profile of haskap berry (*Lonicera caerulea* L.) was investigated using RP-HPLC. Five anthocyanins (ANCs) were quantified: cyanidin 3,5-di-glucoside (4.27 % of the total ANCs), cyanidin 3-glucoside (89.39 %), cyanidin 3-rutinoside (2.07 %), pelargonidin 3-glucoside (0.83 %), and peonidin 3-*O*-glucoside (3.44 %). Freezing did not significantly affect the content of individual ANCs, while frozen storage resulted in significant reductions (16.00-24.50 %). Thawing the frozen berries in the microwave oven retained the highest content of different ANCs. The highest degradation, however, occurred while thawing at room temperature. Extracting juice from the berries significantly reduced the content of individual ANCs. Drying the berries to 25 % moisture content at 60, 100, and 140 °C reduced the individual ANCs by 73.85-76.19, 78.46-80.95 and 90.77-95.40 %, respectively. The overall stability of the five ANCs during storage and processing is summarized by the following trend (from most to least stable): peonidin 3-*O*-glucoside > pelargonidin 3-glucoside > cyanidin 3,5-diglucoside > cyanidin 3-rutinoside > cyanidin 3-glucoside.

Key words: haskap berry, storage, processing, anthocyanins, stability

1. Introduction

The haskap berry (*Lonicera caerulea* L.) has been recently introduced as a commercial crop to the North American market. Some varieties and cultivars are currently available in Canada and USA (Bors *et al.*, 2012). These berries are either consumed fresh or processed into juice, pastries, jams, ice cream, and dried berries (Celli *et al.*, 2014). Haskap berries have attracted attention for their distinct profile of phenolic phytochemicals (Jurikova *et al.*, 2012). They are particularly rich in anthocyanins (ANCs) with varied health benefits (Paredes-Lopez *et al.*, 2010). The total anthocyanin content (TAC) of haskap berry was found to be up to 13.00 mg cyanidin 3-glucoside (C-3-G) equivalents per g fresh weight (FW) (Bakowska *et al.*, 2007; Fan *et al.*, 2011; Rupasinghe *et al.*, 2012). This berry has much higher antioxidant potential than that reported for blueberry, blackberry, raspberry, bilberry, strawberry, sea buckthorn and black currant (Rop *et al.*, 2011; Raudsepp *et al.*, 2013; Celli *et al.*, 2014).

ANCs are the most important water-soluble phytochemicals in nature (Harborne, 1998). They are responsible for the distinguished colors of several fruits and vegetables. They are distinctive from the other flavonoids by their ability to form flavylium cations (Fig. 1) (Mazza, 2007). ANCs consist of an aglycon base or flavylium ring (anthocyanidin), sugars, and may contain acylating groups (Bueno *et al.*, 2012). From the several anthocyanidins found in nature, only cyanidin, delphinidin, petunidin, peonidin, pelargonidin, and malvidin (Fig. 1) are of importance in human nutrition (Harborne, 1998; Jaganath & Crozier, 2010; Bueno *et al.*, 2012). Food ANCs play important roles in preventing various diseases including cancer, diabetes, cardiovascular diseases, and obesity. They are also associated with improving immunity and night vision, retarding aging and reducing the risk of degenerative disorders (Jing, 2006; Nikkhah *et al.*, 2008). These beneficial effects of ANCs are attributed to their antioxidant, detoxification, anti-proliferation, anti-angiogenic, and anti-inflammatory activities (Miguel, 2011).

Upon seasonal harvest, haskap berries are frozen to be used all over the year for consumption or processing. The most economically-important haskap products are the juice and pressed berries, which are a by-product from juice extraction and subsequently sold as a dried berry product. Despite the varied functions and health benefits of ANCs, they are very labile and undergo significant breakage and structural changes during storage and processing (Ochoa et al., 1999; Lohachoompol et al., 2004; Sadilova et al., 2006). Due to their highly reactive nature, ANCs readily degrade to colorless or brown compounds. Loss of ANCs is also accelerated by the presence of oxygen and enzymes, and during the high temperature processing (Jackman et al., 1987).

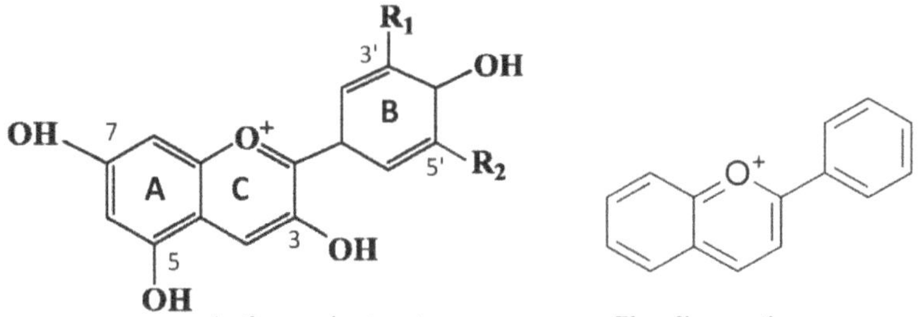

Anthocyanidin	R_1	R_2	Color
Peralgonidin	H	H	Orange
Cyanidin	OH	H	Orange-red
Delphinidin	OH	OH	Bluish-red
Peonidin	OCH_3	H	Orange-red
Petunidin	OCH_3	OH	Bluish-red
Malvidin	OCH_3	OCH_3	Bluish-red

Figure 1. Structures of major ANCs (adapted from Jing, 2006)

The effect of storage, thawing, extraction and drying conditions on the retention of haskap berry ANCs has been investigated in our laboratory. The total anthocyanin content (TAC) was reduced in different varieties by 39.31–59.24 and 36.87–56.57 % upon frozen storage for 6 months at-18 °C and-32 °C, respectively (Khattab et al., 2015a). The reduction in TAC was 32.14–53.25, 28.55–51.45 and 18.92–47.22 % when the frozen fruits were thawed at room temperature (25±2 °C), fridge (4 °C) and in the microwave oven, respectively (Khattab et al., 2015b). In a recent study (Khattab et al., 2016a), we have investigated the effect of juice extraction and drying conditions on the stability of haskap berry ANCs. The juice extraction process significantly reduced the TAC by 48.18 % in the pressed residues as compared to that of the whole berries. Furthermore, the TAC decreased significantly at different drying temperatures with a strong positive correlation between the drying temperature and the degradation rate.

Our previous studies on haskap berries did not report on the levels of individual ANCs during storage and processing. It is important to determine the anthocyanin profiles for different haskap berry products and the effect of storage and processing conditions, as this will help develop strategies to improve the nutritional content of these products. Therefore, this study examines the effect of the processing chain including freezing, frozen storage, thawing, juice extraction, and drying conditions on the stability and retention of specific individual ANCs contained in haskap berry fractions.

2. Materials and Methods

2.1 Materials

2.1.1 Haskap Berries

Haskap berries (Lonicera caerulea L.); variety Indigo Gem (26 kilograms) were obtained from LaHave Natural Farms, Blockhouse, Nova Scotia, Canada. Upon receiving the fruits, they were analyzed for their anthocyanin profile. The fruits were then frozen and stored at −18 °C. Half the berries were thawed and analyzed the next day to study the effect of the freezing process. The other half was kept frozen for 6 months to study the effect of frozen storage, thawing methods and drying conditions.

2.1.2 Chemicals and Phenolic Standards

All chemicals used for this research were of analytical and HPLC grades and were procured from Sigma Aldrich (Oakville, Ontario, Canada) and Fisher Scientific (Ottawa, Ontario, Canada). Authentic anthocyanin standards were obtained from Sigma Aldrich, Canada.

2.2 Experimental Procedures

2.2.1 Fruit Pressing and Juice Extraction

The juice extraction was carried out as previously described (Khattab *et al.*, 2016b), using a lab-scale manual multi-fruit juice extractor (F. Dick 9060600, 6L; Friedrich DICK, Deizisau, Germany). In this study (Fig. 2) the frozen berries were allowed to thaw at different conditions including room temperature (25 ± 2 °C for 12 h), and in the refrigerator (4 °C for 22 h) and microwave oven (1000 Watts for 0.29 h) according to Khattab *et al.* (2015b). The berries thawed at room temperature were loaded to the juice extractor (in 3 kg batches) and manually pressed. The pressing was done until obtaining 70 % of the original fruit weight as juice. The pressed berries were osmotically treated by mixing with sucrose (20 % of their weight), and left to infuse for 24 h. The mixture was then gently pressed using the juice extractor to drain the liquid part (syrup) without drastically affecting or rupturing the berries. The leftover berries were analyzed and dried.

2.2.2 Drying Process

The drying process was carried out according to Khattab *et al.* (2016a) using a lab-scale hot-air drying oven (Isotemp® Oven, Model 630F, Fisher Scientific, USA) at 60, 100, and 140 °C. The drying continued up to 48 h and the time needed to reach 25 % moisture content was recorded.

2.2.3 Moisture Content

Moisture content of the haskap berry samples was determined using a hot-air drying oven (Isotemp® 630F, Fisher Scientific, USA) at $103\pm 2°$ C and atmospheric pressure until constant mass was reached (ISO, 2009).

2.2.4 HPLC Analysis of Anthocyanins from Haskap Berries

Samples were extracted with 80 % acidified methanol and prepared for HPLC analysis according to Khattab *et al.* (2015a). The anthocyanin profiles of fresh, frozen, pressed, and dried haskap berries were analyzed according to Khattab *et al.* (2015c) using the reversed-phase DAD-HPLC (Agilent 1100 Series, Agilent Technologies, Hewlett-Packard, Waldbronn, Germany). Chromatograms were acquired at 520 nm and data were analyzed using the Agilent ChemStation software (version A10.02).

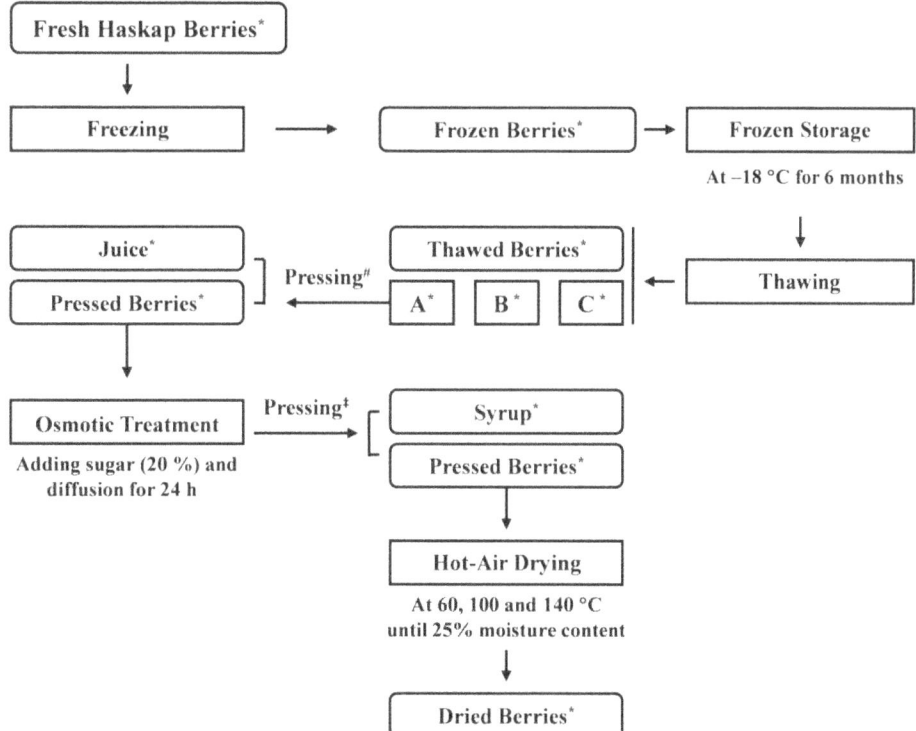

Figure 2. Schematic flowchart of the experimental work. Method of thawing is represented by A: Room temperature; B: Fridge temperature, C: Microwave oven. [#] Pressing until 70 % juice yield; [‡] Pressing to drain the remaining liquid; * HPLC profiling was conducted for these samples

2.2.5 Statistical Analysis

All experiments were done in triplicates and data were analyzed using a one factor analysis of variance (ANOVA). Tukey-Kramer mean separation tests were done for multiple comparisons with SigmaStat software (version 3.5). The significance was accepted at $p \leq 0.05$.

3. Results and Discussion

3.1 HPLC Profiling of Anthocyanins from Fresh Haskap Berries

The chromatogram of haskap berry extract from fresh haskap berries is illustrated in Fig. 3. The contents of the identified ANCs are shown in Table 1.

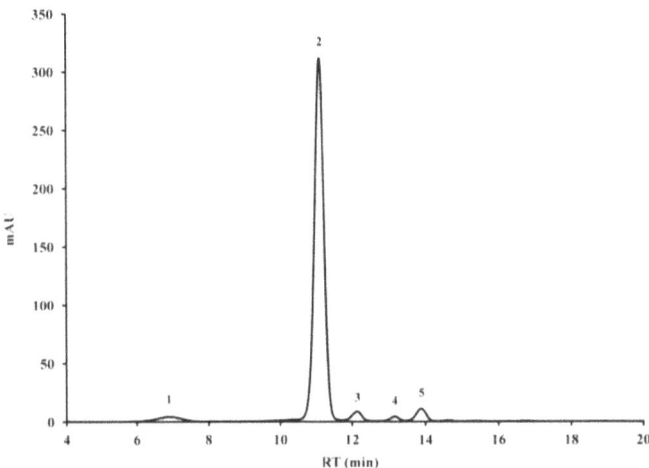

Figure 3. HPLC chromatogram of fresh haskap berries at 520 nm. 1: cyanidin 3,5-di-glucoside; 2: cyanidin 3-glucoside; 3: cyanidin 3-rutinoside; 4: pelargonidin 3-glucoside; 5: peonidin 3-O-glucoside

The TAC of the fresh fruit was 7.26 ± 0.24 mg/g FW. Five ANCs were identified in the whole fruit including cyanidin 3,5-di-glucoside (4.27 % of the TAC), cyanidin 3-glucoside or C-3-G (89.39 %), cyanidin 3-rutinoside (2.07 %), pelargonidin 3-glucoside (0.83 %), and peonidin 3-O-glucoside (3.44 %). The chemical structures of these ANCs are shown in Fig 4. These results are in agreement with those reported by Chaovanalikit *et al.* (2004) and Andersen & Jordheim (2007) where C-3-G was reported to dominate in berries of most *Lonicera* species. The distribution of ANCs in haskap berries was reported to be C-3-G (79–88 %), cyanidin-3-rutinoside (1–11 %), cyanidin-3,5-diglucoside (2.2–6.4 %), peonidin 3-O-glucoside (2.8–4.5 %), peonidin 3-rutinoside (0.3–1.3 %) and pelargonidin 3-glucoside (0.2–1.0 %) (Chaovanalikit *et al.*, 2004).

3.2 Effect of Freezing and Frozen Storage on the Anthocyanin Profile of Haskap Berries

The results of the present study (Table 1) indicate that the freezing process had no significant effect on the content of individual ANCs of haskap berries. The total content of all ANCs was reduced by only 1.79 % as compared to the fresh fruit. In the food industry, a storage temperature of -18 °C effectively reduces the chemical and biological spoilage of foods and extends their shelf life. However, freezing causes cell rupture and division allowing reactions between enzymes and their substrates (Tomás-Barberán & Espín, 2001).

Table 1. Effect of freezing and frozen storage (-18 °C for six months) on the anthocyanin profile of haskap berries

	Fresh	Frozen	Frozen stored	Reduction upon frozen storage (%)
ANC$_1$	0.31 ± 0.00^a	0.30 ± 0.00^a	0.25 ± 0.01^b	19.35
ANC$_2$	6.49 ± 0.23^a	6.37 ± 0.20^a	4.90 ± 0.08^b	24.50
ANC$_3$	0.15 ± 0.00^a	0.15 ± 0.00^a	0.12 ± 0.00^b	20.00
ANC$_4$	0.06 ± 0.00^a	0.07 ± 0.00^a	0.05 ± 0.00^b	16.67
ANC$_5$	0.25 ± 0.01^a	0.24 ± 0.00^a	0.21 ± 0.01^b	16.00

ANC$_1$: Cyanidin 3,5-di-glucoside; ANC$_2$: Cyanidin 3-glucoside; ANC$_3$: Cyanidin 3-rutinoside; ANC$_4$: Pelargonidin 3-glucoside; ANC$_5$: Peonidin 3-O-glucoside. Values are means of duplicate analyses $\pm$ standard deviation (SD). Values in the same row with similar superscript letters are not significantly different ($\rho \leq 0.05$).

Therefore, ANCs may be degraded during freezing and more extensively during thawing due to their interaction with oxidative enzymes. The effects of freezing and frozen storage on anthocyanin content and phenolic profile of different kinds of berries have been investigated by other researchers (Bushway et al., 1992; de Ancos et al., 2000; Häkkinen et al., 2000; Mullen et al., 2002). According to Selman (1992), the process of freezing itself does not alter the nutritive value of the product being frozen. Upon freezing raspberries at –30 °C within 3 hours of picking, Mullen et al. (2002) found no significant differences either in the levels of the individual ANCs or in the TAC of the fresh and frozen raspberries. This might be because ANCs in frozen fruits become more easily extractable due to degradation of cell structures during frozen storage over time. The enhanced extractability might have surmounted any ANCs degradation that might have occurred during the freezing process. In some cases, ANCs have been even reported to increase during freezing (de Ancos et al., 2000).

The effect of frozen storage at –18 °C for 6 months on the individual ANCs from haskap berries is shown in Table 1. The content of the five ANCs significantly decreased upon storage. The highest reduction (24.50 %) was recorded for C-3-G (the most abundant ANC), while the smallest decrease was that of peonidin 3-O-glucoside (16.00 %). This agrees with de Ancos et al. (2000) who found that C-3-G demonstrated a more significant degradation during frozen storage compared to the other ANCs found in raspberries. Structural analysis showed that less free hydroxyl groups and more methoxy groups in the B-ring of the aglycon improve anthocyanin stability (Liu et al., 2014).

Our results showed that cyanidins (C-3-G, cyanidin 3,5-diglucoside, and cyanidin 3-rutinoside) with two OH groups in the B-ring exhibited lower stability than pelargonidin 3-glucoside with one OH group. Peonidin 3-O-glucoside, with one hydroxyl group (OH) and one methoxy group (OCH_3) showed the highest stability among the five ANCs investigated. These results agree with other investigators. Chaovanalikit and Wrolstad (2004) reported that frozen storage at –23 °C for 6 months caused more than 75 % degradation in the ANC content of cherries, while storage at –70 °C resulted in better stability. The effect of frozen storage at –25 °C on the stability of individual ANCs from pomegranate juice was investigated by Mirsaeedghazi et al. (2014). The 5 major ANCs; C-3-G, cyanidin 3,5-diglucoside, delphinidin 3-glucoside, pelargonidin 3-glucoside and pelargonidin 3,5-diglucoside were decreased by 4.8, 3.5, 4.6, 6.0 and 3.4 %, respectively after 20 days of storage.

3.3 Effect of Thawing Conditions on the Anthocyanin Profile of Haskap Berries

Freezing techniques affect how the food thaws and its subsequent structural and compositional changes. Quick freezing retains cell integrity than slower freezing due to the smaller intracellular ice crystals formed. Cell integrity is further influenced by thawing regimes as quick thawing better retains fruit quality (Delgado & Rubiolo, 2005). The freezing and thawing chain ruptures the cells allowing reactions between enzymes and their substrates. Anthocyanins, therefore, may degrade during thawing due to their interaction with oxidative enzymes like polyphenol oxidases and peroxidases that have been reported to be active even at lower temperatures (Chisari et al., 2007).

In the present study, we compared the effect of thawing methods (room temperature, refrigerator, and microwave oven) on the anthocyanin profile of frozen-stored haskap berries. The content of individual ANCs of haskap berries as affected by thawing methods is shown in Table 2. All thawing methods reduced the ANCs of haskap berries with significant differences among them. In agreement with the present study, the quality of frozen food has been reported as being more affected by the thawing process than by the freezing itself (Kim et al., 2011). Thawing has a major impact on the food quality, as the compounds normally kept apart in the intact cell can mix and react with each other (Kmiecik et al. 1995).

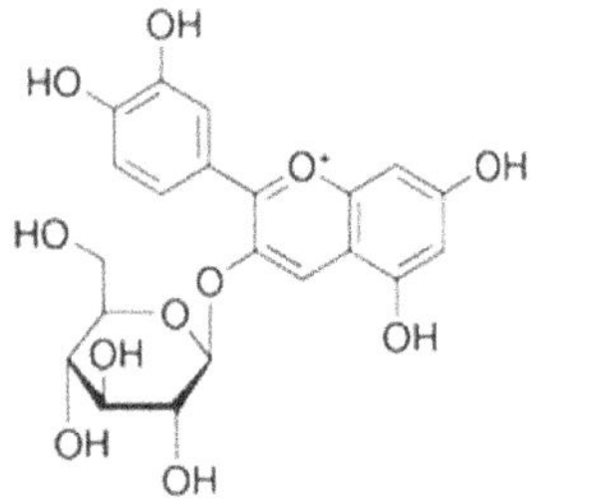

Cyanidin-3-glucoside ($C_{21}H_{21}O_{11}$)
(PubChem CID: 12303203)

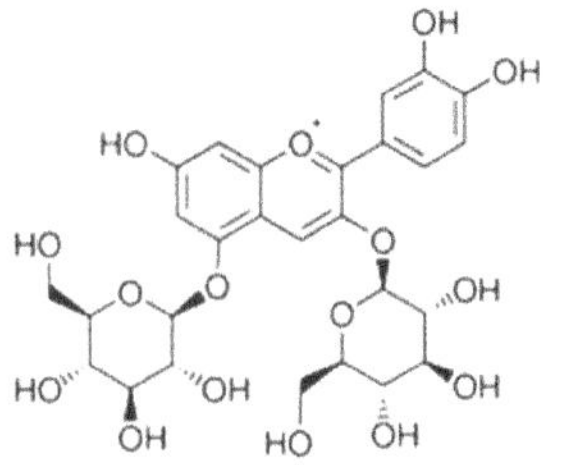

Cyanidin 3,5-di-glucoside ($C_{27}H_{31}O_{16}$)
(PubChem CID: 441688)

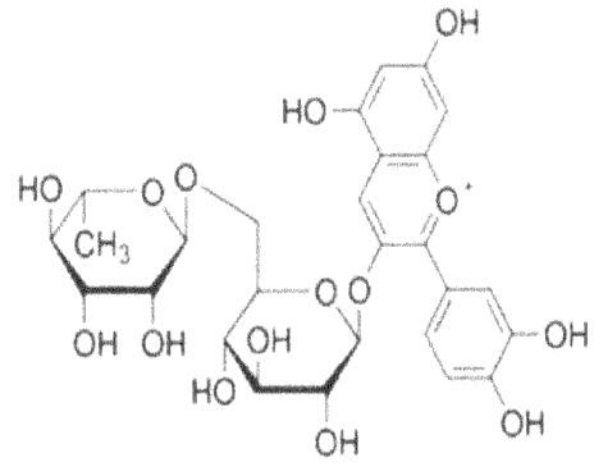

Cyanidin-3-rutinoside ($C_{27}H_{31}O_{15}$)
(PubChem CID: 44256716)

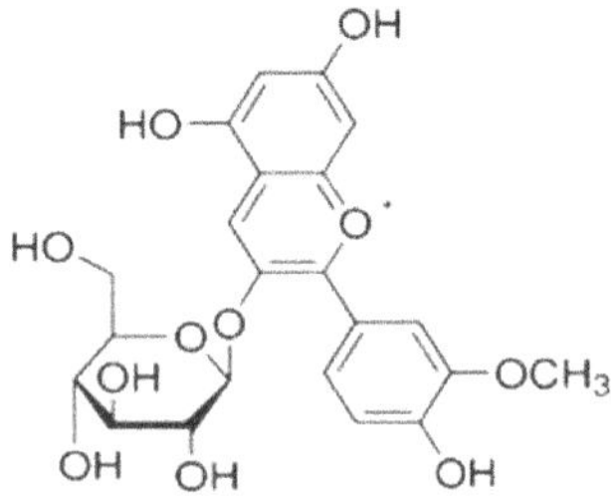

Peonidin-3-O-glucoside ($C_{22}H_{23}O_{11}$)
(PubChem CID: 443654)

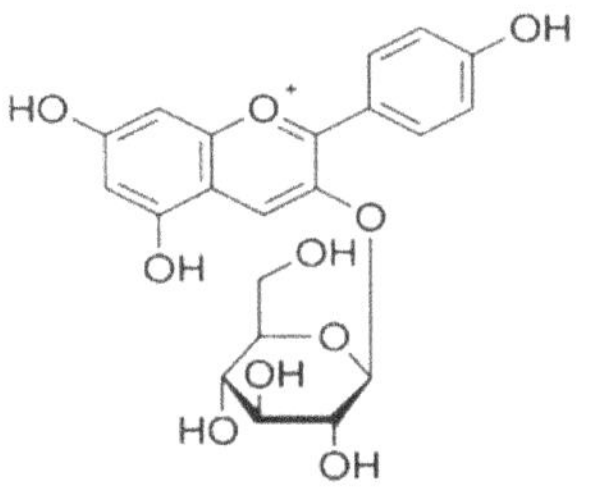

Pelargonidin-3-glucoside ($C_{21}H_{21}O_{10}$)
(PubChem CID: 443648)

Figure 4. Chemical structures of haskap berry ANCs identified in this study

The reductions in Table 2 may be attributed to the hydrolytic reactions that convert anthocyanin glycosides to chalcones, which intuitively degrade into aldehydes and phenolic acids (Kamiloglu et al., 2015). It is also possible that enzymes might have played a role in the reduction of ANCs (Howard et al., 2010). The highest reduction occurred when thawing at room temperature followed by that in the refrigerator, while microwave thawing caused the least reduction. The higher ANCs retention during the microwave thawing might be attributed to the shorter time taken (17 min) as compared to 12 and 22 h in the room temperature and refrigerator thawing, respectively. Using microwave, thawing time was reduced by seven times compared to convective thawing at atmospheric temperature when appropriate conditions were used (Tong et al., 1993). Thawing at lower temperature (refrigerator), despite taking significantly longer time, retained more ANCs compared to room temperature thawing.

These results agree with Oszmiański et al. (2009) where considerable ANC losses were reported after thawing strawberries stored frozen for several months. Moreover, the ANC contents of frozen fruits were found to depend on their thawing methods. The differences of ANC contents between bilberries thawed at 2-4°C and fruit thawed at room temperature (18-20°C) were approximately 10% (Kmiecik et al., 1995).

Table 2. Effect of thawing conditions on the content of individual anthocyanins (mg/g) in haskap berries

	Frozen-Stored berries	Thawed berries		
		Room temperature	Refrigerator temperature	Microwave oven
ANC^1	0.25±0.01^a	0.19±0.00^b	0.21±0.00^b	0.24±0.00^a
ANC$_2$	4.90±0.08^a	3.58±0.00^d	4.08±0.00^c	4.31±0.29^b
ANC$_3$	0.12±0.00^a	0.09±0.01^a	0.10±0.01^a	0.11±0.04^a
ANC$_4$	0.05±0.00^a	0.04±0.01^a	0.05±0.01^a	0.05±0.08^a
ANC$_5$	0.21±0.01^a	0.17±0.01^b	0.19±0.01^a	0.21±0.08^a

ANC$_1$: Cyanidin 3,5-di-glucoside; ANC$_2$: Cyanidin 3-glucoside; ANC$_3$: Cyanidin 3-rutinoside; ANC$_4$: Pelargonidin 3-glucoside; ANC$_5$: Peonidin 3-O-glucoside. Values are means of duplicate analysis ± standard deviation (SD). Values in the same row with similar superscript letters are not significantly different ($\rho \leq 0.05$).

Our results showed that the reductions in individual ANCs were 19.05-26.94, 0.00-16.73 and 0.00-12.04 % after thawing at room, and using the refrigerator and microwave oven, respectively. A similar reduction trend was seen as that observed for berries in frozen storage where C-3-G showed the highest reduction while the lowest reductions were noticed in peonidin 3-O-glucoside and pelargonidin 3-glucoside. This is supported by other studies that report that C-3-G is one of the most reactive ANCs during processing (Rommel *et al.*, 1990; Boyles *et al.*, 1993; Garcia-Viguera *et al.*, 1998). Furthermore, Fleschhut *et al.* (2006) reported that an increase in hydroxyl groups in the B-ring of the anthocyanin nucleus results in reduced stability. They found that cyanins seemed to be less stable than petunins and peonidins indicating that methylation of hydroxyl groups in B-ring increases the stability of ANCs.

3.4 Effect of Juice Extraction on the Anthocyanin Profile of Haskap Berries

The results of the present study (Table 3) indicate that both juice and syrup showed significantly lower ANC content (1.61 and 1.00 mg/g FW, respectively) than that of the thawed berries (4.07 mg/g FW). The syrup showed significantly lower content than that found for the juice fraction, which might be attributed to high extractability of these water soluble compounds. These results are not surprising as it is known that ANCs suffer significant degradation and structural changes during processing (Sadilova *et al.*, 2006). Extraction of fruit juice causes major ANC losses yielding significantly lower ANC contents in the obtained juice as compared to the corresponding fruit used.

Table 3. Effect of juice extraction (until 70 % juice yield) on the content of individual anthocyanins (mg/g) in haskap berry products

	Thawed berries[*]	Juice	Syrup	Pressed berries	
				FW	DW
ANC$_1$	0.19±0.00[a]	0.07±0.00[b]	0.04±0.01[c]	0.19±0.01[a]	0.66±0.01
ANC$_2$	3.58 ±0.00[a]	1.43±0.20[b]	0.89±0.08[c]	3.97±0.08[a]	13.49±0.18
ANC$_3$	0.09±0.01[a]	0.04±0.00[b]	0.02±0.00[b]	0.12±0.00[a]	0.37±0.00
ANC$_4$	0.04±0.01[a]	ND	ND	ND	ND
ANC$_5$	0.17±0.01[a]	0.07±0.00[b]	0.05±0.01[b]	0.19±0.01[a]	0.65±0.01

ANC$_1$: Cyanidin 3,5-di-glucoside; ANC$_2$: Cyanidin 3-glucoside; ANC$_3$: Cyanidin 3-rutinoside; ANC$_4$: Pelargonidin 3-glucoside; ANC$_5$: Peonidin 3-O-glucoside; FW: fresh weight; DW: dry weight. Values are means of duplicate analysis ± standard deviation (SD). [*] Thawed at room temperature. [**] The content of the five ANCs of the thawed berries were 1.21, 25.91, 0.71, 0.28 and 1.08, respectively based on the DW base. Values in the same row with similar superscript letters are not significantly different (ρ≤0.05).

The content of ANCs in the juices affects their storage stability and shelf life, for example, only 11–15% of the original ANCs were detected in the commercial juices at their expiry date, after storage for 35–49 weeks at room temperature (Hellstrom *et al.*, 2013). It is known that manufacturing processes lead to anthocyanin degradation and color alteration in berries (Hager *et al.*, 2008). Hellstrom *et al.* (2013) attributed the lower anthocyanin content in commercial berry juices to the severe production processes applied industrially. Furthermore, processing blueberries into purees caused 43 % loss in total ANCs, compared to the levels in fresh fruit (Brownmiller *et al.*, 2008).

In other studies, it was found that the stability of individual ANCs in food systems depended greatly on their chemical structure (Jackman *et al.*, 1987) and different ANCs had different degradation kinetics in juices (Hellstrom *et al.*, 2013). Hydroxyl, methoxyl, sugar, and acylated sugar substituent groups have pronounced effects on the stability of the ACNs. Diglycosidic substitution gives more stability than monoglycosidic substitution (Mazza & Miniati 1993). Moreover, acylation of the ANC molecule improves its stability by preventing it from hydration (Brouillard, 1981). Pelargonidin 3-glucoside totally degraded and was not detected in the juice, syrup or pressed berries. This is might be due to its marginal content in the initial thawed berries. For the other ANCs, the same reduction trend was observed as seen during the storage and thawing where C-3-G suffered the highest reduction (47.94 %) followed by cyanidin 3-rutinoside (47.89 %) and cyanidin 3,5-di-glucoside (45.45 %). The least reduction, however, was observed for peonidin 3-O-glucoside (39.81 %).

3.5. Effect of Drying Conditions on the Anthocyanin Profile of Haskap Berries

The effect of hot-air drying at different temperatures is shown in Table 4. The drying time taken to reach a moisture content of 25 % was 16.0, 5.6 and 2.5 h at 60, 100 and 140 °C, respectively. Upon drying to this moisture content, the reductions in the individual ANCs of the dried berries were 73.85-79.99, 78.46-82.73 and 90.77-100.00 % at the three drying temperatures, respectively. The HPLC chromatograms of pressed and dried haskap berries at different temperatures are illustrated in Fig. 5. The TAC decreased by 71.85 and 88.30 % after 8 h of drying at 60 and 100 °C, respectively. Even at lower temperature (60 °C), the degradation of ANCs continued with drying time to reach more than 95.00 % after 48 h where all ANCs were significantly degraded with pelargonidin 3-glucoside, cyanidin 3,5-di-glucoside and C-3-G being the most affected. However, ANCs were completely degraded after 32 h of drying at 100 °C.

Table 4. Anthocyanin profile of haskap berries dried at 60, 100 and 140 °C to 25 % moisture content

	Pressed berries	Dried berries		
		60 °C	100 °C	140 °C
ANC_1	0.63 ± 0.01^a	0.15 ± 0.00^b	0.12 ± 0.00^c	ND
ANC_2	13.49 ± 0.18^a	2.70 ± 0.06^b	2.33 ± 0.09^c	0.62 ± 0.03^d
ANC_3	0.40 ± 0.00^a	0.09 ± 0.01^b	0.07 ± 0.01^b	0.02 ± 0.04^c
ANC_4	ND	ND	ND	ND
ANC_5	0.65 ± 0.01^a	0.17 ± 0.01^b	0.14 ± 0.01^b	0.06 ± 0.08^c

ANC_1: Cyanidin 3,5-di-glucoside; ANC_2: Cyanidin 3-glucoside; ANC_3: Cyanidin 3-rutinoside; ANC_4: Pelargonidin 3-glucoside; ANC_5: Peonidin 3-O-glucoside; ND: not detected; Values are means of duplicate analysis ± standard deviation (SD). Values in the same row with similar superscript letters are not significantly different ($\rho \leq 0.05$).

Logarithmic anthocyanin degradation with an arithmetic increase in temperature has been frequently reported (Drdak & Daucik, 1999; Rhim, 2002). The high temperatures blanching (95 °C for 3 min in combination with pasteurisation) involved in processing blueberries into purees resulted in 43 % loss in total ANCs (Brownmiller et al., 2008). In addition, ANCs were significantly decreased as a result of jam and marmalade processing of black carrots (Kamiloglu et al., 2015). After 20 weeks of jam and marmalade storage, ANCs were significantly higher at 4 °C that at 25 °C.

Drying at 140 °C (Table 4), however, had a significant effect on the ANCs. Both pelargonidin 3-glucoside and cyanidin 3,5-diglucoside were completely degraded upon drying to 25% moisture content (2.5 h). The other ANCs were reduced by 90.77 to 95.40 % with C-3-G and peonidin 3-O-glucoside being the highest and least degraded ones, respectively. After 5 h of drying at this temperature, only C-3-G was detected in the samples but no other ANCs. These results agree with Drdak and Daucik (1999) and Rhim (2002). Garcia-Viguera et al. (1999) reported anthocyanin losses of 10 to 80 % during jam processing. Moreover, C-3-G and pelargonidin 3-glucoside in blackberry and strawberry puree were significantly reduced by thermal treatment at 70 °C for 2 min (Patras et al., 2009). Furthermore, the content of total ANCs of dehydrated potato flakes decreased by 23-45 % during the dehydration process at 100-150 °C (Nayak, 2011).

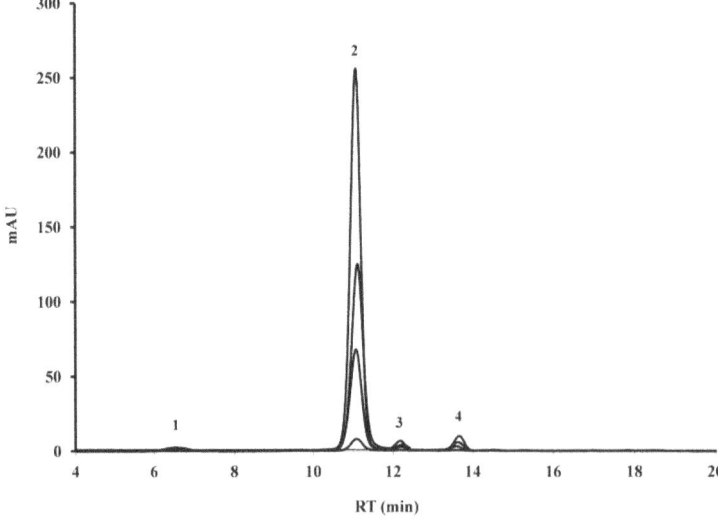

Figure 5. HPLC chromatograms of pressed and dried haskap berries (at different drying temperatures until 25 % moisture content). Chromatograms from top to bottom: pressed berries; dried berries at 60 °C; dried berries at 100 °C; dried berries at 140 °C. 1: cyanidin 3,5-di-glucoside; 2: cyanidin 3-glucoside; 3: cyanidin 3-rutinoside; 4: peonidin 3-O-glucoside

The reduction of ANCs upon drying might be attributed to the heat labile nature of ANCs. The exposure to oxygen at higher temperatures might have also contributed to their degradation (Welcha *et al.*, 2008). Sadilova *et al.* (2006) observed that only 50 % of ANCs were retained after heating elderberry for 3 h at 95 °C. Similar losses in raspberry purees were reported by Ochoa *et al.* (1999). This reduction might also be attributed to the osmotic treatment during which more ANCs might have leached out into the osmotic solution. Lohachoompol *et al.* (2004) found that the reduction in the TAC was 41 % in the dried blueberries and increased to 49 % when drying was preceded with osmotic treatment. The loss of ANCs was further attributed to several factors, including residual enzyme activity or condensation reactions of ANCs with other phenolics at higher temperatures (Jackman *et al.*, 1987; Brownmiller *et al.*, 2008). Fracassetti *et al.* (2013) found that storage of freeze-dried wild blueberry powder for 49 days at 25, 42, 60, and 80 °C reduced single and total ANCs at all temperatures. The reduction in ANCs depended on the temperature and occurred slowly up to 3% at day 14 at 25 and 42 °C, whereas it was faster, reaching 60 and 85 % after three days at 60 and 80 °C, respectively.

The stability of ANCs is influenced by the aglycon B-ring substituents and the presence of additional hydroxyl groups decreases the aglycon stability in neutral media. Furthermore, mono- and diglycosides derivatives are more stable than the non-glycosylated aglycons (Castañeda-Ovando *et al.*, 2009). Our results showed that among the three cyanidins investigated in this study, cyanidin 3,5-diglucoside (with two sugar moieties) showed the highest stability followed by cyanidin 3-rutinoside and C-3-G.

The reductions in individual ANCs were 73.85-76.19, 78.46-80.95 and 90.77-95.40 % upon drying to 25 % moisture content at 60, 100 and 140 °C, respectively. This excludes pelargonidin 3-glucoside (the least abundant anthocyanin in haskap berries) which was completely degraded at both 100 and 140 °C. Cyanidin 3,5-diglucoside also disappeared in the samples dried at 140 °C. Peonidin 3-*O*-glucoside, however, was the most stable anthocyanin at different temperatures. Drying haskap berries at 60 °C is recommended for better retention of ANCs. Drying at this temperature was recommended by Garba and Kaur (2014) who investigated the influence of hot air drying (40-60 °C) on the TAC of black carrot and found that the optimum retention of ANCs was attained from drying at 60 °C. In the study by Zoric *et al.* (2014), the effect of heating temperature (80-120 °C) and processing time (5-50 min) on the stability of ANCs in freeze-dried sour cherry pastes was explored. They found that C-3-G was the most unstable among ANCs.

4. Conclusion

ANCs were significantly affected by frozen storage and subsequent thawing by different methods. Microwave thawing revealed the highest ANC retention. Drying significantly reduced ANCs and higher drying temperatures resulted in higher degradation. All ANCs were significantly reduced during the frozen storage, thawing, juice extraction and drying. Cyanidins were more degradable than pelargonidin than peonidin. Microwave thawing and lower-temperature storage and processing are recommended for better retention of haskap berry ANCs. Understanding the structure-stability relations and behavior of ANCs during storage and processing will help haskap and other berry processors to design high-quality berry products with improved nutritional/functional properties.

Acknowledgements

This work was funded by the Natural Sciences and Engineering Research Council of Canada (NSERC). LaHave Natural Farms, Blockhouse, Nova Scotia, Canada is sincerely acknowledged for providing the haskap berries used in this study.

References

Andersen, M., & Jordheim, M. (2007). *The Anthocyanins. Chemistry, Biochemistry and Applications*, 4th ed., *CRC Press*: Boca Raton, FL, USA, 471-553.

Bakowska, A., Marianchuk, M., & Kolodziejczyk, P. (2007). Survey of bioactive components in Western Canadian berries. *Canadian Journal of Physiology and Pharmacology, 85*, 1139-1152.

Bors, B., Thomson, J., Sawchuk, E., Reimer, P., Sawatzky, R., & Sander, T. (2012). Haskap breeding and production-final report (pp. 1–142). Saskatchewan Agriculture: Regina.

Boyles, M. J., & Wrolstad, R. E. (1993). Anthocyanin composition of red raspberry juice: influence of cultivar, processing, and environmental factors. *Journal of Food Science, 58*, 1135-1141. http://dx.doi.org/10.1111/j.1365-2621.1993.tb06132.x

Brouillard, R. (1981). Origin of the exceptional color stability of the Zebrina anthocyanin. *Phytochemistry, 20,*

143-145. http://dx.doi.org/10.1016/0031-9422(81)85234-X

Brownmiller, C., Howard, L. R., Prior, R. L. (2008). Processing and storage effects on monomeric anthocyanins, percent polymeric colour, and antioxidant capacity of processed blueberry products. *Journal of Food Science*, H72-H79. http://dx.doi.org/10.1111/j.1750-3841.2008.00761.x

Bueno, J. M., Saez-Plaza, P., Ramos-Escudero, F., Jimenez, A. M., Fett, R. & Asuero, A. G. (2012). Analysis and antioxidant capacity of anthocyanin pigments. Part II: chemical structure, color, and intake of anthocyanins. *Critical Reviews in Analytical Chemistry, 42*, 126-151. http://dx.doi.org/10.1080/10408347.2011.632314

Bushway, A. A., Bushway, R. J., True, R. H., Work, T. M., Bergeron, D., Handley, D. T., & Perkins, L. B. (1992). Comparison of the physical, chemical and sensory characteristics of five rapsberry cultivars evaluated fresh and frozen. *Fruit Varieties Journal, 46*, 229-234.

Castañeda-Ovando, A., Pacheco-Hernández, M. L., Páez-Hernández, M. E., Rodríguez, J. A., & Galán-Vidal, C. (2009). Chemical studies of anthocyanins: a review. *Food Chemistry, 113*, 859-871. http://dx.doi.org/10.1016/j.foodchem.2008.09.001

Celli, G. B., Ghanem, A., & Brooks, M. S. (2014). Haskap Berries (*Lonicera caerulea* L.) - a Critical Review of Antioxidant Capacity and Health-Related Studies for Potential Value-Added Products. *Food and Bioprocess Technology, 7*, 1541-1554. http://dx.doi.org/10.1007/s11947-014-1301-2

Chaovanalikit, A, Thompson M. M., & Wrolstad, R. E (2004). Characterization and quantification of anthocyanins and polyphenolics in blue honeysuckle (*Lonicera caerulea* L.). *Journal of Agricultural & Food Chemistry, 52*, 848-852. http://dx.doi.org/10.1021/jf030509o

Chisari, M., Barbagallo, R. N., & Spagna, G. (2007). Characterisation of polyphenol oxidase and peroxidase and influence on browning of cold stored strawberry fruit. *Journal of Agricultural and Food Chemistry, 55*, 3469-3476. http://dx.doi.org/10.1021/jf063402k

De Ancos, B., Ibanez, E., Reglero, G., & Cano, M. P. (2000). Frozen storage effects on anthocyanins and volatile compounds of raspberry fruit. *Journal of Agricultural and Food Chemistry, 48*, 873-879. http://dx.doi.org/10.1021/jf990747c

Delgado, A. E., & Rubiolo, A. C. (2005). Microstructural changes in strawberry after freezing and thawing processes. *Lebensmittelwissenschaft und -Technologie, 38*, 135-142. http://dx.doi.org/10.1016/j.lwt.2004.04.015

Drdak, M., & Daucik, P. (1999). Changes of elderberry (*Sambucus nigra*) pigments during the production of pigment concentrates. *Acta Alimentaria, 19*, 3-7.

Fan, Z., Wang, Z., & Liu, J. (2011). Cold-field fruit extracts exert different antioxidant and antiproliferative activities in vitro. *Food Chemistry, 129*, 402-407. http://dx.doi.org/10.1016/j.foodchem.2011.04.091

Fleschhut, J., Kratzer, F., Rechkemmer, G., & Kulling, S. (2006). Stability and biotransformation of various dietary anthocyanins *in vitro*. *European Journal of Nutrition, 45*, 7-18. http://dx.doi.org/10.1007/s00394-005-0557-8

Fracassetti, D., Del Bo, C., Simonetti, P., Gardana, C., Klimis-Zacas, D., & Ciappellano, S. (2013). Effect of Time and Storage Temperature on Anthocyanin Decay and Antioxidant Activity in Wild Blueberry (*Vaccinium angustifolium*) Powder. *Journal of Agricultural and Food Chemistry, 61*, 2999-3005. http://dx.doi.org/10.1021/jf3048884

Garba, U., & Kaur, S. (2014). Effect of drying and pretreatment on anthocyanins, flavenoids and ascorbic acid content of black carrot (*Daucus carrota* L.). *Journal of Global Biosciences, 3*, 772-777.

Garcia-Viguera, C., Zafrilla, P., Artés, F., Romero, F., Abellán, P., & Tomas-Barberan, F. A. (1998). Colour and anthocyanin stability of red raspberry jam. *Journal of the Science of Food and Agriculture, 78*, 565-573. http://dx.doi.org/10.1002/(SICI)1097-0010(199812)78:4<565::AID-JSFA154>3.0.CO;2-P

Garcia-Viguera, C., Zafrilla, P., Romero, P., Abellan, P., Artes, F., & Tomas-Barberan, F. A. (1999). Color stability of strawberry jam as affected by cultivar and storage temperature. *Journal of Food Science, 64*, 243-247. http://dx.doi.org/10.1111/j.1365-2621.1999.tb15874.x

Hager, T. J., Howard, L. R., & Prior, R. L. (2008). Processing and storage effects on monomeric ANCs, percent polymeric color, and antioxidant capacity of processed blackberry products. *Journal of Agricultural and Food Chemistry, 56*, 689-695. http://dx.doi.org/10.1021/jf071994g

Häkkinen, S. H., Kärenlampi, S. O., Mykkänen, H. M., & Törrönen, A. R. (2000). Influence of domestic processing and storage on flavonol contents in berries. *Journal of Agricultural and Food Chemistry, 48*, 2960-2965. http://dx.doi.org/10.1021/jf991274c

Harborne, J. B. (1998). Phenolic Compounds in Phytochemical Methods-a Guide to Modern Techniques of Plant Analysis, 3rd ed., *Chapman & Hall*, New York, pp. 66-74.

Hellstrom, J., Mattila, P., & Karjalainen, R. (2013). Stability of anthocyanins in berry juices stored at different temperatures. *Journal of Food Composition and Analysis, 31*, 12-19. http://dx.doi.org/10.1016/j.jfca.2013.02.010

Howard, L. R., Castrodale, C., Brownmiller, C., & Mauromoustakos, A. (2010). Jam processing and storage effects on blueberry polyphenolics and antioxidant capacity. *Journal of Agricultural and Food Chemistry, 58*, 4022-4029. http://dx.doi.org/10.1021/jf902850h

International Organization of Standardization (ISO) (2009). Cereals and cereal products - determination of moisture content - reference method. ISO 712.

Jackman, R. L., Yada, R. Y., Tung, M. A., & Speers, R.A. (1987). Anthocyanins as food colorants-a review. *Journal of Food Biochemistry, 11*, 201-247. http://dx.doi.org/10.1111/j.1745-4514.1987.tb00123.x

Jaganath I. B. & Crozier A. (2010). Dietary flavonoids and phenolic compounds. In Plant Phenolics and Human Health: Biochemistry, Nutrition, and Pharmacology (edited by Cesar G. Fraga). *John Wiley & Sons, Inc.*, Hoboken, New Jersey.

Jing, P. (2006). Purple corn anthocyanins: chemical structure, chemopreventive activity and structure/function relationships. PhD thesis; The Ohio State University, U.S.A., pp. 5-90.

Jurikova, T., Rop, O., Mlcek, J., Sochor, J., Balla, S., Szekeres, L., Hegedusova, A., Hubalek, J., Adam, V., & Kizek, R. (2012). Phenolic Profile of Edible Honeysuckle Berries (*Genus Lonicera*) and Their Biological Effects-a review. *Molecules, 17*, 61-79. http://dx.doi.org/10.3390/molecules17010061

Kamiloglu, S., Pasli, A. A., Ozcelik, B., Camp, J. V., & Capanoglu, E. (2015). Colour retention, anthocyanin stability and antioxidant capacity in black carrot (*Daucus carota*) jams and marmalades: Effect of processing, storage conditions and in vitro gastrointestinal digestion. *Journal of Functional Foods, 13*, 1-10. http://dx.doi.org/10.1016/j.jff.2014.12.021

Khattab, R., Celli, G. B., Ghanem, A., & Brooks, M. S. (2015a) Effect of frozen storage on polyphenol content and antioxidant activity of haskap berries (*Lonicera caerulea* L.). *Journal of Berry Research, 5*, 231-242. http://dx.doi.org/10.3233/JBR-150105

Khattab, R., Celli, G. B., Ghanem, A., & Brooks, M. S. (2015b). Effect of thawing conditions on polyphenol content and antioxidant activity of frozen haskap berries (*Lonicera caerulea* L.). *Current Nutrition & Food Science, 11*, 223-230. http://dx.doi.org/10.2174/1573401311666150519235845

Khattab, R., Brooks, M. S., & Ghanem, A. (2015c). Phenolic analyses of haskap berries (*Lonicera caerulea* L.): spectrophotometry versus high performance liquid chromatography. *International Journal of Food Properties, 19*, 1708-1725. http://dx.doi.org/10.1080/10942912.2015.1084316

Khattab, R., Ghanem, A., & Brooks, M. S. (2016a). Quality of Dried Haskap Berries (*Lonicera caerulea* L.) as Affected by Prior Juice Extraction, Osmotic Treatment and Drying Conditions. *Drying Technology*, http://dx.doi.org/10.1080/07373937.2016.1175472

Khattab, R., Ghanem, A., & Brooks, M. S. (2016b). Effect of Juice Extraction Methods on the Physicochemical Characteristics of Haskap Berry (*Lonicera caerulea* L.) Products. *Current Nutrition & Food Science, 12*, 220-229. http://dx.doi.org/10.2174/1573401312666160608122248

Kim, T. H., Choi, J. H., Choi, Y. S., Kim, H. Y., Kim, S. Y., Kim, H. W., & Kim, C. J. (2011). Physicochemical properties of thawed chicken breast as affected by microwave power levels. *Food Science and Biotechnology, 20*, 971-977. http://dx.doi.org/10.1007/s10068-011-0134-2

Kmiecik, W., Jaworska, G., & Budnik A. (1995). Influence of thawing methods of frozen berries fruits on their quality. *Rocz. PZH* 46, 2, 135-143 [in Polish].

Liu, Y., Zhang, D., Wu, Y., Wang, D., Wei, Y., Wu, J., & Ji, B. (2014). Stability and absorption of anthocyanins from blueberries subjected to a simulated digestion process. *International Journal of Food Science & Nutrition, 65*, 440-448. http://dx.doi.org/10.3109/09637486.2013.869798

Lohachoompol, V., Srzednicki, G., & Craske, J. (2004). The change of total ANCs in blueberries and their antioxidant effect after drying and freezing. *Journal of Biomedicine & Biotechnology, 5*, 248-252. http://dx.doi.org/10.1155/S1110724304406123

Mazza, G. J. (2007). Anthocyanins and heart health. *Annali dell'Istituto Superiore di Sanità, 43*, 369-374.

Mazza, G., & Miniati, E. (1993). Anthocyanins in Fruits, Vegetables and Grains. Boca Raton, FL: CRC Press, Inc.

Miguel, M. G. (2011). Anthocyanins: Antioxidant and/or anti-inflammatory activities. *Journal of Applied Pharmaceutical Science, 1*, 7-15.

Mirsaeedghazi, H., Emam-Djomeh, Z., & Ahmadkhaniha, R. (2014). The effect of frozen storage on the anthocyanins and polyphenolic components of pomegranate juice. *Journal of Food Science and Technology, 51*, 382-386. http://dx.doi.org/10.1007/s13197-011-0504-z

Mullen, W., Stewart, A. J., Lean, M. E. J., Gardner, P., Duthie, G. G., & Crozier, A. (2002). Effect of freezing and storage on the phenolics, ellagitannins, flavonoids, and antioxidant capacity of red raspberries. *Journal of Agricultural and Food Chemistry, 50*, 5197-5201. http://dx.doi.org/10.1021/jf020141f

Nayak, B. (2011). Effect of Thermal Processing on the Phenolic Antioxidants of Colored Potatoes. Academic dissertation, Department of Biological Systems Engineering, Washington State University.

Nikkhah, E., Khayami, M., & Heidari, R. (2008). In vitro screening for antioxidant activity and cancer suppressive effect on blackberry (*Morus nigra*). *Iranian Journal of Cancer Prevention, 1*, 167-172.

Ochoa, M. R., Kesseler, A. G., Vullioud, M. B., & Lozano, J. E. (1999). Physical and chemical characteristics of raspberry pulp: storage effect on composition and color. *LWT–Food Science and Technology, 149*, 149-153. http://dx.doi.org/10.1006/fstl.1998.0518

Oszmianski, J., Wojdylo, A., & Kolniak, J. (2009). Effect of L-ascorbic acid, sugar, pectin and freeze-thaw treatment on polyphenol content of frozen strawberries. *LWT - Food Science and Technology, 42*, 581-586. http://dx.doi.org/10.1016/j.lwt.2008.07.009

Paredes-Lopez, O., Cervantes-Ceja, M. L., Vigna-Perez, M., & Hernandez-Perez, T. (2010). Berries: improving human health and healthy aging, and promoting quality life-a review. *Plant Foods for Human Nutrition, 65*, 299-308. http://dx.doi.org/10.1007/s11130-010-0177-1

Patras, A., Brunton, N. P., Gormely, T. R., & Butler, F. (2009). Impact of high pressure processing on antioxidant activity, ascorbic acid, ANCs and instrumental colour of blackberry and strawberry puree. *Innovative Food Science and Emerging Technologies, 10*, 308-313. http://dx.doi.org/10.1016/j.ifset.2008.12.004

Raudsepp, P., Anton, D., Roasto, M., Meremäe, K., Pedastsaar, P. & Mäesaar, M. (2013). The antioxidative and antimicrobial properties of the blue honeysuckle (*Lonicera caerulea* L.), Siberian rhubarb (*Rheum rhaponticum* L.) and some other plants, compared to ascorbic acid and sodium nitrite. *Food Control, 31*, 129-135. http://dx.doi.org/10.1016/j.foodcont.2012.10.007

Raudsepp, P., Anton, D., Roasto, M., Meremäe, K., Pedastsaar, P. & Mäesaar, M. (2013). The antioxidative and antimicrobial properties of the blue honeysuckle (*Lonicera caerulea* L.), Siberian rhubarb (*Rheum rhaponticum* L.) and some other plants, compared to ascorbic acid and sodium nitrite. *Food Control, 31*, 129-135. http://dx.doi.org/10.1016/j.foodcont.2012.10.007

Rein, M. (2005). Copigmentation Reactions and Color Stability of Berry Anthocyanins. Academic dissertation, Department of Applied Chemistry and Microbiology, University of Helsinki.

Rhim, J. W. (2002). Kinetics of thermal degradation of anthocyanin pigment solutions driven from red flower cabbage. *Food Science and Biotechnology, 11*, 361-364.

Rommel, A., Heatherbell, D. A., & Wrolstad, R. E. (1990). Red raspberry juice and wine: effect of processing and storage on anthocyanin pigment composition, color and appearance. *Journal of Food Science, 55*, 1011-1017. http://dx.doi.org/10.1111/j.1365-2621.1990.tb01586.x

Rop, O., Řezníček, V., Mlček, J., Juríková, T., Balík, J., & Sochor, J. (2011). Antioxidant and radical oxygen species scavenging activities of 12 cultivars of blue honeysuckle fruit. *Horticultural Science, 38*, 63-70.

Rupasinghe, H. P. V., Yu, L. J., Bhullar, K. S., & Bors, B. (2012). Haskap (Lonicera caerulea): A new berry crop with high antioxidant capacity. *Canadian Journal of Plant Science, 92*, 1311-1317. http://dx.doi.org/10.4141/cjps2012-073

Sadilova, E., Stintzing, F. C., & Carle, R. (2006). Thermal degradation of acylated and nonacylated ANCs. *Journal of Food Science, 71*, C504-C512. http://dx.doi.org/10.1111/j.1750-3841.2006.00148.x

Selman, J. D. (1992). Vitamin retention during blanching of vegetables. Presented at the Royal Society of Chemistry Symposium 'Vitamin Retention in Cooking and Food Processing', 24 November, London, UK: 137-147.

Tomás-Barberán, F. A., & Espín, J. C. (2001). Phenolic compounds and related enzymes as determinants of quality in fruits and vegetables. *Journal of the Science of Food and Agriculture, 81*, 853-876. http://dx.doi.org/10.1002/jsfa.885

Tong, C. H., Lentz, R. R., & Lund, D. B. (1993). A microwave oven with variable continuous power and a feedback temperature controller. *Biotechnology Progress, 9*, 488-496.
http://dx.doi.org/10.1021/bp00023a600

Welcha, C. R., Wub, Q. & James E. Simonb, J. E. (2008). Recent Advances in Anthocyanin Analysis and Characterization. *Current Analytical Chemistry, 4*, 75-101. http://dx.doi.org/10.2174/157341108784587795

Zoric, Z., Dragovic-Uzelac, V., Pedisic, S., Kurtanjek, Z., Garofulic, I. E. (2014). Heat-Induced Degradation of Antioxidants in Marasca Paste. *Food Technology & Biotechnology, 52*, 101-108.

3

Meat Colour Stability and Fatty Acid Profile in Commercial Bison and Beef

Galbraith, J. K.[1], Aalhus, J. L.[2], Juárez, M.[2], Dugan, M. E. R.[2], Larsen, I. L.[2], Aldai, N.[3], Goonewardene, L. A.[4] & Okine, E. K.[4]

[1] Alberta Agriculture and Rural Development, 5712-48 Avenue T4V 0K1, Camrose, Alberta Canada

[2] Agriculture and Agri-Food Canada, 6000 C&E Trail, T4L 1W1, Lacombe, Alberta, Canada

[3] Food Science and Technology, Faculty of Pharmacy, Universidad del País Vasco/Euskal Herriko Unibertsitatea, 01006 Vitoria-Gasteiz, Spain

[4] Department of Agricultural, Food and Nutritional Science, University of Alberta, Edmonton, Alberta, Canada, T6G 2P5, Canada

Correspondence: Juárez, M., Agriculture and Agri-Food Canada, 6000 C&E Trail, T4L 1W1, Lacombe, Alberta, Canada. E-mail: Manuel.Juarez@agr.gc.ca

Abstract

Commercial bison meat has been found to discolour more rapidly than beef in retail display. The influence of fat content, fat composition and vitamin E on the colour stability of commercially produced bison and beef were examined. *Longissimus* samples from grain-fed beef (n = 20) and grass-fed bison (n = 14) were analyzed for fat content, fatty acid composition, vitamin E levels, pigments, TBARS and retail stability. Intramuscular fat content was lower and richer in PUFA in bison ($P < 0.01$) compared to beef. Pigment and TBARS levels in bison were significantly higher ($P < 0.01$), leading to higher ($P < 0.01$) metmyoglobin levels. Regression analysis results showed that differences in total fat content and fatty acid composition were the most responsible factors for early discolouration of commercial bison meat compared to commercial beef. In conclusion, total fat and fatty acid composition can be manipulated to improve the colour stability of bison meat.

Keywords: *Bison bison*, myoglobin, lipid, retail, tocopherol

1. Introduction

Bison (*Bison bison*) are raised in North America for their meat and other products. In 2012, in the Canadian Province of Alberta, over 12,000 bison were slaughtered in inspected abattoirs (Steenbergen, 2013). Bison meat composition has been found to be nutrient dense, with high proportion of protein (Galbraith et al., 2006; Marchello & Driskell, 2001). Bison meat is sold as both frozen and fresh product and can be found increasingly at mainstream grocery chains. Bison meat has been reported to discolour more rapidly than beef (Dhanda et al., 2002; Janz et al., 2000; Pietrasik et al., 2006) and the reason for this has not been determined. Meat colour is important because it is used by consumers as an indicator of freshness and wholesomeness (Mancini & Hunt, 2005). Structurally, beef and bison have identical myoglobin which displays no difference in primary structure, kinetics of oxidation, and thermostability (Joseph et al., 2010). Therefore, the differing rate of discolouration of bison cannot be attributed to differences in the structure and biochemistry of myoglobin. Early browning in bison meat was also not attributable to a difference in microflora, but rather pigment oxidation (Janz et al., 2001). In Canada, commercial beef is usually grain-fed, while a large proportion of Canadian bison is grass-fed. Thus, a combination of inter-species intrinsic differences, production system and dietary factors may be responsible for the different discolouration rates observed between commercially produced beef and bison meat.

Polyunsaturated fatty acids (PUFA) in phospholipid membranes are susceptible to oxidative breakdown resulting in changes to the colour, smell and taste of the meat (Wood et al., 2008). Differences in susceptibility of meat to oxidation have been linked to the heme iron content (Rhee et al., 1996). Heme iron has been proposed as an initiator and promoter of lipid oxidation in raw meats and H2O2-activated metmyoglobin has been seen to promote lipid oxidation in model systems (Decker et al., 2000). High levels of iron have been found in raw bison meat compared to those typically found in beef (Galbraith et al., 2006; Marchello & Driskell, 2001). The relatively rapid deterioration of colour quality of bison muscle compared with beef may be related to the significantly higher content of both total PUFA's (Rule et al., 2002) and total iron. These characteristics may make bison meat more susceptible to a reduction in display life because of oxidation-related changes in appearance. Additionally, feeding vitamin E to steers has been found to increase lipid and oxymyoglobin stability in several muscles (Chan et al., 1996).

The purpose of this study was thus to examine the influence of fatty acid profile, vitamin E levels and pigment on the oxidative and colour stability of fresh commercial grain-fed beef and grass-fed bison in a retail display environment in order to understand the origin of the different rate of discolouration observed in these two types of meat.

2. Method and Methods

2.1 Animals and Slaughter

A total of 20 feedlot British×Continental composite steers were fed a commercial diet containing approximately 8% grass hay and up to 80% steam-rolled barley. Animals were finished to a target backfat depth of 8-9 mm. Fourteen intact male bison from three commercial farms were also slaughtered at the AAFC Lacombe Meat Research Centre abattoir. Bison were grass finished on native grass pasture. All animals were stunned and exsanguinated in accordance with the principles and guidelines established by the Canadian Council on Animal Care (CCAC, 1993).

The average carcass weight of beef was 328 ± 5.44 whereas the average carcass weight of bison was 208 ± 8.83 Kg. Following the overnight chill, at approximately 24 h *post-mortem*, pH was measured as described by Juárez et al. (2011) and carcasses were knife-ribbed at the grade site (between the 12th and 13th rib for beef and between the 11th and 12th rib for bison). After being exposed to atmospheric oxygen for 20 min, carcasses were assessed for grade by certified graders (CFIA, 2003). The left *longissimus* muscle (grade site) was pulled from the carcasses, labelled and trimmed. One steak was removed for subsequent fatty acid and α-tocopherol determination. The remainder of the muscle was labelled, bagged and aged until 7 d *post-mortem* in a cooler at 2°C.

Following the 6 d ageing period, the stored loin muscle was removed from the cooler and two steaks (25 mm thick) were collected closest to the grade site. The first steak was placed into a polystyrene tray with a dri-loc pad, over-wrapped with oxygen permeable film (8000 ml m^{-2} 24 h^{-1} vitafilm choice wrap; Goodyear Canada Inc., Toronto, ON, Canada) and put into a retail display case (Hill Refrigeration of Canada Ltd., Barrie, ON, Canada) at 1°C for retail evaluation at 0, 1, 2 and 3 d under fluorescent room lighting (GE deluxe cool white), supplemented with incandescent lighting directly above the display case (GE clear cool beam 150 W/120 V spaced 91.5 cm apart) to provide an intensity of 1076 lux at the meat surface for 12 h per day. The second steak was cut in half and one half was immediately prepared for determination of thiobarbituric reactive (TBAR) substances (0 d in retail), as described by Nielsen et al. (1997). The remaining half was placed on pre-labelled polystyrene tray with a dri-loc pad, over-wrapped with oxygen permeable film and put into the retail display case for an additional 3 d, before determining final TBAR values.

2.2 Lipid Analyses

Lipid extraction was performed using 2:1 chloroform: methanol and with the same solvent to sample ratio as reported by Folch et al. (1957). Lipids were methylated using 1.5 N methanolic hydrochloric acid as described by Kramer et al. (1997). Fatty acid methyl esters (FAMEs) thus obtained were dissolved and purified using Supelclean LC-Si solid phase extraction tubes (Supelco, Bellefonte, PA, USA). The sample was then analyzed using a Varian CP-3800 GC (Varian Chromatography Systems, Walnut Creek, CA, USA) with Model 1079 injector and a flame ionization detector (25 psi and hydrogen as carrier gas) and a Varian CP-Sil88 – 100 m column. The temperature program used included an initial temperature of 45°C held for 4 min, to 175°C at 13°C / min and held for 27 min, to 215°C at 4°C / min and held for 35 min as outlined by Cruz-Hernandez et al. (2004). Fatty acid concentrations were reported individually or by lipid class as percentage of the total fatty acids identified.

2.3 Pigment Content

Pigment was determined in duplicate using a modified procedure from Trout (1991) and evaluated using a spectrophotometer (Pharmacia Ultraspec 3000 Model 80-2106-20; McKinley Scientific, Sparta, NJ, USA) at wavelengths of 730 nm and 409 nm.

2.4 Tocopherol Levels

Muscle tissue levels of tocopherol were determined using normal phase HPLC with tocopherol acetate as an internal standard as described by Katsanidis and Addis (1999), using am isocratic high performance method and avoiding saponification (in order to protect sensitive homologs) and adapted for fluorescence detection by Hewavitharana et al. (2004).

2.5 Retail Stability

Treatment samples were placed into the retail display case controlling for known temperature gradients within the retail case. On each specific day in retail objective colour measurements (CIE L^*[brightness], a^*[red-green axis], b^*[yellow-blue axis] values; (CIE 1978) were collected, and converted to hue (H_{ab}=arctan[b^*/a^*]) and chroma (C_{ab}=[$a^{*2}+b^{*2}$]$^{0.5}$), in triplicate across the face of each steak using a Minolta CR-300 with Spectra QC-300 Software (Minolta Canada Inc., Mississauga, ON, Canada). Spectral reflectance readings were also collected at the same time to calculate the relative contents of MetMb, Mb and MbO$_2$ as described by Krzywicki (1979). Following the objective colour measurements, steaks were subjectively evaluated for retail appearance, lean colour score, percent surface discolouration, colour of discolouration, amount of marbling and marbling colour by five trained raters using an 8-point hedonic (1=extremely undesirable and 8=extremely desirable), 8-point descriptive (1=white and 8=extremely dark red), 7-point descriptive (1=0% and 7=100% discolouration), 7-point descriptive (1=no browning and 7=black), 6-point descriptive (1=devoid and 6=abundant) and 5-point descriptive (1=white and 5=red) scales, respectively.

2.6 Retail Stability

Differences between species (bison and beef) were determined for retail evaluation data on day 0, 1, 2 and 3 using a repeated measures design with PROC MIXED (SAS 2003). The fixed effects were species, day and their interaction and the experimental unit was the individual animal (species). The model of best fit was determined using the Bayesian Information Criterion (BIC) where the lower BIC indicated a better fit (Wang and Goonewardene 2004).

A comparison of bison to beef for tocopherol, pigment and all fatty acids using PROC MIXED LS means and standard errors for the dependant variable were determined.

PROC STEPWISE was used (SAS 2003) to examine the effect of inherent tissue levels of total fat, omega-3 (n-3), omega-6 (n-6), the n-6:n-3 ratio, saturated fatty acids (SFA), monounsaturated fatty acids (MUFA), PUFA, pigments and vitamin E on the change in the percent of surface discolouration and MetMb content between d 0 and d 3 in retail display. When a factor was significant ($P<0.15$) it was included in the model contributing to the overall model R^2 value.

3. Results and Discussion

3.1 Fatty Acid Composition

No differences were observed for pH values among groups at 24 h (5.71; $P>0.05$; data not shown) Intramuscular fat content (Table 1) was much lower in meat from commercial bison than in meat from commercial beef ($P<0.01$). Fat content in commercial Canadian bison cuts has been reported to range between 0.31 and 1.08 % (Galbraith et al. 2006). On the other hand, Marchello et al. (1998) reported higher values in meat from bison fed concentrate diets (1.6-2.4 %). Total fat content influences the fatty acid composition of meat, as low concentrations of total lipids in muscle lead to a relative increase in the proportion of membrane phospholipids, which have higher proportions of PUFA (Wood et al. 2008).

The percentages of all the individual fatty acids and indices were also different ($P<0.01$) between commercial beef and bison (Table 1). Significant differences were observed for SFA, MUFA, PUFA, n-3, and n-6 fatty acids. Overall, SFA was lower in bison due to lower amounts ($P<0.01$) of 16:0 and 14:0. The long chain SFA 18:0 was significantly higher in bison (18.40 mg / 100 mg fat compared to 12.52 mg / 100 mg fat in beef; $P<0.01$). Furthermore, α-linolenic (18:3n-3) levels were more than 7 times higher in bison than beef (Table 1). It is well known that diet can alter the fatty acid composition in bison (Rule et al. 2002) and beef (Laborde et al. 2002; Nuernberg et al. 2008). However, changes to diet have less of an effect in a ruminant animal compared to a monogastric, due to bio-hydrogenation of dietary fatty acids in the rumen (Scollan et al. 2006).

The fatty acid composition of muscle affects its oxidative stability during retail display. The PUFA's in phospholipids are susceptible to oxidative breakdown at this stage (Wood et al. 2008). PUFA levels were over three times higher in commercial bison than commercial beef (Table 1). In a previous study, *longissimus* muscle samples from bison contained more PUFA than either Hereford or Brahman cattle (26.2, 20.8, and 21.1 mg / 100 mg fat, respectively; Larick et al. 1989). Authors also attributed an increased off-flavour and aftertaste in bison to the increased levels of PUFA. In the present study, commercial beef n-3, n-6 and n-6:n-3 were 0.94 mg / 100 mg fat, 4.21 mg / 100 mg fat and 4.52 whereas commercial bison n-3, n-6 and n-6:n-3 were 6.16 mg / 100 mg fat, 13.25 mg / 100 mg fat and 2.11, respectively (Table 1). Thus, such differences among the different production systems may account for the differences in oxidative stability of the bison meat and beef during retail display.

Table 1. Fatty acid content of *longissimus thoracis* muscle of entire male bison and beef steers

	Beef	Bison	SEM	*P* value
Number of animals	20	14		
Total Fat, %	5.56	1.00	0.39	<0.01
Fatty Acids, mg 100 mg fat^{-1}				
14:0	2.68	1.27	0.17	<0.01
16:0	27.2	18.9	0.70	<0.01
*c*9-16:1	3.92	1.43	0.16	<0.01
17:0	1.12	1.41	0.07	<0.01
18:0	12.5	18.4	0.47	<0.01
Σtrans 18:1	3.50	2.95	0.55	0.03
*c*9-18:1	37.5	30.2	0.94	<0.01
18:2n-6	3.00	9.54	0.80	<0.01
*Σ*CLA	0.59	0.78	0.05	<0.01
18:3 n-3	0.38	2.90	0.11	<0.01
20:3 n-6	0.28	0.38	0.04	0.04
20:4 n-6	0.80	3.17	0.31	<0.01
20:5 n-3	0.14	1.15	0.09	<0.01
22:4 n-6	0.13	0.16	0.02	0.04
22:5 n-3	0.35	1.72	0.11	<0.01
22:6 n-3	0.07	0.41	0.03	<0.01
SFA	44.0	40.6	1.00	<0.01
MUFA	49.5	38.3	0.97	<0.01
PUFA	6.43	21.11	1.44	<0.01
n-3	0.94	6.16	0.32	<0.01
n-6	4.21	13.25	1.14	<0.01
n-6:n-3	4.52	2.11	0.18	<0.01

[a-b] Least squares means in the same row with different letters differ (*P*<0.05).

3.2 Pigment

Mb is the principle protein responsible for meat colour, and is the predominant meat pigment (Mancini & Hunt, 2005). Oxygenation occurs when Mb is exposed to oxygen and is characterized by the development of a bright cherry-red colour. The loss of oxygen from oxymyoglobin and an electron from ferrous ion producing the MetMb, are the changes that occur to alter the absorption properties and the complementary colour which turns from bright red to dark-red and further to brown (Kanner, 1994). Pigment levels in grass-fed bison can be seen to be significantly higher (*P* < 0.01) than in grain-fed beef (Figure 1). The iron atom in the centre of the heme ring (in Mb) can form six bonds, the sixth of which can reversibly bind to ligands which dictate muscle colour (Mancini & Hunt, 2005). While structure of Mb between bison and beef have been found to be identical (Joseph et al., 2010), the increased levels of pigment found in commercial bison compared to commercial beef in the present study combined with high levels of iron found in bison muscle tissue in previous studies (Galbraith et al., 2006) could contribute to poor colour stability in fresh bison meat during retail display.

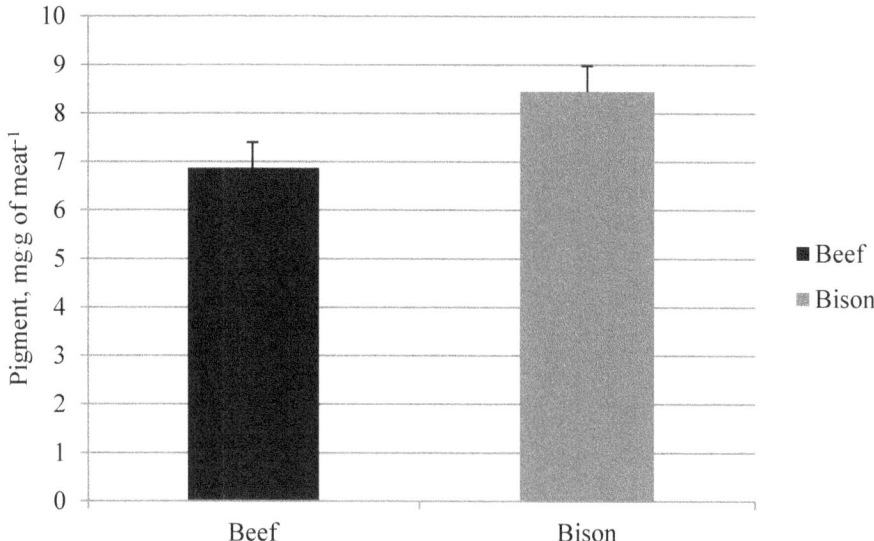

Figure 1. Pigment levels (mg/g meat) in bison and beef tissue

3.3 Retail Evaluation

Metmyoglobin levels in commercial bison were significantly higher than commercial beef on all retail display days (Table 2) confirming the early browning in bison reported previously (Dhanda et al., 2002; Janz et al., 2000; Pietrasik et al., 2006). Commercial beef MetMb level on day 3 was equivalent to the level in commercial bison on day 0 and, by day 3, the levels in bison were twice that of beef. Over the same time, MbO$_2$ decreased (P <0.01) slightly in commercial beef (0.78 to 0.71) and substantially in the commercial bison (0.70 to 0.49) (Table 2). Hence, it appears that colour stability of commercial bison was already compromised on entry into the retail case after 6 d of ageing.

Table 2. Retail performance for beef (n=20) and bison (n=14) steaks (*P value for interaction*)

	Beef (n=20)				SEM	Bison (n=14)				SEM	*P* value		
	0	1	2	3		0	1	2	3		Species(S)	Retail(R)	S×R
Objective Retail Measurements													
L*	42.2[a]	41.2[b]	41.3[b]	41.4[b]	0.43	37.1[c]	35.2[d]	34.5[d]	36.4[d]	0.52	<0.01	0.03	0.03
Chroma	23.9[a]	23.3[a]	23.14[a]	22.9[a]	0.45	18.6[b]	17.5[c]	15.2[d]	14.1[e]	0.54	<0.01	<0.01	<0.01
Hue	36.2[f]	36.1[f]	36.6[cef]	37.2[bce]	0.66	32.4[d]	35.9[cf]	39.0[b]	42.2[a]	0.79	<0.01	<0.01	<0.01
Metmyoglobin	0.15[f]	0.18[e]	0.19[de]	0.20[d]	0.01	0.21[d]	0.28[c]	0.34[b]	0.40[a]	0.01	<0.01	<0.01	<0.01
Myoglobin	0.07[d]	0.08[b]	0.08[b]	0.09[b]	0.00	0.10[bc]	0.09[b]	0.11[ac]	0.11[a]	0.01	<0.01	<0.01	0.05
Oxymyoglobin	0.78[a]	0.74[b]	0.73[bc]	0.71[c]	0.01	0.70[c]	0.63[d]	0.55[e]	0.49[f]	0.01	<0.01	<0.01	<0.01
Subjective Retail Measurements													
Retail Appearance	7.66[a]	7.01[b]	6.69[c]	6.01[d]	0.17	7.40[ab]	5.26[e]	3.96[f]	3.26[g]	0.20	<0.01	<0.01	<0.01
Lean Colour Score	5.01[c]	4.95[c]	5.02[c]	5.04[c]	0.07	6.00[d]	6.49[c]	6.93[b]	7.19[a]	0.09	<0.01	<0.01	<0.01
% Surface Discolouration	1.02[f]	1.33[ef]	1.63[de]	2.03[d]	0.19	1.26[ef]	3.59[c]	4.77[b]	5.33[a]	0.23	<0.01	<0.01	<0.01
Colour of Discolouration	1.04[e]	1.16[de]	1.45[cd]	1.72[c]	0.11	1.11[e]	2.13[b]	2.70[a]	2.83[a]	0.13	<0.01	<0.01	<0.01
Marbling Score	3.59	3.73	3.80	3.78	0.09	2.29	2.34	2.40	2.43	0.10	<0.01	0.088	0.57
Marbling Colour	1.00[b]	1.00[b]	1.00[b]	1.00[b]	0.04	1.23[a]	1.17[a]	1.00[b]	1.00[b]	0.04	<0.01	0.022	<0.01

[a-f] Least squares means in the same row with different letters differ (*P*<0.05).

There was a significant interaction for objective colour measurements ($P<0.05$) over time in retail (Table 2). By day 1, the subjective retail measurements were showing commercial bison to be less resilient in the retail environment. Retail appearance, lean colour score, and percent surface discolouration were significantly different for commercial bison than commercial beef, with bison scoring less favourably on all of these measurements (Table 2).

3.4 Vitamin E

Vitamin E is a term that encompasses a number of tocopherols and trienols that have antioxidant properties (Brigelius-Flohé & Traber, 1999). The efficacy with which they exert their biological action is very low in the case of the tocotrienols, whereas tocopherols, and in particular α-tocopherol, are much more active and potent and account for almost all the vitamin E activity of living tissues (Berges, 1999). Tocopherols constitute a series of benzopyranols that occur in plant tissues and vegetable oils and are powerful lipid-soluble antioxidants (Christie, 2010). In the current study, vitamin E was significantly higher in grass-fed bison meat ($P<0.01$) than in grain-fed beef (3.47 μg / g meat compared to 2.20 μg / g meat; Figure 2). However, levels of α-tocotrienols were higher ($P<0.01$) in beef than in bison. Elevated tissue levels of antioxidants have been reported to have a stabilizing effect on meat colour. Larraín et al. (2008) reported beef fed a corn diet were found to have α-tocopherol of 188 μg / 100 g meat, which is also lower than the levels reported for both commercial bison and beef in the present study. Arnold et al. (1992) reported a vitamin E supplementation (500 IU / head / d) could extend display life of beef by 2.5 days based on a threshold value of 15% MetMb, which corresponded to first detection of discolouration. In the present study, MetMb levels started in the display case at d 0 at 15 % for commercial beef and 21% for commercial bison (Table 2).

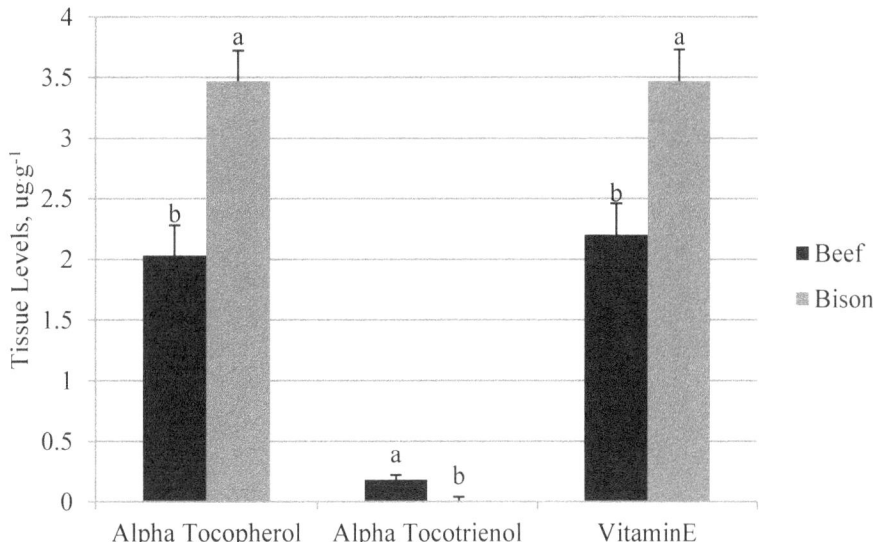

Figure 2. Tissue levels (μg/g meat) of α-tocopherol, α-tocotrienol and total vitamin E in beef and bison

3.5 TBARS

Lipid oxidation in muscle tissues can promote Mb oxidation causing discolouration and rancid odours and flavours (Scollan et al., 2006). TBAR values above about 0.5 mg malonaldehyde / kg meat are considered critical since at this level lipid oxidation products, which produce a rancid odour and taste, detectable to consumers, are present (Wood et al., 2008). In the present study (Figure 3), both commercial beef and bison had TBAR values exceeded this critical level by day 3 of retail display (0.54 and 0.73, respectively). Despite a numerically higher value after 3 d of retail display in the commercial bison compared to commercial beef, there was no significant differences ($P = 0.28$) due to the large range in values. In a study where TBAR values were measured under different packaging regimes, bison meat was found to form higher TBAR values during storage compared to beef (Pietrasik et al., 2006). In this study, authors used storage intervals of 1 d, 1 wk and 2 wks, which were longer than the current study. Thus, had the current study extended the time in retail, significant differences in TBAR levels between commercial bison and beef would likely have emerged.

In a previous study, a positive correlation was found between metmyoglobin accumulation and the production of lipid oxidation products (Faustman & Cassens, 1990). Basically, the rate of discolouration of meat is related to the effectiveness of oxidation processes and enzymatic reducing systems in controlling metmyoglobin levels in meat (Faustman & Cassens, 1990).

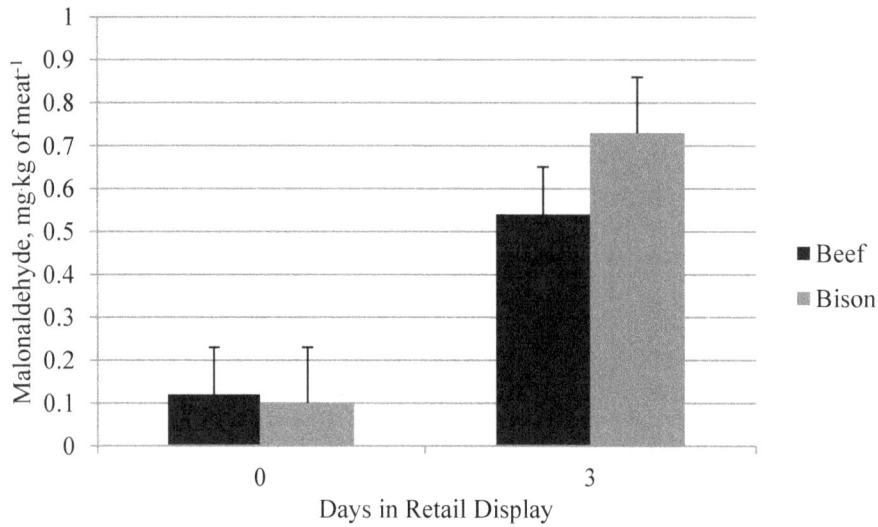

Figure 3. Tissue levels of malonaldehyde in bison and beef after 0 and 3 days retail display

3.6 Changes to Appearance Traits

The current study shows that inherent traits in the muscle tissue of bison influence the lack of retail colour stability. Prolonged *ante-mortem* stress in bison has been related to preharvest depletion of glycogen, with both pH and lactate levels indicating a more rapid glycolysis while carcass temperatures are still high (Galbraith et al., 2009). These metabolic processes may lead to higher drip loss and lower colour stability. In all retail display days commercial bison had higher ($P<0.01$) MetMb (Table 2), and on display days 1, 2 and 3 it had a higher ($P<0.01$) percentage of discolouration than the commercially produced beef steaks. Retail appearance was also rated as less desirable ($P<0.01$) in bison for retail d 1, 2 and 3 (Table 2).

3.7 Regression Analysis

As pointed-out in section 1, the reason for a rapid discolouration of meat bison compared to beef has not been clear. These types of meat have identical structure and biochemistry of myoglobin. Additionally, commercially produced bison meat is grass-fed whereas commercially produced beef is grain-fed in Canada. The main purpose of this study was, therefore, to identify the parameters that are important for the different rate of discolouration of the commercially produced bison meat and beef. For this reason, a regression analysis was performed to estimate the contribution of different factors. The results indicated that the change (Δ) in the percentage of surface discolouration from d 0 to d 3 in retail was most influenced by the inherent n-3 levels in the tissue, followed by the n-6 and the n-6:n-3 ratio ($R^2 = 0.72$). This shows that 72% of the explained variability was due to the differences in the fatty acid composition. Using the *b* values and the intercept, an equation to describe the influence that the independent variables were having on the dependant variables was derived. For instance, the equation for changes in percentage of surface discolouration from d 0 to d 3 in retail was derived as follows:

$$\Delta \% \text{ discolouration} = -0.271 - 1.029(\text{n-3}) + 0.217(\text{n-6}) - 0.159(\text{n-6:n-3}) \ (P<0.05)$$

In addition to this, the following equation was also derived for changes in MetMb ($R^2 = 0.56$):

$$\Delta \text{ MetMb} = -0.230 + 0.017(\text{total fat}) - 0.021(\text{n-3}) + 0.010(\text{n-6}) + 0.013(\text{n-6:n-3}) \ (P< 0.05)$$

This relationship showed that as levels of total fat increase and levels of n-3 and n-6 decrease, an increased change in MetMb content will occur between d 0 and d 3. Although included in the statistical analyses, the levels of vitamin E or pigment content, as well as SFA, MUFA and PUFA levels, were not included in any of the stepwise regression models.

These equations show the inherent tissue traits of commercial bison were strongly linked to undesirable changes in fresh meat colour in a retail display environment. Commercial bison had lower total fat, lower n-6 and n-6:n-3 levels and higher n-3 levels than beef and, consequently, showed a higher change in the percentage of surface discolouration and MetMb content between d 0 and d 3 in retail. Based on these results, it could be suggested that feeding bison with grain during finishing can help to stabilize the colour of bison meat by increasing the n:6 and decreasing n:3. A recent study (Tuner et al., 2014) revealed that bison meat from grain-fed bison had a lower proportion of n:3 compared to grass-fed bison. Of course, while an increase in n:6 and a decrease in n:3 in grain-fed bison can improve the colour stability of bison meat, this strategy will have a negative impact on the health of consumers. As previously mentioned, PUFA are susceptible to oxidative breakdown resulting in changes to the colour (Wood et al., 2008) and these differences have been linked to the heme iron content (Rhee et al., 1996). Rule et at. (2002) hypothesized that the rapid deterioration of colour quality of bison could be linked to its higher PUFA content. The results of the present study seem to corroborate this idea. However, further research is required to determine the appropriate proportion of n:6 and n:3 in order to improve the colour stability of commercially produced bison meat without affecting the human health and fat quality. Furthermore, specific packaging strategies should be evaluated in order to extend the shelf-life of bison meat.

4. Conclusions

The expected differences in colour stability between commercial beef and commercial bison meat were evident in the present study. Further analyses revealed that the differences in total fat content and lipid composition (n-3, n-6 and n-6:n-3 levels) between the two types of meat were among the main reasons for the early deterioration of bison meat, compared to beef. Vitamin E levels, although higher in bison, did not protect against this process. Summing up, this study revealed that there are factors that are inherent within the two commercially produced types of meat that impact meat colour, but there are also factors that can be manipulated in order to improve the colour stability of bison meat.

References

Arnold, R. N., Scheller, K. K., Arp, S. C., Williams, S. N., Buege, D. R., & Schaefer, D. M. 1992. Effect of long- or short-term feeding of alpha-tocopheryl acetate to Holstein and crossbred beef steers on performance, carcass characteristics, and beef color stability. *Journal of Animal Science, 70*(10), 3055-65.

Berges, E. 1999. Importance of vitamin E in the oxidative stability of meat: Organoleptic qualities and consequences. In J. Brufau & A. Tacon, (eds.), *Feed manufacturing in the Mediterranean region: Recent advances in research and technology* (pp. 347-363). Zaragoza : CIHEAM.

Brigelius-Flohé, R., & Traber, M. G. (1999). Vitamin E: Function and metabolism. *FASEB Journal, 13*(10), 1145-1155.

CCAC. (1993). Canadian Council on Animal Care. Guide to the care and use of experimental animals. Vol 1, 2nd ed ED Olfert, BM Cross, AA McWilliam, eds CCAC, Ottawa, ON.

CFIA. (2003). *Canadian Food Inspection Agency*. Guide to food labelling and advertising. Retrieved Feb. 14, 2011, from http://www.inspection.gc.ca/english/fssa/labeti/guide/ch7be.shtml.

Chan, W. K. M., Hakkarainen, K., Faustman, C., Schaefer, D. M., Scheller, K. K., & Liu, Q. (1996). Dietary vitamin E effect on color stability and sensory assessment of spoilage in three beef muscles. *Meat Science, 42*(4), 387-399. http://dx.doi.org/10.1016/0309-1740(95)00055-0

Christie, W. M. (2010). The lipid library- tocopherols and tocotrienols. Retrieved July 29, 2010, from http://lipidlibrary.aocs.org/Lipids/tocol/file.pdf

CIE. (1978). Recommendations on uniform color spaces – color difference equations – psychometric color terms. Commission Internationale de l'Eclairage (publication no. 15, supplement no. 2, pp. 8-12). Paris, France: CIE.

Cruz-Hernandez, C., Deng, Z., Zhou, J., Hill, A. R., Yurawecz, M. P., Delmonte, P., … Kramer, J. K. G. (2004). Methods to analyze conjugated linoleic acids (CLA) and trans-18:1 isomers in dairy fats using a combination of GC, silver ion TLC-GC, and silver ion HPLC. *Journal of AOAC International, 87*(2), 545-562.

Decker, E., Faustman, C., & Lopez-Bote, C. J. (2000). *Antioxidants in muscle foods: nutritional strategies to improve quality*. New York.

Dhanda, J. S., Pegg, R. B., Janz, J. A. M., Aalhus, J. L., & Shand, P. J. (2002). Palatability of bison semimembranosus and effects of marination. *Meat Science, 62*(1), 19-26.

http://dx.doi.org/10.1016/S0309-1740(01)00222-4

Faustman, C., & Cassens, R. G. (1990). The biochemical basis for discoloration in fresh meat: A review. *Journal of Muscle Foods, 1*(3), 217-243. http://dx.doi.org/10.1111/j.1745-4573.1990.tb00366.x

Folch, J., Lees, M., & Sloane Stanley, G. H. (1957). A simple method for the isolation and purification of total lipides from animal tissues. *The Journal of biological chemistry, 226*(1), 497-509.

Galbraith, J., Robertson, W., Aalhus, J., Schaefer, A., Cook, N., Larsen, I., & Okine, E. (2009). Meat quality of bison slaughtered in a mobile or stationary abattoir. Proc. 55th International Conference of Meat Science and Technology, Copenhagen, Denmark.

Galbraith, J. K., Hauer, G., Helbig, L., Wang, Z., Marchello, M. J., & Goonewardene, L. A. (2006). Nutrient profiles in retail cuts of bison meat. *Meat Science, 74*(4), 648-654. http://dx.doi.org/10.1016/j.meatsci.2006.05.015

Givens, D. I., Kliem, K. E., & Gibbs, R. A. (2006). The role of meat as a source of n - 3 polyunsaturated fatty acids in the human diet. *Meat Science, 74*(1), 209-218. http://dx.doi.org/10.1016/j.meatsci.2006.04.008

Hewavitharana, A. K., Lanari, M. C., & Becu, C. (2004). Simultaneous determination of Vitamin E homologs in chicken meat by liquid chromatography with fluorescence detection. *Journal of Chromatography A, 1025*(2), 313-317. http://dx.doi.org/10.1016/j.chroma.2003.10.052

Insani, E. M., Eyherabide, A., Grigioni, G., Sancho, A. M., Pensel, N. A., & Descalzo, A. M. (2008). Oxidative stability and its relationship with natural antioxidants during refrigerated retail display of beef produced in Argentina. *Meat Science, 79*(3), 444-452. http://dx.doi.org/10.1016/j.meatsci.2007.10.017

Janz, J. A. M., Aalhus, J. L., Price, M. A., & Greer, G. G. (2001). Bison meat quality research: and investigation of spray chilling, very fast chilling, and the relationship of grade factors to meat quality. A final report to the Peace Country Bison Association

Janz, J. A. M., Aalhus, J. L., Price, M. A., & Schaefer, A. L. (2000). The influence of elevated temperature conditioning on bison (Bison bison bison) meat quality. *Meat Science, 56*(3), 279-284. http://dx.doi.org/10.1016/S0309-1740(00)00054-1

Joseph, P., Suman, S. P., Li, S., Beach, C. M., Steinke, L., & Fontaine, M. (2010). Characterization of bison (Bison bison) myoglobin. *Meat Science, 84*(1), 71-78. http://dx.doi.org/10.1016/j.meatsci.2009.08.014

Juárez, M., Dugan, M. E. R., Aalhus, J. L., Aldai, N., Basarab, J. A., Baron, V. S., & McAllister, T. A. (2011). Effects of vitamin E and flaxseed on rumen-derived fatty acid intermediates in beef intramuscular fat. *Meat Science, 88*(3), 434-440. http://dx.doi.org/10.1016/j.meatsci.2011.01.023

Kanner, J. (1994). Oxidative processes in meat and meat products: Quality implications. *Meat Science, 36*(1-2), 169-189. http://dx.doi.org/10.1016/0309-1740(94)90040-X

Katsanidis, E., & Addis, P. B. (1999). Novel HPLC analysis of tocopherols, tocotrienols and cholesterol in tissue. *Free Radical Biology and Medicine, 27*, 1137-1140. http://dx.doi.org/10.1016/S0891-5849(99)00205-1

Kramer, J. K. G., Fellner, V., Dugan, M. E. R., Sauer, F. D., Mossoba, M. M., & Yurawecz, M. P. (1997). Evaluating acid and base catalysts in the methylation of milk and rumen fatty acids with special emphasis on conjugated dienes and total trans fatty acids. *Lipids, 32*(11), 1219-1228. http://dx.doi.org/10.1007/s11745-997-0156-3

Krzywicki, K. (1979). Assessment of relative content of myoglobin, oxymyoglobin and metmyoglobin at the surface of beef. *Meat Science, 3*(1), 1-10. http://dx.doi.org/10.1016/0309-1740(79)90019-6

Laborde, F. L., Mandell, I. B., Tosh, J. J., Buchanan-Smith, J. G., & Wilton, J. W. (2002). Effect of management strategy on growth performance, carcass characteristics, fatty acid composition, and palatability attributes in crossbred steers. *Canadian Journal of Animal Science, 82*(1), 49-57. http://dx.doi.org/10.4141/A01-022

Larick, D. K., Turner, B. E., Koch, R. M., & Crouse, J. D. (1989). Influence of Phospholipid Content and Fatty Acid Composition of Individual Phospholipids in Muscle from Bison, Hereford and Brahman Steers on Flavor. *Journal of Food Science, 54*(3), 521-526. http://dx.doi.org/10.1111/j.1365-2621.1989.tb04641.x

Larraín, R. E., Schaefer, D. M., Richards, M. P., & Reed, J. D. (2008). Finishing steers with diets based on corn, high-tannin sorghum or a mix of both: Color and lipid oxidation in beef. *Meat Science, 79*(4), 656-665. http://dx.doi.org/10.1016/j.meatsci.2007.10.032

Mancini, R. A., & Hunt, M. C. (2005). Current research in meat color. *Meat Science, 71*(1), 100-121.

http://dx.doi.org/10.1016/j.meatsci.2005.03.003

Marchello, M. J., & Driskell, J. A. (2001). Nutrient composition of grass- and grain-finished bison. *Great Plains Research, 11*(1), 65-82.

Marchello, M. J., Slanger, W. D., Hadley, M., Milne, D. B., & Driskell, J. A. (1998). Nutrient Composition of Bison Fed Concentrate Diets. *Journal of Food Composition and Analysis, 11*(3), 231-239. http://dx.doi.org/10.1006/jfca.1998.0583

Nielsen, J. H., Sørensen, B., Skibsted, L. H., & Bertelsen, G. (1997). Oxidation in pre-cooked minced pork as influenced by chill storage of raw muscle. *Meat Science, 46*(2), 191-197. http://dx.doi.org/10.1016/S0309-1740(97)00016-8

Nuernberg, K., Fischer, A., Nuernberg, G., Ender, K., & Dannenberger, D. (2008). Meat quality and fatty acid composition of lipids in muscle and fatty tissue of Skudde lambs fed grass versus concentrate. *Small Ruminant Research, 74*(1-3), 279-283. http://dx.doi.org/10.1016/j.smallrumres.2007.07.009

Pietrasik, Z., Dhanda, J. S., Shand, P. J., & Pegg, R. B. (2006). Influence of injection, packaging, and storage conditions on the quality of beef and bison steaks. *Journal of Food Science, 71*(2), S110-S118. http://dx.doi.org/10.1111/j.1365-2621.2006.tb08913.x

Rhee, K. S., Anderson, L. M., & Sams, A. R. (1996). Lipid oxidation potential of beef, chicken, and pork. *Journal of Food Science, 61*(1), 8-12. http://dx.doi.org/10.1111/j.1365-2621.1996.tb14714.x

Rule, D. C., Broughton, K. S., Shellito, S. M., & Maiorano, G. (2002). Comparison of muscle fatty acid profiles and cholesterol concentrations of bison, beef cattle, elk, and chicken. *Journal of Animal Science, 80*(5):1202-1211.

SAS. (2003). SAS® user's guide: Statistics. SAS for Windows, version 9.1. SAS Institute, Inc, Cary, NC.

Scollan, N., Hocquette, J. F., Nuernberg, K., Dannenberger, D., Richardson, I., & Moloney, A. (2006). Innovations in beef production systems that enhance the nutritional and health value of beef lipids and their relationship with meat quality. *Meat Science, 74*(1), 17-33. http://dx.doi.org/10.1016/j.meatsci.2006.05.002

Steenbergen, J. (2013). Business Development – Elk/Bison, Alberta Agriculture and Rural Development.

Trout, G. R. (1991). A rapid method for measuring pigment concentration in porcine and other low pigmented muscles. Proc. 37th International Congress of Meat Science and Technology, Kulmbach, Germany.

Turner, T. D. (2005). Evaluation of the effect of dietary forages and concentrate levels on the fatty acid profile of bison tissue MSc. University of Saskatchewan. Canada.

Wang, Z., & Goonewardene, L. A. (2004). The use of MIXED models in the analysis of animal experiments with repeated measures data. *Canadian Journal of Animal Science, 84*(1), 1-11. http://dx.doi.org/10.4141/A03-123

Wood, J. D., Enser, M., Fisher, A. V., Nute, G. R., Sheard, P. R., Richardson, R. I., … Whittington, F. M. (2008). Fat deposition, fatty acid composition and meat quality: A review. *Meat Science, 78*(4), 343-358. http://dx.doi.org/10.1016/j.meatsci.2007.07.019

4

Combining Modified Atmosphere Packaging and Nisin to Preserve Atlantic Salmon

Dong Han[1], Inyee Han[2] & Paul Dawson[2]

[1]Auburn University, Auburn, AL, USA

[2]Department of Food, Nutrition and Packaging Sciences, Clemson University, Clemson, SC, USA

Correspondence: Paul Dawson, Department of Food, Nutrition and Packaging Sciences, Clemson University, Clemson, SC 29634, USA. E-mail: pdawson@clemson.edu

Technical Contribution No. 6460 of the Clemson University Experiment Station.

Abstract

Preservation effects of modified atmosphere package combined with nisin on fresh Atlantic salmon were evaluated. Farm-raised Atlantic salmon were purchased from the local market and packaged using either 19 % CO_2: 70 % N_2: 11 % O_2, 38 % CO_2: 51 % N_2: 11 % O_2, and under atmospheric air (with and without nisin at 400 IU/g) resulting in a total of 6 treatments. The microbiological (aerobic plate count, psychrotrophic bacteria, and lactic acid bacteria) and the total volatile basic nitrogen analyses were evaluated on Day 0, 2, 4, 7 and 10. Package headspace and sensory evaluation were also conducted on Day 0, 2 and 4. The presence of CO_2 effectively inhibited the growth of all three types of bacteria while nisin significantly inhibited the growth of aerobic microorganisms with less impact on lactic acid bacteria. The TVB-N test indicated that CO_2 delayed the spoilage of Atlantic salmon while nisin had a lesser but measurable impact on Atlantic salmon shelf-life. The experiments support the potential for combining modified atmosphere package and nisin as an effective method to limit the spoilage of Atlantic salmon compare to traditional preservation methods.

Keywords: Atlantic salmon, modified atmosphere packaging, nisin, shelf life

1. Introduction

Modified atmosphere packaging (MAP) has become a popular preservation technology (McMillin, 2008). The basis of MAP is a sealed food package with an altered headspace gas mixture. Both microbiological and chemical reactions continue during shipping and storage often altering the package gas headspace. Several researchers (Fagan et al., 2004; Wang et al., 2008; Economou et al., 2009; Tsironi et al., 2010) have studied the effect of the MAP on fresh seafood product shelf life. Other researchers (Jayasingh, et al., 2002; Sivertsvik et al., 2003; Fagan et al., 2004; Wang et al., 2008; Economou et al., 2009) have demonstrated that higher concentrations of CO_2 extend the microbiological shelf-life of fresh seafood. However, high concentrations of CO_2 can also lead to quality loss.

Salmon spoilage is multifaceted and one of the main factors causing spoilage is the growth of microorganisms (Rasmussen et al., 2002). In MAP preserved salmon, O_2 levels and temperature are lower than ambient conditions and these different environments have an impact on the growth of microorganisms. Beside microbiological growth, other chemical indexes can reflect the spoilage level or freshness of Atlantic salmon. For example, total volatile basic nitrogen (TVB-N) (Dhaouadi et al., 2007) and trimethylamine (TMA) have been used for this purpose (Gökoğlu et al., 2004; Erkan and Özden, 2008). TVB-N is an indicator of total nitrogen and during the storage of salmon, the reduction of trimethylamine oxide (TMAO) results in the production of trimethylamine (TMA) (Pena-Pereira et al., 2010). The measurement of TVB-N and TMA has been employed as spoilage indicators for decades and the ratio of TVB-N to TMA (TVB-N/TMA) has been an index of seafood freshness (Mitsubayashi et al., 2004; Howgate, 2010).

Nisin is a well-known bacteriocin that inhibits the growth of gram-positive bacteria and is generally recognized as safe (GRAS) as a food preservative (de Arauz et al., 2009). Researchers have confirmed that nisin (Jofré et al., 2008; Lu et al., 2010; Shirazinejad et al., 2010) and MAP (Raju et al., 2003; Economou et al., 2009; Tsironi et al., 2010; López‑Mendoza et al., 2007,) can extend the shelf life of seafood when used individually. However, little has been published on combining MAP with nisin to extend the shelf life of fresh Atlantic salmon. Therefore the objective of this research was to determine the effect of nisin combined with MAP on Atlantic salmon shelf life.

2. Material and Methods

2.1 Salmon

Fresh, sliced farm-raised Atlantic salmon fillets were transported on ice (2:1 ice:salmon ratio)to the laboratory. Salmon was purchased as skinless and boneless fillets from a local grocery store and held under super-chilled conditions (-4±2°C) until experiments (up to 3 days). Frozen fillets were thawed at 4±2°C for 12 hours. These fillets were cut into portions of approximately 150 g and then further cut a 25 g portion and a 10 g portion of the 150 g fillet. Thirty nine portions of 150 g Atlantic salmon samples were prepared and 3 were tested immediately. Processing and packaging procedures were conducted under strict hygienic conditions.

2.2 Nisin Activity

Commercial nisin (Nisaplin®, 10^6 IU/g) was purchased from the manufacturer (Danisco, New Century, KS, USA). Nisaplin® activity was determined in triplicate using a zone of inhibition assay. Nisaplin® sample was diluted 1:500 in sterile water, then a series of two-fold dilutions were tested against *Lactobacillus plantarum* ATCC 14917 by spotting 10 µL of each dilution on the surface of the MRS agar medium seeded uniformly with the suspension of *L. plantarum*. After 48 h incubation at 37°C under 5% CO_2, the plates were examined for inhibition zones. The activity of nisin in Arbitrary Units per mL (AU/mL) was expressed as the reciprocal of the highest dilution showing a clear inhibition zone for each triplicate sample. The activity of nisin was expressed in AU/mg based on the weight of the nisin compounds used in serial dilution. Stock nisin (10^6 IU/mL solution was then prepared by dissolving Nisaplin® in sterile water.

2.3 Modified Atmosphere Packaging

Final gas mixture ratios in packages were 19% CO_2 : 70% N_2 : 11% O_2 and 38% CO_2 : 51% N_2 : 11% O_2. Previous researchers have tested 60:40 CO_2:N_2 ratios and studies from our group with other meat types found lower CO_2 with N_2 and O_2 mixtures were successful in extending shelf life (Naas et al., 2013; Kalleda et al., 2013).Three salmon portions (10, 25 and 115 g) were placed in foam trays (C976 Sealed Air Cryovac, Duncan, SC) (8¾ * 6¾ * 15/8 ″). The nisin solution was spread on the surface of fillets using a sterile spreader to obtain a final nisin concentration of 400 IU/g. The six MAP -nisin combination treatments were; air, 19:70:11% CO_2:N_2:O_2 or 38:51:11% CO_2:N_2:O_2 each either without or with 400 IU/g nisin. A Ross Jr™ preformed tray MAP machine (Model No. S-3180, Robert Reiser & Co. Inc., Canton, MA 02021) was used for packaging all salmon. Gases used in the package were pure mixtures of CO_2, O_2 and N_2 (National Specialty Gases, Durham, NC 27713). A vacuum pressure of 150 mbar, gas pressure of 765 mbar, seal time of 2.1 sec, knife temperature of 143°C and seal temperature of 141°C were pre-set on the MAP machine. The trays were sealed using lid stock film (Lid 1050) (18.5″wide) to achieve a gas:product ratio of 5:1. All samples were refrigerated (2±4°C C) until the analyses. At the same time, 7 empty packages were also sealed with the 3 gas mixtures and stored in the same environment as salmon samples to monitor the gas composition of empty packages.

2.4 Analyses - Headspace Gas Analysis

There were 4 major analyses (headspace gas analyses, microbiologic enumeration, TVB-N titration test, sensory testing) employed to evaluate freshness of salmon samples. All analyses were conducted on samples from each of the 6 treatments as described in Table 1 on days 0, 2, 4, 7 and 10, except for package headspace gas mixtures which were taken on days 0, 2 and 4. The gas mixture in the headspace of an airtight food package was monitored to determine how gas composition changed relative to the spoilage of the salmon. A gas chromatograph (series 200, Gow-Mac Inst.Co., Bethlehem, PA) fitted with CTR-1 gas analysis column (catalog no.8700, Alltech, Sanjose, CA) and TCD (thermal conductivity detector) was used to determine the package headspace gases (O_2, CO_2, N_2). An integrator (Hewlett Packard, Wilmington, DE) was used to plot chromatograms and calculate gas percentages from peak areas. A 0.05 mL package headspace gas sample was analyzed at each sampling interval by injecting a needle (syringe type) through a gas-tight septum placed onto the package film surface. The chromatograph was calibrated using standardized gas mixtures verified by manufacturer (Air Products and Chemicals Inc., Allentown, PA).

2.5 Microbiological Enumeration

Twenty-five g portion of fish samples were aseptically removed from trays, placed in sterile stomacher bags (model 400, 6041/STR, Seward Limited, London, UK) and homogenized for 2 min at 230 rpm in a laboratory blender (Model 400, Seward™, FL, USA 33330) containing pre-added 225 mL pre-chilled sterile peptone-physiological saline solution (0.1% peptone + 0.85% NaCl) (Difco™, Bactopeptone, Becton, Dickison

& Company, MD, USA 21152). Then decimal serial dilutions were prepared from this homogenate in the same chilled sterile diluent. Culture medium for aerobic microorganisms was Plate Count Agar (PCA) (Difco[TM], Bactopeptone, Becton, Dickison & Company, MD, USA 21152). The plates were incubated for 48 h at 37°C and the population of psychrotrophic bacteria was determined by a spread plate counting method with PCA with 1% NaCl and incubated at 4 °C for 7 days. For estimation of potential lactic acid bacteria (LAB), diluted samples were plated on deMan, Rogosa, and Sharpe (MRS) agar and incubated at 37°C for 72h in a CO_2 incubator with a continuous CO_2 flow. (Difco[TM], Bactopeptone, Becton, Dickinson & Company, MD, USA 21152) Prior to data analyses of microbiological data, bacterial populations were converted to logarithmic values (CFU/g).

2.6 Total Volatile Basic Nitrogen

Total volatile basic nitrogen (TVB-N) tests were prepared by homogenizing 10 g of fish from trays with 100 ml water in a laboratory blender (Model 400, Seward[TM], FL, USA 33330) for 1 min at 230 rpm. Then the salmon-water mixture was centrifuged at 3000 rpm for 5 min with the centrifuge at 10,000 x g (J-26 XPI, Beckman Coulter, Inc., CA, USA 92821). 10 ml of the supernatant was placed into a distillation tube, followed by 10mL of 1% (w/v) magnesium oxide suspension. Vacuum-distillation was conducted using a vertical distillation unit (Model RV 10 digital, IKA ® Works, Inc., NC, USA 28405) and the distillate was placed into 20mL of 2% (v/v) aqueous boric acid solution with the 7-8 drops of indicator solution. After five minutes, the distillation was ended and titrated. The titration was conducted using 0.005 mol/L sulphuric acid solution. The indicator solution was a mixture of 0.2% methyl red ethanol solution and 0.1% methylene blue solution added immediately before titration. The titration endpoint was a color change from green to blue/purple.

2.7 Sensory Test

All the assessors were trained using pre-spoiled salmon series. 5 samples were prepared for each of 7 days of training and held at 20°C for different time periods to accelerate spoilage. 3 terms (general appearance, color, and odor) were used on a 5 point scale for the 5 samples so that assessors agreed on the different levels of salmon spoilage.

At least 7 assessors were involved in each day's sensory test. On each day of testing, packages were opened and each of the six samples was tested immediately by the trained panel using the 5-point scale. The minimal scoring difference was set at 0.5. After the grading, panelists were asked to decide if the samples were acceptable or not for consumption.

2.8 Five-Level Sensory Scale (Modified From Matis, 2016)

General Appearance:

level. 1: Firm texture with natural and fresh fish fillet appearance.

level. 2: Slight drip loss and minor reduction on firmness and appearance

level. 3: Soft texture and obvious reduction in appearance

level. 4: Extreme soft texture and critical appearance reduction

level. 5: Totally spoiled salmon texture and appearance

Color:

level. 1: Flesh-colored salmon tissue with almost no effect of spoilage.

level. 2: Minor change of flesh-colored salmon tissue.

level. 3: Obvious change from flesh-colored to deeper red colors.

level. 4: Extreme change of flesh-color deep red color.

level. 5: Totally spoiled salmon dark red color

Odor:

level. 1: Flesh seafood-like smell

level. 2: Slight fish odor

level. 3: Obvious change from flesh odor to fishy odor.

level. 4: Strong fishy odor.

level. 5: Completely spoiled fishy odor

2.9 Statistical Analysis

All analyses (headspace gas analyses, microbiological enumeration, TVB-N tests and sensory tests) were replicated three times in separate trials with different lots of salmon. Analysis of variance was conducted for each parameter to determine if there was a significant effect ($p \leq 0.05$) due to the treatments. When the treatment was determined to be significant for a parameter based on an analysis of variance, the Least Significant Difference (LSD) test was used to separate the means ($p \leq 0.05$) (SAS, Version 9.0, 2004).

3. Results

3.1 Nisin Activity Detection

An activity level of 1.6×10^6 AU/ml was confirmed for the commercial nisin sample. Nisin was first detected from its production by *Lactococcus lactis* subsp. *lactis* in 1928, and has become a popular bacteriocin with widespread commercial use (Ross et al., 2002). Nisin effectively inhibits the growth of Gram-positive bacteria, such as *Micrococcus*, *Lactococcus*, *Staphylococcus*, *Lactobacillus* and *Listeria* (Arauz et al., 2009). The gram-positive bacteria strain used to determine nisin activity was *Lactobacillus plantarum* ATCC 14917. The results verified that growth of *Lactobacillus plantarum* ATCC 14917 can be inhibited by nisin. Also, the commercial nisin sample was found to retain high activity throughout use in the study. Nisin doesn't generally inhibit the growth of gram-negative bacteria, fungi and virus (Arauz et al., 2009). This group of microorganisms can cause spoilage of food products which limits nisin preservation effects. Nisin has been reported to restrict some pathogenic bacterial growth. When *L. monocytogenes* was inoculated into long-life cottage cheese the number of viable *L. monocytogenes* cells was reduced by one log with nisin (Ferreira and Lund, 1996). *Clostridium sp.* can be susceptible to nisin and spore outgrowth of *Clostridium sp.* is more likely to be restricted than vegetative cell growth by nisin (Delves-Broughton, Blackburn, Evans & Hugenholtz, 1996).

3.2 Headspace Gas Analyses

During the storage of empty MAP sealed packages, there was no change in CO_2 and O_2 percentage concentrations for both 19% CO_2/70% N_2/11% O_2 and 38% CO_2/51% N_2/11% O_2 packages ($P > 0.05$). This result indicates there was an airtight seal and stable gas environment throughout the Atlantic salmon preservation study. Most MAP is formed from one or more of these four materials: polyvinylchloride (PVC), polyethylene terephthalate (PET), polyethylene (PE) and polypropylene (PP), moreover, PE is usually the major component in a MAP film since it provides the hermetic seal. Polyethylene is also considered for the characteristics of anti-fogging ability, peel ability and the ability to seal under less than optimal conditions (Phillips, 1996).

The technique of sealing modified atmospheres in the polymeric film without further exchange of gasses between inside and outside of the package often generates lower O_2 and higher CO_2 concentrations compared to atmospheric conditions to extend food shelf life. These conditions can influence the biological activity (respiration, enzymatic reactions, and microbial growth) within the package which can, in turn, inhibit spoilage (Zhao et al., 1995; Mangaraj et al., 2009).

Initial O_2 concentration ranged from about 11% in MAP packages to 21% for air packages (Table 1). On Day 2, salmon samples packaged in the air with and without nisin had similar O_2 concentrations ($p > 0.05$), however, by Day 4, there was a significant difference ($p \leq 0.05$) of the oxygen level between air/nisin and air/no nisin. Also, the 38% CO_2 MAP treatment inhibited O_2 consumption compared to the 19% CO_2 MAP ($p \leq 0.05$) at day 2. Thus as CO_2 concentration increased, the O_2 consumption rate of microorganisms present on salmon decreased at early sampling times. This implies that the microorganism's respiration rate was slowed with the increase in CO_2. Similar to the contrast between air/nisin and air/no nisin treatments, the 19% CO_2 package with and without nisin on day 4 also limited the metabolism rate of the microorganisms ($p \leq 0.05$). As the primary gas consumed by spoilage microorganisms during growth, reduced oxygen has been recognized as an important factor in MAP. Taylor, Davidson, & Zhong, (2007) reported commercial nisin samples with similar composition as the one used in the current study [2.5% pure nisin, 74.4% sodium chloride, 23.8% denatured milk solids and 1.7% moisture (w/w)] had similar activity (10^6 IU/g). Nisin limited the growth of spoilage microorganisms and reduced total oxygen consumption in both the present and study by Taylor, Davidson & Zhong (2007).

Table 1. The concentration of O_2 of farmed Atlantic salmon packaged in different air packages and nisin stored at 2-4 °C

Treatment (package condition)	Day 0	Day 2	Day 4
	Percentage of Oxygen package headspace		
Air + Nisin(400 IU/g)	20.60 ± 0.26 a	18.08 ± 0.57 a	8.92 ± 0.42 b
Air without nisin (control)	20.60 ± 0.26 a	17.76 ± 0.15 a	5.89 ± 0.44 b
38% CO_2/51% N_2/11% O_2 + Nisin(400 IU/g)	10.75 ± 0.10 a	10.45 ± 0.31 a	6.95 ± 0.33 b
19% CO_2/70% N_2/11% O_2 + Nisin(400 IU/g)	11.01 ± 0.18 a	8.14 ± 0.28 b	6.5 ± 0.46 b
38% CO_2/51% N_2/11% O2 without nisin	10.75 ± 0.10 a	9.62 ± 0.23 a	6.26 ± 0.24 b
19% CO_2/70% N_2/11% O_2 without nisin	11.01 ± 0.18 a	8.18 ± 0.39 b	4.16 ± 0.69 cb

a-c means within rows with different letters are significantly different (p≤0.05). n=3

3.3 Microbiologic Enumeration

In fresh food product preservation, especially fresh seafood, and meat ingredients, microorganism growth is the primary cause of spoilage (Hozbor et al., 2006). Enumeration of particular microorganisms' groups is important to understand factors affecting spoilage.

3.3.1 Aerobic Plate Count Cultured at 37 °C

The average starting population of aerobic microorganisms on salmon was 3.25 ± 0.06 log CFU/g. The total aerobic bacteria for the 2 MAP treatments differed from air-packed salmon (p≤0.05) on day 2 (Figure 1).

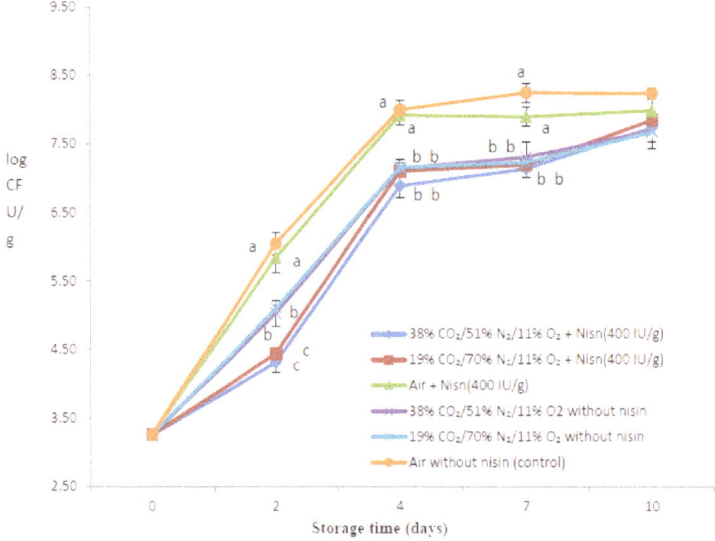

Figure 1. Total aerobic bacterial populations of farmed Atlantic salmon stored at 2-4 °C packaged in various modified atmosphere packaging with and without nisin (n=4) a-c means with a different letter are significantly different (p≤0.05)

Furthermore, the 2 MAP treatments with nisin had a lower population of aerobes on salmon on day 2 compared to the MAP treatments without nisin. No difference in aerobic bacteria population was observed between salmon packaged in CO_2 concentrations of 19% and 38% (p>0.05). On days 2, 4 and 7 the population of bacteria on salmon packaged in the air was greater than MAP-packaged salmon (p≤0.05). By day 10, MAP salmon treatments had reached ~ 8 log CFU/g aerobic, however, air-packed salmon reached this level by Day 4. Ibrahim Sallam (2004) used 7 log CFU/g aerobic plate count population as the spoilage indicator in fish products and following this standard, MAP-packaged salmon had approximately a 1-2 day longer shelf life compared to salmon packaged in air.

3.3.2 Psychrotrophic Bacteria Count

Similar to aerobic microorganisms, no difference in psychrotrophic bacteria population was observed between salmon packaged in CO_2 concentrations of 19% and 38% (p>0.05) (Figure 2). Air-packaged salmon had higher psychrotrophic bacteria populations (about 1 log CFU/g) than the other four treatments at day 2 (p≤0.05). Thus, CO_2 inhibited the growth of psychrotrophic bacteria while nisin had no observable inhibition of psychrotrophic bacteria (p>0.05). As storage continued, psychrotrophic bacteria population differences due to the various treatments diminished (p>0.05).

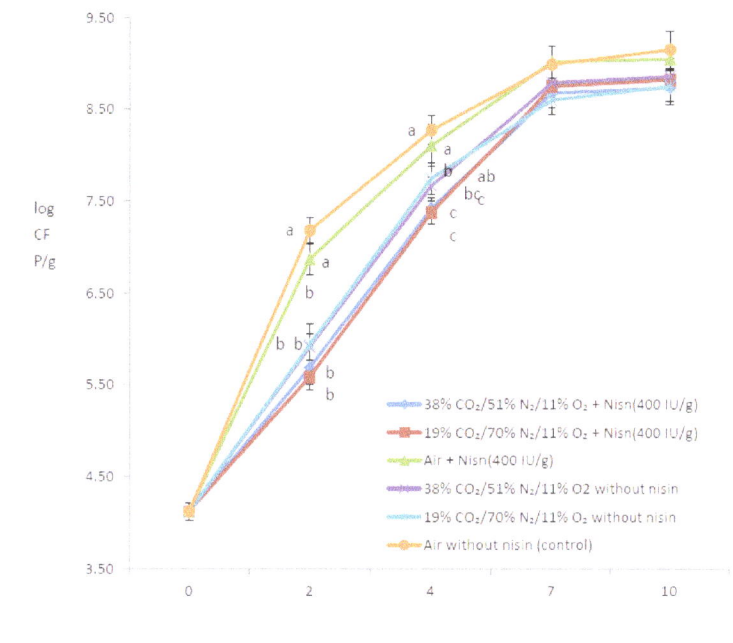

Storage time (days)

Figure 2. Psychrotrophic bacterial populations of farmed Atlantic salmon stored at 2-4 °C packaged in various modified atmosphere packaging with and without nisin (n=6). a-c means within rows with a different letter are significantly different (p≤0.05)

Due to the storage of fresh fish at refrigerated temperatures, psychrotrophic bacteria dominate the spoilage of refrigerated fish (Sivertsvik et al., 2002). Gram-negative psychrotrophic bacteria often dominate in the spoilage of refrigerated fresh food products and thus may reduce the efficiency for nisin to extend shelf life since nisin alone has a little inhibitory effect on Gram-negative bacteria. Nisin can target vegetative cells acting at the cytoplasmic membrane causing pores resulting in cell degradation (Bauer and Dicks, 2005). However, nisin is usually ineffective against gram-negative bacteria due to the presence of the lipopolysaccharide layer (LPS) (Millette et al., 2004).

3.3.3 Lactic Acid Bacterial Count

Initial lactic acid bacteria (LAB) populations were 1.25 ± 0.17 log CFU/g and compared to MAP packaged salmon, air packaged samples without nisin had higher populations at day 2 and 4 than any CO_2 packaged salmon (p≤0.05)(Figure 3). The 38% CO_2 preserved salmon inhibited LAB population compared to 19% CO_2 regardless of nisin application at day 2. As storage continued, LAB population differences due to the various treatments dissipated. Interestingly, there may be a synergistic effect of LAB and nisin on Gram-negative bacteria since LAB has been shown to disrupt the outer Gram-negative membrane (Alakomi et al, 2000), possibly making Gram negative bacteria more susceptible to nisin.

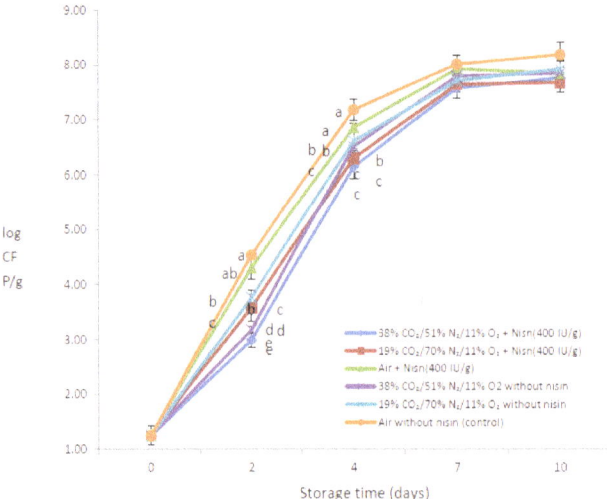

Figure 3. Lactic acid bacterial populations of farmed Atlantic salmon stored at 2-4 °C packaged in various modified atmosphere packaging with and without nisin (n=6). a-e means within rows with a different letter are significantly different (p≤0.05)

3.3.4 TVB-N Titration Test

The initial concentration of total volatile basic nitrogen (TVB-N) titration was 2.89 ± 0.01 mg/100 g. On all 4 sampling days, TVB-N concentration was higher in air-packed salmon compared to MAP packed salmon (Figure 4) ($p \leq 0.05$). The only difference in TVB-N between air/nisin and air/non-nisin packed salmon was on day 2 ($p \leq 0.05$). No significant TVB-N difference was detected between the 4 CO_2 MAP treatments ($p > 0.05$). The application of CO_2 delayed Atlantic salmon spoilage as determined by TVB-N. A significant difference was detected between the CO_2 MAP and the air packed samples on all four sampling days ($p < 0.05$). No significant difference in the production of TVB-N was found between the 38% and 19% CO_2 ($p > 0.05$). Nisin had no impact on the TVB-N concentration except for air packed samples on day 2 ($p \leq 0.05$). Application of nisin in the aerobic atmosphere has been extensively researched but nisin under anaerobic conditions has not received as much attention. Nisin was not as effective as CO_2 headspace in inhibiting the general spoilage of packaged Atlantic salmon. The increase in TVB-N shows a similar trend as the bacterial growth of the psychrotrophic bacteria. Previous research has verified a strong connection between the generation of TVB-N and the spoilage of refrigerated fresh products (Arashisar, Hisar, Kaya & Yanik, 2004, Ojagh, Rezaei, Razavi, & Hosseini, 2010)

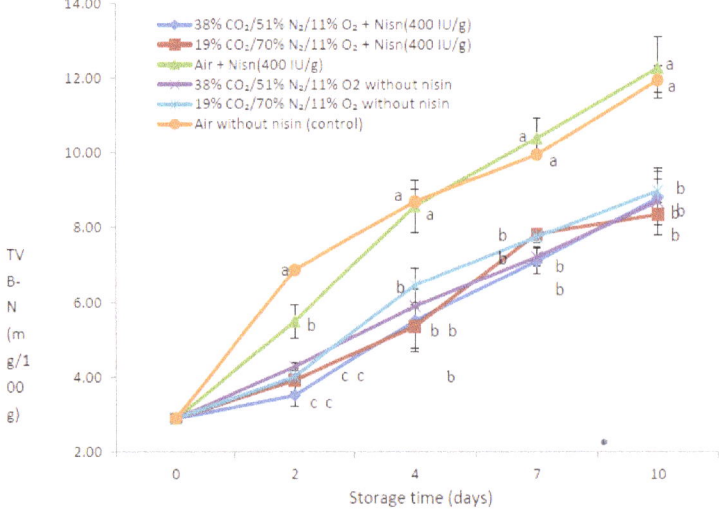

Figure 4. TVB- N of farmed Atlantic salmon stored at 2-4 °C packaged in various modified atmosphere packaging with and without nisin (n=6). a-c means within rows with a different letter are significantly different (p≤0.05)

Lactic acid bacteria are recognized as one of the major spoilage groups in MAP foods and controlling LAB can impact shelf-life. The higher package CO_2 concentration resulted in a slower growth of LAB ($p \leq 0.05$). Although LAB is recognized as a spoilage bacteria that can grow at low oxygen levels, the present of CO_2 successfully slowed the growth of LAB in the current experiment. Sivertsvik et al. (2002) concluded that CO_2 decreases the growth of microorganism especially aerobic bacteria and that this inhibition could not be explained only by the limited O_2 nor the pH changes caused by the CO_2. Other research has revealed that the presence of oxygen can inhibit the growth of LAB (Ibrahim Sallam, 2007). Sivertsvik et al. (2002) theorized that the effect of CO_2 on microorganisms could be described by changes in cell membrane function which effect nutrient absorption, the deactivation of enzymes, degradation of membranes and changes in the proteins. The results show that nisin was not as effective as CO_2 in limiting the growth of LAB.

3.5 Sensory Test

While sensory analysis can be affected by microbiological status, microbial safety is a priority during shipping and storage of fish.. No significant difference ($p > 0.05$) were detected for the first three evaluation terms (appearance, color, and odor) due to the 6 treatments. On the acceptability, a significant difference was observed that air packaged sample without nisin application displayed the lowest acceptable rate on day 4 ($p \leq 0.05$). Thus all the CO_2 or nisin -treated Atlantic salmon samples were more acceptable than control salmon samples. Fish tissue usually contains 60-80% (w/w) water (Ghaly et al., 2010) and the drip loss can cause the major decrease of fresh fish's appearance. The release of water from the fish tissue resulting from microbial growth can reduce the general appearance and overall quality of Atlantic salmon. Alfnes et al. (2006) mentioned the color of salmon is an essential factor of freshness. Ottestad et al. (2011) reported that the color of Atlantic salmon is dependent on the relationship between oxygen and myoglobin. No sensory color difference ($p > 0.05$) was observed in the current study. Inside and surface volatile and non-volatile amines (Bulushi et al., 2009) are the major spoilage amine produced by the blooming of all kinds of microorganisms. Although differences between TVB-N test results were observed, there was no significant difference ($p > 0.05$) in the sensory odor evaluation. However, a difference in salmon acceptability ($p \leq 0.05$) was observed between the control (no nisin packaged in air) and the other 5 storage treatments of Atlantic salmon.

4. Conclusion

Combining MAP with nisin can impact the spoilage rate of Atlantic salmon, however, little synergistic effect between MAP and nisin were observed in this study. In all three microbiological test methods (aerobic plate count, psychrotrophic bacteria, and LAB), the present of CO_2 effectively inhibited the growth of these microorganisms. However, the inhibition effectiveness difference between the medium concentration of CO_2 (38%) and low concentration of CO_2 (19%) was only found in LAB. Furthermore, nisin only significantly inhibited the growth of aerobic microorganisms. CO_2 can efficiently limit the spoilage of Atlantic salmon as measured by the TVB-N test but no difference was observed between 38% and 19% CO_2. The sensory evaluation found no differences in appearance, color, and odor. But more assessors tended to reject the non-nisin and non-CO_2 packed samples compared to the other treatments. Modified atmosphere packaging alone may be a more cost-effective than adding nisin to shelf life extension of fresh salmon.

References

Alakomi, H.L., Skytta, E., Saalrela, M., Mattila-Sandholm, T., Latva-Kala, K. & Helander, I.M. 2000. Lactic acid permeabilizes Gram-negative bacteria by disrupting the outer membrane. *Applied and Environmental Microbiology 66*(5), 2001-2005. http://dx.doi: 10.1128/AEM.66.5.2001-2005.2000

Alfnes, F., Guttormsen, A. G., Steine, G., & Kolstad, K. (2006). Consumers' willingness to pay for the color of salmon: a choice experiment with real economic incentives. *American Journal of Agricultural Economics, 88*(4), 1050-1061. http://dx.doi.org/10.1111/j.1467-8276.2006.00915.x

Arashisar, Ş., Hisar, O., Kaya M., & Yanik, T. (2004). Effects of modified atmosphere and vacuum packaging on microbiological and chemical properties of rainbow trout (*Oncorhynchus mykiss*) fillets. *International Journal of Food Microbiology, 97*(2), 209-214. http://dx.doi.org/10.1016/j.ijfoodmicro.2004.05.024

Bauer, R., & Dicks, L. M. T. (2005). Mode of action of lipid II-targeting lantibiotics. *International Journal of Food Microbiology, 101*(2), 201-216. http://dx.doi.org/10.1016/j.ijfoodmicro.2004.11.007

Bulushi, I. A., Poole, S., Deeth, H. C., & Dykes, G. (2009). Biogenic amines in fish: roles in intoxication, spoilage, and nitrosamine formation-a review. *Critical reviews in Food Science and Nutrition, 49*(4), 369-377. http://dx.doi.org/10.1080/10408390802067514

Davies, A. R., (1995). New methods of food preservation. (Chapter 14) In G.W. Gold (ED.) *Advances in modified-atmosphere packaging* (pp. 304-320). Springer US

De Arauz, L. J., Jozala, A. F., Mazzola, P. G., & Penna, T.C. (2009). Nisin biotechnological production and application: a review. *Trends in Food Science & Technology, 20*(3), 146-154. http://dx.doi.org/10.1016/j.tifs.2009.01.056

Delves-Broughton, J., Blackburn, P., Evans, R. J., & Hugenholtz, J. (1996). Applications of the bacteriocin, nisin. *Antonie van Leeuwenhoek, 69*(2), 193-202. http://dx.doi.org/10.1007/BF00399424

Devlieghere, F., Debevere, J., & Van Impe, J. (1998). Effect of dissolved carbon dioxide and temperature on the growth of Lactobacillus sake in modified atmospheres. *International Journal of Food Microbiology, 41*(3), 231-238. http://hdl.handle.net/1854/LU-348287

Dhaouadi, A., Monser, L., Sadok, S., & Adhoum, N. (2007). Validation of a flow-injection-gas diffusion method for total volatile basic nitrogen determination in seafood products. *Food Chemistry, 103*(3), 1049-1053. DOI: http://dx.doi.org/10.1016/j.foodchem.2006.07.066

Erkan, N., & Özden, Ö. (2008). Quality assessment of whole and gutted sardines (Sardina pilchardus) stored in ice. *International Journal of Food Science and Technology, 43*(9), 1549-1559. http://dx.doi.org/10.1111/j.1365-2621.2007.01579.x

Ferreira, M., & Lund, B. M. (1996). The effect of nisin on Listeria monocytogenes in culture medium and long‑life cottage cheese. *Letters in Applied Microbiology, 22*(6), 433-438. http://dx.doi.org/10.1111/j.1472-765X.1996.tb01197.x

Ghaly, A. E., Dave, D., Budge, S., & Brooks, M. S. (2010). Fish spoilage mechanisms and preservation techniques: review. *American Journal of Applied Sciences, 7*(7), 859-863. http://dx.doi.org/10.3844/ajassp.2010.859.877

Gökoğlu, N., Cengız, E., & Yerlıkaya, P. (2004). Determination of the shelf life of marinated sardine (*Sardina pilchardus*) stored at 4° C. *Food Control, 15*(1), 1-4. http://dx.doi.org/10.1016/S0956-7135(02)00149-4

Hozbor, M. C., Saiz, A. I., Yeannes, M. I., & Fritz, R. (2006). Microbiological changes and its correlation with quality indices during aerobic iced storage of sea salmon (*Pseudopercis semifasciata*). *LWT-Food Science and Technology, 39*(2), 99-104. http://dx.doi.org/10.1016/j.lwt.2004.12.008

Ibrahim Sallam, K. (2007). Antimicrobial and antioxidant effects of sodium acetate, sodium lactate, and sodium citrate in refrigerated sliced salmon. *Food control, 18*(5), 566-575. http://dx.doi.org/10.1016/j.foodcont.2006.02.002

Jayasingh, P., Cornforth, D. P., Brennand, C. P., Carpenter, C., & Whittier, D. R. (2002). Sensory evaluation of ground beef stored in high‑oxygen modified atmosphere packaging. *Journal of Food Science, 67*(9), 3493-3496. http://dx.doi.org/10.1111/j.1365-2621.2002.tb09611.x

Kalleda, R. K., Han, I., Toler, J., Chen, F., Kim, H. J., & Dawson, P. L. (2013). Shlef life extension of shrimp (white) using modified atmosphere packaging. *Polish Journal of Food and Nutrition Sciences, 63*(2), 87-94. http://dx.doi.org/10.2478/v10222-012-0071-7.

López-Mendoza, M. C., Ruiz, P., & Mata, C. M. (2007). Combined effects of nisin, lactic acid and modified atmosphere packaging on the survival of Listeria monocytogenes in raw ground pork. *International Journal of Food Science and Technology, 42*(5), 562-566. http://dx.doi.org/10.1111/j.1365-2621.2006.01275.x.

Mangaraj, S., Goswami, T. K., & Mahajan, P. V. (2009). Applications of plastic films for modified atmosphere packaging of fruits and vegetables: a review. *Food Engineering Reviews, 1*(2), 133-158. http://dx.doi.org/10.1007/s12393-009-9007-3

Matis. (2016). Chill fish from catch to consumer. Chill-on (EU Integrated project), Chill add-on (Icelandic project) and Thermal modelling of chilling and transport of fish (Icelandic project) and also on regulations from The Icelandic Food and Veterinary Authority. Vínlandsleið 12, 113 Reykjavík. http://www.kaeligatt.is/english. Accessed 11-10-16

McMillin, K. W. (2008). Where is MAP going? A review and future potential of modified atmosphere packaging for meat. *Meat Science, 80*(1), 43-65. http://dx.doi.org/10.1016/j.meatsci.2008.05.028

Millette, M., Smoragiewicz, W., & Lacroix, M. (2004) Antimicrobial potential of immobilized *Lactococcus lactis* subsp. *lactis* ATCC 11454 against selected bacteria. *Journal of Food Protection, 67*(6), 1184-1189.

Ojagh, S. M., Rezaei, M., Razavi, S. H., & Hosseini, S. M. H (2010). Effect of chitosan coatings enriched with cinnamon oil on the quality of refrigerated rainbow trout. *Food Chemistry, 120*(1), 193-198. http://dx.doi.org/10.1016/j.foodchem.2009.10.006

Naas, H. Martinez-Dawson, R., Hand I., & Dawson, P. L. (2013). Effect of combining nisin with modified atmosphere packaging on inhibition of Listeria monocytogenes in ready-to-eat turkey bologna. *Poultry Science, 92*, 1930-1935. http://dx.doi.org/10.3382/ps.2012-02141

Ottestad, S., Sørheim, O., Heia, K., Skaret, J., & Wold, J. P. (2011). Effects of storage atmosphere and heme state on the color and visible reflectance spectra of salmon (*Salmo salar*) fillets. *Journal of Agricultural and Food Chemistry, 59*(14), 7825-7831. http://dx.doi.org/10.1021/jf201150x

Phillips, C. A. (1996). Review: modified atmosphere packaging and its effects on the microbiological quality and safety of produce. *International Journal of Food Science and Technology, 31*(6), 463-479. http://dx.doi.org/10.1046/j.1365-2621.1996.00369.x

Raju, C. V., Shamasundar, B. A., & Udupa, K. S. (2003). The use of nisin as a preservative in fish sausage stored at ambient (28±2 °C) and refrigerated (6±2 °C) temperatures. *International Journal of Food Science and Technology, 38*(2), 171-185. http://dx.doi.org/10.1046/j.1365-2621.2003.00663.x

Rasmussen, S. K. J., Ross, T., Olley, J., & McMeekin, T. (2002). A process risk model for the shelf life of Atlantic salmon fillets. *International Journal of Food Microbiology, 73*(1), 47-60. http://dx.doi.org/10.1016/S0168-1605(01)00687-0

Ringø, E., & Gatesoupe, F. J. (1998). Lactic acid bacteria in fish: a review. *Aquaculture, 160*(3), 177-203.

Scott, V. N., & Taylor, S. L. (1981). Temperature, pH, and spore load effects on the ability of nisin to prevent the outgrowth of Clostridium botulinum spores. *Journal of Food Science, 46*(1), 121-126. http://dx.doi.org/10.1111/j.1365-2621.1981.tb14544.x

Sivertsvik, M., Jeksrud, W. K., & Rosnes, J. T. (2002). A review of modified atmosphere packaging of fish and fishery products-significance of microbial growth, activities and safety. *International Journal of Food Science and Technology, 37*(2), 107-127. http://dx.doi.org/10.1046/j.1365-2621.2002.00548.x

Sivertsvik, M., Rosnes, J. T., & Kleiberg, G. H. (2003). Effect of modified atmosphere packaging and superchilled storage on the microbial and sensory quality of Atlantic salmon (Salmo salar) fillets. *Journal of Food Science, 68*(4), 1467-1472. http://dx.doi.org/10.1111/j.1365-2621.2003.tb09668.x

Taylor, T. M., Davidson, P. M., & Zhong, Q. (2007). Extraction of nisin from a 2.5% commercial nisin product using methanol and ethanol solutions. *Journal of Food Protection, 70*(5), 1272-1276.

Zhao, Y., Wells, J. H., & McMillin, K. W. (1995). Dynamic changes of headspace gases in CO_2 and N_2 packaged fresh beef. *Journal of Food Science, 60*(3), 571-575.

5

Fad Diets: Lifestyle Promises and Health Challenges

Jomana Khawandanah[1] & Ihab Tewfik[1]

[1]Department of Life Sciences, Faculty of Science and Technology, University of Westminster, London, UK

Correspondence: Jomana Khawandanah, Department of Life Sciences, Faculty of Science and Technology, University of Westminster, 115 New Cavendish Street, London W1W 6UW, United Kingdom. E-mail: j_khawandanah@hotmail.com

Abstract

Chronic excess of dietary intake combined with reduced energy expenditure increase the positive energy balance. This transition in behaviour contributes significantly to prevalence of obesity, impairment of health, reduction in quality of life and increases health-care costs. While obesity has turned into a public health threat, with the government failing to reverse this growing trend, good number of people is undertaking fad diets with the hope to lose weight fast and easy. Furthermore, media and peers contribute to the popularity of fad diets as they put pressure to individuals who desire a certain body image, which leads to low self-esteem and perhaps eating disorders. Despite the fact that fad diets may appeal as simple way to lose weight, recent studies have shown that such diets in the long term are unsustainable and can bring adverse side effects to health. Consideration of the reviewed literature suggests that long-life changes in diet and lifestyle might be the best approaches to maintain a healthy weight in the long term. Overweight individuals should consult nutrition professions before adopting any fad diets to minimise the health risks and psychological impacts.

Keywords: Atkins diet, dieting, eating disorder, fad diets, obesity, overweight, public health, weight loss, yo-yo dieting

1. Introduction

1.1 The Epidemic of Obesity

Since 1980 the number of people suffering from obesity has doubled on a worldwide level. According to the latest figures published by World Health Organisation (WHO) (2015) almost 2 billion adults (39%) were overweight (Body Mass Index, BMI≥25kg/m^2), with 600 million of these (13%) being obese (BMI≥30kg/m^2). Both being overweight and obese is characterised by excess fat mass in the body (WHO, 2015). The main factors causing obesity involve excessive calorie intake, unhealthy eating habits like consuming processed food, sedentary lifestyle, but also medical conditions (hypothyroidism) or genetics such as the Prader-Willi syndrome (National Health Services [NHS], 2014a).

There is plenty of evidence showing that obesity is one of the major public health threats also to the UK; according to NHS statistics in 2011, 65% of men and 58% of women over 16 years old are overweight or obese (Health and Social Care Information Centre [HSCIC], 2013). These figures pose a threat to the society, which can be translated into non-communicable diseases, morbidity and mortality if not treated, with a high economic burden to NHS.

There is no doubt that obesity can lead to various health problems. Common public health consequences associated with abnormal fat deposits include cardiovascular diseases, especially heart attack and stroke, diabetes, cancer as well as musculoskeletal disorders, mainly osteoarthritis (Weight-control Information Network [WIN], 2014; WHO, 2015). On the other hand, the mechanical stress induced by obesity can result in disabilities caused by several conditions such as shortness of breath, sleep apnoea, osteoarthritis and low back pain (Visscher & Seidell, 2001).

Given the current trend it seems like government actions to tackle obesity are failing. No country has managed to show any supporting evidence for improvements. Policymakers do not seem to be working quickly to identify those reasons and yet they expect fast outcomes. Intervention plans based on a broader approach that deals with agriculture, product manufacture and education might be a starting point to reverse those trends. Table 1 below shows the complexity of factors that contribute to obesity interventions failure.

1.2 Health Practices: Healthy Eating and Physical Activity

There is no doubt that adjusting the daily nutritional intake and eating habits can have a major impact on weight management. In order to lose weight, the daily energy expenditure has to be greater than the consumed energy intake; therefore reducing the daily calories can help towards this goal. Healthy eating together with being physically active should be the right choice for obese individuals if they want to achieve long-term weight loss (NHS, 2014b). However, current lifestyles and easy access to cheap junk food has resulted in unhealthy eating behaviours and a routine with minimal physical activity (NHS, 2013).

Table 1. Failure and factors shaping obesity in developed countries (Source: Lang & Rayner, 2007)

Focus of failure	Factors shaping obesity				
	Domains		Transitions		
	Body	Mind	Diet	Physical activity	Culture
Markets	• Highlight and over- supply particular taste receptors (sweet and fat) • Invest in technical fixes and single-factor solutions	• Appeal to pleasure • Build brand value over nutritional value • Exploit vulnerable groups (e.g. children and low income)	• Produce an excess of inappropriate, energy- dense foods cheaply • Offer only limited investments in workforce training	• Promote fossil-based fuels • Glamorise private motor transport rather than expenditure of food-as-energy	• Market and mould mass consciousness • Barrage consumers with energy-dense food and drink as entertainment
Governments	• Adopt inconsistent modes of protection (interventions on sexual protection but not nutrition) • Are unwilling to modernise public health scope and capacity	• Limit health education to become a minor partner of market information, generating asymmetry of information flow and education	• Subsidise overproduction of fat and sugar compared with micronutrient-rich foods • Emphasize food safety while semi-abandoning nutrition • De-emphasize nutrition and food education	• Oversee decline of physical activity (transport, public spaces, sports facilities) • Prioritise car use in retail and transport planning	• Permit genderised and inadequate food literacy and skills • Promote rights of individualized choice • Facilitate media transmission by paid marketing • Confuse citizenship with marketplace meritocracy (everyone is equal in the market)
Consumers	• Disconnect appetite from need and satiety	• Adopt distorted images of body acceptability • Accept temporality (short-termism) of choice	• Eat a price-led rather than nutrition-led diet • Respond individually rather than en masse to identity crises about meaning and values	• Bow to the ubiquity of the non-energy-expending material world (e.g. in travel to work/shop/school) • Are disinclined to build exercise into daily life	• Consume rather than expend energy as the norm of consumer culture • Participate in physical activity by proxy (TV sports) • Accept inequalities or indulge in victim-blaming

Individuals tend to ignore that the way we approach or manage the process of change, and potentially weight loss, plays an important role. As discussed by Strecher et al. (1995), setting goals can have beneficial effects in health behaviour change and maintenance interventions. Cognitive behavioural therapy (CBT) strategies include specific goal setting, self-monitoring, feedback and reinforcement from outside sources, boosting the confidence in succeeding as well as the use of incentives.

2. Fad Diets

As a consequence, people are more susceptible to adopt various fad diets that claim to aid in losing weight very fast. As stated in CDC's "Healthy Weight – It's not a diet, It's a lifestyle!" a fad diet is any weight loss plan that promises quick results and is usually a temporary nutritional change (Centers for Disease Control and Prevention [CDC], 2014). These diets are considered unhealthy as they provide individuals with less calories and nutrients.

2.1 History of Fad Diets

Fad diets are known for centuries. Since ancient times, it was reported that Greeks and Romans had used them; however, at that time it was more about a healthy and active lifestyle. It was Victorians later who actually adopted fad diets. According to Foxcroft (2011) in her book 'Calories & Corsets: A History of Dieting Over 2 000 Years', the word diet originates from the Greek word *diaita,* which represents a way of life including mental and physical health. It was in the 19th century that people started dieting for aesthetic purposes.

One of the most famous dieter of all time was Lord Byron, who in 1820 made the Vinegar and Water Diet very popular (Foxcroft, 2011). A century later, the Grapefruit Diet was created, where eating grapefruit with each meal was suggested as part of a low calorie diet plan. Interestingly, the Lucky Strike cigarette company launched the known Cigarette Diet based on the appetite suppressing effects of nicotine. Later in 1963, Jean Nidetch founded Weight Watchers and in 1970 the 'sedative' Sleeping Beauty Diet became famous (Rotchford, 2013). The last few decades, fad diets, such as The Atkins and Dukan Diets, became well known, based on high-protein and low-carbohydrate intake (Hughes, 2012). Other examples include the Zone Diet, suggesting a certain ratio of fat protein and carbohydrates, the South Beach Diet, which is a lighter version of Atkins Diet and the Master Cleanse, a diet based mostly in liquid food (Rotchford, 2013). Figure 1 demonstrates the most important diets throughout history.

2.2 How to Identify a Fad Diet

A fad diet is usually described as a weight loss plan that guarantees quick weight loss and dramatic results with no much effort. There are various types of fad diets that can be recognized; however, all of them share some common characteristics (Bastin, 2004). All fad diets promise fast weight loss (more than 1 kg a week) without giving away fatty, rich-in-calories food and without the need to regularly exercise. Most fad diets limit the range of food types included in the meal plan and do not reassure a balanced and healthy diet. They usually propose 'miracle' foods that need to be consumed in abnormal quantities and help fat burning with a minimal effort. Some focus on consuming large quantities of one food type that could result in intestinal disturbance, bloating, bad breath and nutritional imbalances (Bastin, 2004).

Fad diets are usually promoted by 'before and after' images of successful examples of people that have followed the particular diet or by 'experts' in the field of nutrition. However, no health warnings about possible consequences of adopting these fad diets on individuals with chronic diseases are included in the advertisements. Most fad diets are usually based in no or limited research and can lead to serious health impacts (Bastin, 2004).

2.3 Popular Fad Diets

As mentioned before, there is a wide range of proposed fad diets over the last centuries, which are summarised in Tables 2 and 3. They can be categorised into several main groups including low-/no- carbohydrate, high-carbohydrate/high-fibre as well as the liquid formula diets (Bastin, 2004; British Dietetic Association [BDA], 2014).

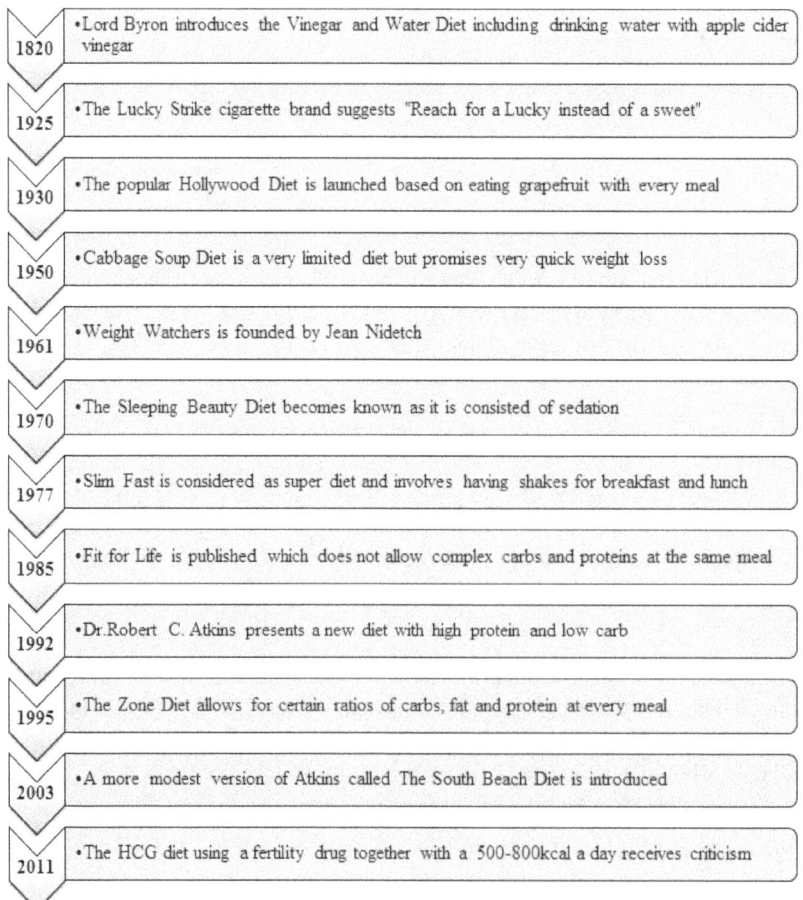

Figure 1. History of dieting over time including the most popular fad diets (adjusted from Rotchford, 2013)

When following a low- or no-carbohydrate diet, a high intake of protein and/or fat is recommended. Example fad diets of this type include the Atkins diet, the Dukan diet, the South Beach diet and the Grapefruit diet. These diets are particularly popular as due to the low intake of energy through carbohydrates, a rapid weight loss mainly due to water loss occurs immediately. The success of such a 'ketogenic' diet together with high protein consumption promotes great weight loss by increasing satiety, which makes it easier to adhere (Geissler & Powers, 2005). However, it is common that the weight is gained back once the diet is discontinued as the body tries to 'correct' the water imbalance (Bastin, 2004) (Figure 2). Although the diet's claimed weight loss compared to conventional energy-restricted diets has been demonstrated in controlled trials (Foster et al., 2003), its long-term effects are still being investigated by the scientific community.

These diets can be quite dangerous due to the ketone formation in excess amounts as a result of incomplete fat breakdown and dehydration. Ketones are stored in blood and can even lead to death if the diet is continued long-term. The Recommended Dietary Allowance (RDA) for carbohydrates is 130 grams per day for both adults and children as an average minimum intake of glucose used by the brain for normal function (Food and Nutrition Board, Institute of Medicine, 2005).

Table 2. Types of fad diets (adjusted from Bastin, 2004; BDA, 2014)

Diet type	Known examples
Low carbohydrates (<100g/day)	• Atkins Diet Revolution
	• South Beach Diet
Extremely low fat (<20% kcal from fat/day)	• Pritikin Diet
	• Pasta Diet
Combination	• Fit for Life
	• Zone Diet
Very low kcalorie (<800 kcal/day)	• Cambridge Diet
	• Rotation Diet
Novelty (certain nutrients or foods)	• Beverly Hills Diet
	• Junk Food Diet
Formula	• Slim Fast
	• Last Chance Diet
Pre-measured	• Jenny Craig
	• Nutri-System
Detox	• The Master Cleanse
High fat	• Ketogenic Diet
High protein	• Dukan Diet
	• Bodybuilder Diet

Table 3. Main categories of fad diets as suggested by the BDA

High-protein Low-Carbohydrate	Moderate-fat Low-carbohydrate	Low-fat Very-high-carbohydrate	Very-low-calorie
Atkins	Jenny Craig	Ornish	Bernstein
Dukan	Nutri-System	The New Pritikin program	Lighter Life
South Beach	Weight Watchers	LEARN	Slim Fast
Zone			

On the other hand, high-carbohydrate/high-fibre-diets are also popular, including the Pritikin diet plan and Save-Your-Life diet. This type of diets provide with low levels of both proteins and fats, which can lead to reduced immunity and problematic wound healing (Bastin, 2004). Furthermore, the liquid formula diets contain very limited calorie intake (often 400-500 calories) and promise to supply with the necessary nutrients. An example of a very low-calorie diet is the Cambridge diet. These diets should be adopted with caution as the self-prescribed starvation be linked with health risks and serious illnesses such as anaemia, reduction of vitamins and minerals supply, fatigue as well as weakness and dizziness. Lastly, diets suggesting the consumption of only one food group like the Cider Vinegar and Vitamin B_6 Diet are also popular but can cause malnutrition and reduced renal function (Bastin, 2004).

3. Causes Leading to Fad Dieting

3.1 The Effect of Peer Pressure

Following a fad diet is often a result of peer pressure. Peer pressure occurs when people of the same age influence an individual's behaviours and decisions by making the person feeling uncomfortable, including the way of thinking of themselves, dressing and eating. Peer pressure originates not only from family and friends but also from the outside environment (Berry, 1999). Criticism of weight and diet by peer members is associated particularly with dieting in teenagers (Cattarin & Thompson, 1994). The ideal body image introduced by peers can lead to feelings of inferiority, low self-esteem and depression. Emotions of guilt and unattractiveness are strongly developed causing individuals to adopt fad diets to make themselves more likable. These individuals appear to be obsessed with their appearance, weight and popularity within their peers that results in inefficient dieting (Berg, 1996).

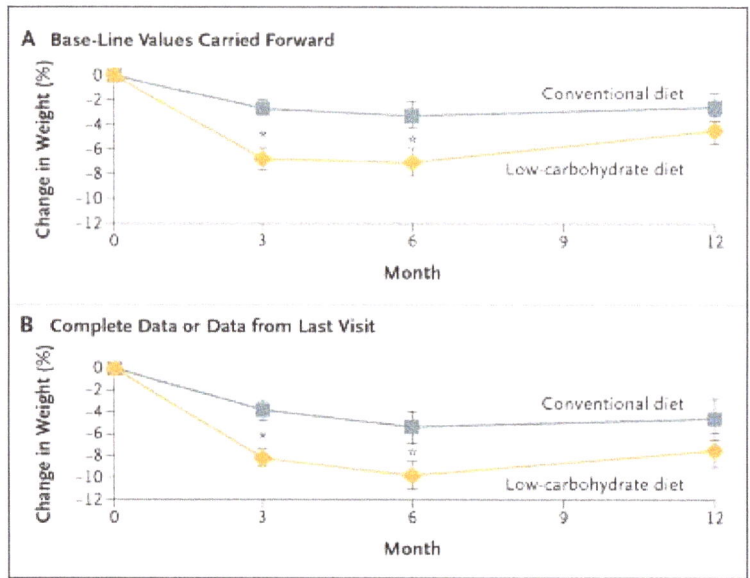

Figure 2. Average weight changes amongst subjects on a low-carbohydrate diet and a low-calorie/high-carbohydrate conventional diet (adapted from Foster et al., 2003)

Furthermore, possible failed attempts to quickly lose weight can lead to depression and 'yo-yo' dieting. Extreme weight loss followed by quick weight gain is associated with many health risk factors such as heart disease, cancer, diabetes, increased in LDL cholesterol as well as reduced muscle and energy. It is believed that these negative health impacts are linked with the stress hormone called cortisol (McNight, 2013). These feelings and continuous dieting can result in unhealthy eating habits, which can continue over someone's lifetime (Berry, 1999). The fear of being fat can also lead to eating disorders from a young age, such as anorexia nervosa and bulimia. In both disorders, sufferers show body image distortions and feelings of anxiety and shame about eating (Human Diseases and Conditions [HDC], 2014).

3.2 The Effect of Media on Body Image and Self-Esteem

Over the last decade, statistics have shown an increase in people suffering from eating disorders in many societies due to the continuous value that they put on being thin. In every aspects of someone's life, such as going for shopping, watching television, reading fashion magazines and following favourite celebrities, the very thin figure is linked with a happy and successful life. Therefore, thousands of teenagers are exposed to the 'ideal', unrealistic image of models, who according to medical standards, try to maintain a 15% below normal weight and meet the criteria for anorexia (Mirror Mirror Eating Disorders [MMED], 2014). Television, movies and social media are full of diet advertisements about food supplements, diet programs and in general chemically based ways to lose weight. Every month new novel diets promise to cause dramatic changes to the appearance of overweight people; probably due to the fact that all previous diets did not work and are rather unhealthy (MMED, 2014). The inevitable effects of food advertising on eating behaviours was studied in elementary-school-aged children which received a snack after watching a cartoon with either a food advertisement or one about other products (Harris, Bargh, & Brownell, 2009). The results showed that children consumed 45% more snacks (28.5 gr) when exposed to the food advertisements than their controls (19.7 gr) (Harris et al., 2009).

Apart from advertisements, television and fashion magazines also include articles or reports about appearance meaning how to look perfect, how to be in shape, how to apply makeup or suggestions about clothing. Magazines are also full of photos that have previously been photo-shopped containing wrinkle-free faces and fat-free bodies which are idealistic and far from reality (Education.com, 2014). A study has found that 69% of teenage girls agreed that the photos in magazines affected their idea of the 'ideal' body image, while 47% said that as a result they desired to lose weight (Field, 2000). The frequency of reading magazine's articles about weight loss and diets increases the possibilities of adopting unhealthy eating and weight control behaviours like skipping meals and fasting, especially among teenagers (Van den Berg, Neumark-Sztainer, Hannan, & Haines, 2007).

4. Weighing Short-Term Pros against Long-Term Cons

Table 4 summarises the advantages and drawbacks of the main categories of fad diets, which are discussed in more detail below.

Table 4. Advantages and disadvantages of fad diets as suggested by the BDA (2014)

Types of diets	Pros	Cons
High-protein diets: *Atkins, Dukan, South Beach, Zone*	• Rapid weight loss • Increase satiety • Improved TG level • Improved serum cholesterol	• Not sustainable • High-fat content • Nutrient deficiency • Detrimental to brain and heart • Increased risk for CHD
Moderate-fat, high-carbohydrate diets: *Jenny Craig, Nutri-System, Weight Watchers*	• Reduced saturated fat intake • Increased consumption of fruit and vegetables • Significant weight loss • Reduction of the risk for diabetes	
Low-fat, very high-carbohydate diets: *Ornish, The New Pritikin Program*	• Possible reduction of cardiovascular disease risks	• Increased TG levels • Decreased HDL-C levels • Micronutrient deficiency
Very low-calorie diets: *Bernstein, Leighter Life, Slim Fast*	• Initiates quick weight loss • Improved quality of life • Long-term benefits in conjunction with exercise	• Enhanced diuresis • Electrolyte loss • Disturbed acid-base balance • Should be used only under medical supervision

4.1 Benefits of Fad Diets

Fad diets are usually known for showing drastic results and some of them can actually offer some benefits. There are plenty of benefits when following a low-carbohydrate diet. Atkins diet is known for the suppression of appetite and its anorectic properties, mainly due to the high protein intake (McClernon, Yancy, Eberstein, Atkins, & Westman, 2007). It results in a rapid weight loss, especially during the first few weeks. In a controlled study of 63 obese patients, Atkins diet resulted in a weight loss of 6.8% body weight after three months compared to 2.7% accomplished by an energy-restricted diet (Foster et al., 2003). Low-carbohydrate diets have also a great potential in eliminating a larger proportion of abdominal fat (Volek et al., 2004). In addition, carbohydrate restriction can be efficient in significantly lowering both the plasma LDL cholesterol and triglyceride following a twelve-week weight loss intervention (Wood et al., 2006). At the same time, HDL levels can be dramatically increased because of the high-fat consumption included in Atkins diet (Brinkworth, Noakes, Buckley, Keogh, & Clifton, 2009). Furthermore, in a study testing the effect of a low-carbohydrate diet in type 2 diabetes, it was reported that most patients improved their glycemic index and reduced or eliminated medication control within six months (Westman et al., 2007). Lastly, Atkins diet has shown to reduce blood pressure short-term, therefore leading to a decreased risk of developing cardiovascular diseases (Gardner et al., 2007).

The proposed meal plans such as in many Detox diets or the Grapefruit diet often consist of lots of fruits and vegetables. These food types are popular for being healthy and beneficial but of low calories at the same time. Fad diet users will therefore supply with many vitamins, minerals and antioxidants. Following these eating habits also helps in promoting physical and psychological health. A person that loses weight firstly reduces the risks linked with obesity and secondly feels lighter, healthier and with increased self-confidence (Health Research Funding [HRF], 2014). Moreover, adopting a fad diet can help someone recognise which characteristics of his/her current diet are responsible for the excess weight. Fad diet programs like the French Woman's diet can also help establish the right portion sizes and avoid unwanted snaking (Cespedes, 2014).

4.2. Health consequences of fad dieting

On the other hand, fad diets do not come without disadvantages. Weight loss occurs too fast, most of the lost weight being water and muscle, not fat tissue. Rapid weight loss can further lead to various health risks such as constipation, low nutrients and energy intake and tiredness, all caused by eating less calories (HRF, 2014).

Fad diets, especially the low-carbohydrate/high-protein diets such as The Dukan diet, involve physiological aspects; they limit the amount of carbohydrate and sugar intake and at the same time promote the consumption of animal protein. As a result, there is a shift from using glucose to using fatty acids and ketones as the main fuel sources; therefore, the glycogen availability increases (Westman et al., 2007). Ketosis could then result in a decrease of appetite but also cause hyperuricemia since ketones compete with urine acid for renal tubular excretion (Denke, 2001). Despite the reputation of the ketogenic diets there is little scientific support of their long-term side effects and sustainability. In order to suggest the adaptation of such diets as a 'life nutritional philosophy', possible adverse effects on health and disease prevention need to be studied.

Studies have shown that high-fat/high-protein diets like The Atkins diet also result in higher risk of heart disease (WHO/Food & Agriculture Organisation Expert Consultation, 2013), colon cancer (Giovannucci et al., 1994), bad breath (Mahon & Escott-Stump, 1996) and sleeping disorders (Fitness, 2014). Anderson, Konz, and Jenkins (2000) showed that the Atkins diet increase the cholesterol levels in the serum, increasing the risk of atherosclerosis and coronary heart disease by greater than 50% if used in long term. Various adverse effects of following a very-low-carbohydrate short-term diet have previously been reported including constipation due to low fibre intake (68% of participants), bad breath (63%), headache (51%), hair loss (10%) and increased menstrual bleeding (3%) (Westman, Yancy, Edman, Tomlin, & Perkins, 2002).

High-protein diets have been linked with a greater production of nitrogen waste products, which add a pressure on the kidneys, especially in dehydration state (Denke, 2001). Another study has shown that after six weeks of such a diet there is an increased acid load to the kidney maximising the possibility of stone formation and a decrease of calcium balance leading to bone loss and osteoporosis (Reddy, Wang, Sakhaee, Brinkley, & Pack, 2002). Consuming two or three times more protein than the recommended daily allowance has been associated with loss of calcium through the urinary tract, which could predispose to bone loss long-term (Eisenstein, Roberts, Dallal, & Saltzman, 2002). Although significant reduction in insulin responses has been observed after a low-carbohydrate diet (Westman et al., 2007), it is believed that insulin sensitivity and resistance may be negatively influenced in the long-term (Riccardi, Giacco & Rivellese, 2004; Shulman, 2000). Moreover, following such a diet for several months can cause an increase in plasma homocysteine (Clifton, Noakes, Foster, & Keogh, 2004), the effect of which has been linked with higher cardiovascular risk (Refsum, Ueland, Nygard, & Vollset, 1998).

On the other hand, low-protein/low-fat diets might increase the risk of inadequate intake of minerals such as calcium and zinc as well as high-quality protein. For example when following the Pritikin diet the recommended fat intake should be less than 10% of the daily energy intake which is close to the lowest limit of the daily requirement for essential fatty acids (Pritikin, 1981). Other nutritional deficiencies, such as iron deficiency, can be associated with growth decrease and anaemia in adolescents (Dietz & Hartung, 1985). Additionally, adopting unbalanced eating habits has been connected with menstrual irregularity or amenorrhea (Kreipe, Strauss, Hodgman, & Ryan, 1989). Table 5 summarises the most popular fad diets, their main components and their physiological implications, the main of which are explained below.

4.3 Psychological Implications

Studies in adults have suggested that constant dieting (yo-yo dieting) is associated with a range of symptoms such as food obsession, constant calorie counting, distractibility, increased emotional responsiveness and fatigue (Polivy, 1996). Chronic dieters also tend to overeat, have low self-esteem as well as suffer from some eating disorders and depression (Polivy, 1996). It is known that these effects are particularly true in children and teenagers, causing complicated consequences in their social and psychological development (Canadian Paediatric Society [CPS], 2004). From previous studies it has been shown that following a structured, multidisciplinary weight loss regime might have a negative effect on the self-esteem of both children (Cameron, 1999) and adolescents (Stice, Cameron, Killen, Hayward, & Taylor, 1999). Current research has linked dieting with higher risk of developing disordered eating as well as a trend to overeat; however, the causes are still unclear or controversial (Polivy, 1996). Lastly, self-directed dieting early in life (9-14 years old) has been associated with weight gain overtime in large-scale, 3-year-long study involving more than 15 000 subjects (Field et al., 2003).

4.4 Safety of Fad Diets

Fad diets promise quick weight loss but are generally thought to be unhealthy and temporary. Weighing the advantages over the drawbacks of fad dieting can be hard but necessary since there is no scientific evidence proving their safety. According to NHS (2013) in 2011 the BDA had warned against many fad diets since they were not based on clinical trials and reportedly did not give rise to long-term results. Also, most of these diets overemphasise one particular food type (for example the Banana or the Rice diet) and are considered dangerous and nutritionally unbalanced. Furthermore, there is not always supplementary scientific evidence to support the proposed claims (Fisher, 2011).

5. Sustainability of Weight Loss by Fad Diets

Furthermore, the strict regime of fad diets makes it very difficult to commit as eating low-calorie foods requires both preparation and lifestyle change (HRF, 2014). It is common that most of dieters regain the weight they have lost - or even more within a few months - a situation that leads to yo-yo dieting. It is believed that it is very hard to maintain a fad diet long-term as the dieters revert back to their previous eating habits at some point. Thomas, Hyde, Karunaratne, Kausman, and Komesaroff (2008) conducted a study on obese subjects and found that they were unable to commit to these diets for various reasons such as unsuitable or costly diet, concordance with their lifestyle and monotony in food choices.

Foster et al. (2003) showed that although following a low-carbohydrate diet produced a larger weight loss compared to the conventional diet at six months period, the difference was not statistically significant after a year. Moreover, these findings are supported by a study published in 2006 from the BBC "diet trials". This randomized controlled study compared the effectiveness of four popular fad diets in the UK over six months including the Dr Atkins' new diet revolution, the Slim-Fast plan, the Weight Watchers pure points plan, and the Rosemary Conley's eat yourself slim diet and fitness plan (Truby et al., 2006). The trial was based in 293 healthy overweight or obese adults, who all succeed a significant loss of both weight and body fat regardless of the fad diet followed. Although the Dr Atkins' new diet revolution resulted in higher weight loss during the first month, there were no statistically significant differences among the diet plans after 6 months. The study also showed that the initial sustainability was not feasible over a long period of time as demonstrated in Figure 3. These findings are also supported by the review conducted by Jebb and Goldberg (1998); however, it was also shown that there was a small portion of participants (3-4% based on 381 subjects) that were still 30% or more below baseline after 5 years and actually managed to maintain their weight loss. Nevertheless, it should be noted that these individuals had incorporated physical activity and meal replacement into their lifestyle (Jebb & Goldberg, 1998).

Table 5. Popular fad diets and their health consequences

Type of diet	Example	Summary/Main components	Health consequences	References
Low-carbohydrate/ High-protein	**Atkins diet**	Less than 50g CHO per day, high consumption of animal protein	Water imbalance, ketosis, appetite suppression, renal dysfunction, nausea, low performance capacity, dehydration, osteoporosis	Atkins (1992)
	Dukan diet	Carb-free, high in protein diet structured in four different phases		Diabetes.co.uk (2011)
High-carbohydrate/ Low-fat	**Pritikin diet**	Low-fat, low-calorie, plant-based foods, mainly fruits and vegetables, fats not exceeding 10% of total daily calories	Inadequate intake of good quality protein, vitamin and mineral deficiencies, coronary heart disease	Pritikin (1981)
Low carbohydrate/ High-fat	**Ketogenic diet**	High-fat, adequate-protein, low-carbohydrate diet changing the way energy is used in the body	Acidosis, hypoglycemia, gastrointestinal distress, dehydration, lethargy, kidney stones, dyslipidemia, decreased bone density	Freeman, Kossoff, & Hartman (2007)
Combination	**Zone diet**	Hormonal control (insulin, glucagon and eicosanoids) via a specific ratio of protein, CHO and fat intake	Vitamin and mineral deficiencies	Sears & Lawren (1995)
One-food	**Grapefruit diet**	Certain combinations of food with grapefruit and grapefruit juice, no sugary/starchy food allowed	Unbalanced nutrition, interfere with certain medication, risk for increase intake of saturated fat, sodium and cholesterol.	Ipatenco (2014)
Formula	**Slim Fast diet**	Dietary supplement based on a drinkable meal replacement once or twice a day, each containing around 240 kcal	Low nutritional intake, low energy, weakness, risk for eating disorders	Stern (2015)

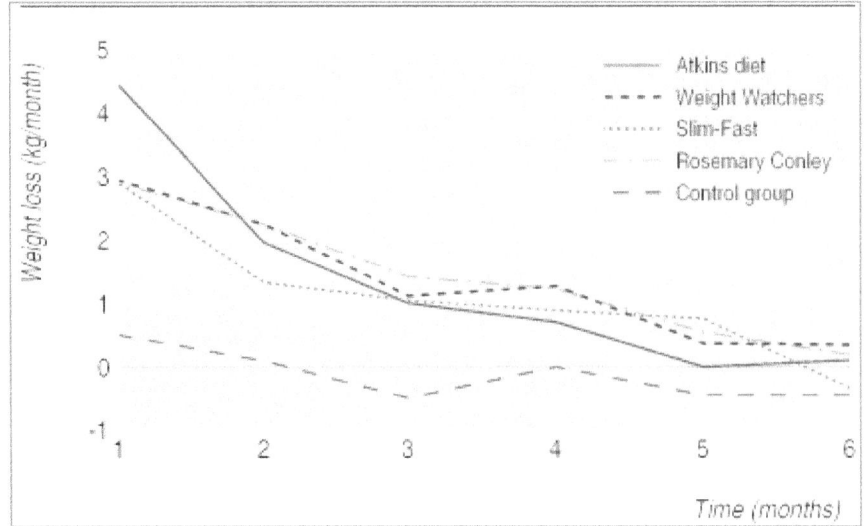

Figure 3. Weight loss over six months after following four commercial weight loss programs in UK (Source: Truby et al., 2006)

6. Fibre-Enriched Diets and Weight Management

As shown above, although following a fad diet could result in rapid weight loss within the first few weeks, the sustainability of the achieved results is very low. It should be mentioned that adjusting daily dietary habits and physical activity is the key to weight management including weight loss and maintenance. To gain the required knowledge and positively progress towards long-term weight loss, one has to go through various stages of adaptation such as commitment to change and self-efficacy. Temporary changes of eating habits only leads to temporary weight loss results and to overcome fad dieting drawbacks, a high-in-fruits-and-vegetables (50%) but low-in-fat-and-sugars (<30%) diet has been recommended by various researchers (National Institutes of Health [NIH], 1998).

It is commonly believed that a high-fruit diet is ideal for weight loss since fruits are known to have unique properties and play an important part of a healthy and balanced diet (He et al., 2004). Fruits are high-in-water, high-in-sugar, low-in-fat and high-in-fibre, have only a few calories as well as essential minerals, vitamins and antioxidants (Martin, Cherubini, Andres-Lacueva, Paniagua, & Joseph, 2002; Swinburn, Caterson, Seidell, & James, 2004). Both soluble and insoluble fibres are found in considerable amounts in fruits (especially in fruits' skins) and have been reported to help with weight management (Pereira, & Ludwig, 2001). Researchers have shown that dietary fibre increases post-consumption satiety and also decreases following hunger (Howarth, Saltzman, & Roberts, 2001). Therefore, by eating fruits with low energy density rather than foods with high energy density, one can eat greater amounts but the same calorie content (CDC, 2015). Howarth et al. (2001) have shown that by increasing the daily fibre intake to an extra 14 gr per day for two days decreased the energy intake by 10% and resulted in ~2 kg weight loss over 4 months. Furthermore, soluble fibre has a beneficial impact in controlling post-meal glycaemic and insulin responses as it affects gastric emptying and how macronutrients are absorbed by the gut (Babio, Balanza, Basulto, Bulló, & Salas-Salvadó, 2010).

Given all the benefits of increased fruit intake mentioned above, researchers have introduced and tested various interventions to prevent or tackle obesity. As an example, Tohill (2005) summarised a large number of epidemiological studies on the relationship of fruit intake and BMI on both children and adults. It has been proven that a high fruit and vegetable intake (two servings of fruits and three servings of vegetables a day) is associated with a decrease in high fat and sugar intake (less than 10 servings of high-fat/high-sugar foods per week), potentially helpful in families with obesity history (Epstein et al., 2001). Moreover, in contrast with fad diets, as illustrated after a 4-year-long intervention, individuals can sustain drastic dietary changes that mainly include high-fibre, low-fat food provided the appropriate support (Lanza et al., 2001).

6.1 Health Benefits of Dietary Fibre

Various observational and experimental studies have shown that there is a positive link between higher fibre intake and lower risk of many health risk factors. Increasing dietary soluble fibre results in the reduction of various biochemical parameters such as total and LDL cholesterol and can be introduced as a small contributor to dietary therapy to lower blood cholesterol (Brown, Rosner, Willett, & Sacks, 1999). Similarly, greater dietary fibre intake has been associated with a lower risk of both cardiovascular disease and coronary heart disease in a total of 22 cohort studies (Threapleton et al., 2013).

Also, better kidney function as well as reduced inflammation is considered one of the health benefits of soluble fibre (Xu et al., 2014). It is known that rich-in-fibre food types contain large amounts of essential minerals such as calcium, potassium and magnesium that can lead in a higher total bone mass providing evidence of a positive association between fruit and vegetable consumption and bone health (New et al., 2000). Lastly, initial controlled trial recently conducted by our research group introducing soluble fibre in the form of a fruit salad in the daily nutritional intake has also highlighted the positive effects of increased dietary fibre in both weight management and better health.

7. Government Public Health Initiatives

In the UK, the National Health Service is trying to combat fad diets and obesity by informing the public on health and nutrition as well as promoting healthier eating habits for weight loss (NHS Choices, 2014). Together with the British Dietetics Association, NHS emphasises the pros and cons of popular fad diets. Also, the Medicines and Healthcare products Regulatory Agency (Medicines and Healthcare products Regulatory Agency [MHRA], 2014) is trying to prevent people from using diet drugs like the homeopathic human chorionic gonadotropin (hCG) by not authorizing it for use in the UK.

Furthermore, within the UK, the National Health Service has initiated the Change4life campaign that promotes the '5-a-day' which includes consuming at least five portions of fruit and vegetables per day as part of a balanced diet (Change4life, 2015). Additionally, in an attempt to tackle obesity, especially in children, the non-profit organization Mind, Exercise, Nutrition… Do it! (MEND) offers weight management programs and family support all over the UK (MEND, 2015). This program provides help designed by health professionals including nutritionists, dietitians, physical activity experts as well as behaviour change specialists through one-to-one sessions with parents and children. However, this type of initiative can last for up to 10 weeks meaning that results can only be effective and sustainable if the families take the recommendations on board.

Due to its nutritional composition, fad diets remain a subject of much controversy since the recommended daily allowance for total energy is 35% of fat, 50% of carbohydrates and 15% of proteins (Scientific Advisory Committee on Nutrition, 2015). Possibly one way forward would be to educate people on healthy practices and enable lifestyle changes. This requires realistic goals and slow weight loss followed by a balanced diet, engagement in physical activity and small portion sizes. Following the eatwell plate where 1/3 of foods should be composed of fruits and vegetables, 1/3 of starch, moderate amounts of dairy and meat products and occasionally small amounts of food high in fat and sugar would be a good start (Food Standards Agency [FSA], 2010) (Figure 4).

8. Conclusion

Fad diets have been popular for decades due to societal and peer pressure to have certain body shape. Most of them limit the range of food types included in the meal plan and do not reassure a balanced and healthy diet. They promise fast weight loss; nevertheless they suffer from many drawbacks such as introducing health risks and low sustainability. Nowadays, there are various types of fad diets available; however, individuals should be aware that there is no scientific evidence proving their safety. Fad diets have been linked with many physiological conditions such as cardiovascular disease, renal dysfunction and osteoporosis as well as psychological implications like eating disorders and depression. A long-term, high-fibre diet will overcome these complications, as the weight loss results are more likely to be sustainable. Increased dietary soluble fibre also provides with various health benefits such as better bone health in elderly populations commonly suffering from osteoporosis.

Figure 4. The eatwell plate (Source: FSA, 2010)

References

Anderson, J. W., Konz, E. C., & Jenkins, D. J. (2000). Health advantages and disadvantages of weight-reducing diets: a computer analysis and critical review. *Journal of American College of Nutrition, 19*, 578-590.

Atkins, R. (1992). *Dr Atkins' new diet revolution*. New York, NY: Avon.

Babio, N., Balanza, R., Basulto, J., Bulló, M., & Salas-Salvadó, J. (2010). Dietary fibre: influence on body weight, glycemic control and plasma cholesterol profile. *Nutrición Hospitalaria, 25*(3), 327-340.

Bastin, S. (2004). *Fad Diets*. Available at: http://www2.ca.uky.edu/hes/fcs/factshts/fn-ssb.119.pdf

Berg, F. (1996). Children in weight crisis. *Healthy Weight Journal, 10*(5), 86-87.

Berry, L. (1999). Media and peer influence on fad diets tried by adolescent females. *University of Wisconsin-Stout*. Available at: http://www2.uwstout.edu/content/lib/thesis/1999/1999berry.pdf

Brinkworth, G. D., Noakes, M., Buckley, J. D., Keogh, J. B., & Clifton, P. M. (2009). Long-term effects of a very-low-carbohydrate weight loss diet compared with an isocaloric low-fat diet after 12 mo. *American Journal of Clinical Nutrition, 90*(1), 23-32. http://dx.doi.org/10.3945/ajcn.2008.27326

British Dietetic Association. (2014). *Food Fact Sheet – Detox Diets*. Available at: https://www.bda.uk.com/foodfacts/detoxdiets

Brown, L., Rosner, B., Willett, W. W., & Sacks, F. M. (1999). Cholesterol-lowering effects of dietary fiber: a meta-analysis. *American Journal of Clinical Nutrition, 69*(1), 30-42.

Cameron, J. W. (1999). Self-esteem changes in children enrolled in weight management programs. *Issues in Comprehensive Pediatric Nursing, 22*(2-3), 75-85. http://dx.doi.org/10.1080/014608699265301

Canadian Paediatric Society. (2004). Dieting in adolescence, *Paediatrics & Child Health, 9*(7), 487-491.

Cattarin, J. A., & Thompson, J. K. (1994). A three-year longitudinal study of body image, eating disturbance and general psychological functioning in adolescent females. *Eating Disorders: Journal of Treatment and Prevention, 2*, 114-125. http://dx.doi.org/10.1080/10640269408249107

Centers for Disease Control and Prevention. (2014). *Healthy Weight - it's not a diet, it's a lifestyle!*. Available at: http://www.cdc.gov/healthyweight/index.html

Centers for Disease Control and Prevention. (2015). *Can eating fruits and vegetables help people to manage their weight?*. Available at: http://www.cdc.gov/nccdphp/dnpa/nutrition/pdf/rtp_practitioner_10_07.pdf

Cespedes, A. (2014). *The pros and cons of fad diet*. Available at: http://www.livestrong.com/article/366562-the-pros-cons-of-fad-diets/

Change4Life (2015). *5 A DAY- tips for getting five portions of fruit and veg each day.* Available at: http://www.nhs.uk/Change4Life/Pages/five-a-day.aspx

Clifton, P., Noakes, M., Foster, P., & Keogh, J. B. (2004). Do ketogenic diets for weight loss lower cardiovascular risk?. *International Journal of Obesity, 28*, S26.

Denke, M. (2001). Metabolic effects of high-protein, low-carbohydrate diets. *American Journal of Cardiology, 88*, 59-61. http://dx.doi.org/10.1016/S0002-9149(01)01586-7

Diabetes.co.uk. (2011). *Dukan Diet.* Available at: http://www.diabetes.co.uk/diet/dukan-diet.html

Dietz, W. H., & Hartung, R. (1985). Changes in height velocity of obese preadolescents during weight reduction. *American Journal of Diseases of Children, 139*, 705-707.

Education.com. (2014). How do magazines affect body image. Available at: http://www.education.com/reference/article/how-magazines-affect-body-image/

Eisenstein, J. I., Roberts, S. B., Dallal, G., & Saltzman, E. (2002). High-protein weight-loss diets: are they safe and do they work? A review of the experimental and epidemiologic data. *Nutrition Reviews, 60*(7 Pt1), 189-200. http://dx.doi.org/10.1301/00296640260184264

Epstein, L. H., Gordy, C. C., Raynor, H. A., Beddome, M., Kilanowski, C. K., & Paluch, R. (2001). Increasing Fruit and Vegetable intake and Decreasing Fat and Sugar Intake in Families at Risk for Childhood Obesity. *Obesity Research, 9*(3), 171-178. http://dx.doi.org/10.1038/oby.2001.18

Field, A. E. (2000). Media influence in self-image: The real fashion emergency. *Healthy Weight Journal, 11*, 88-95.

Field, A. E., Austin, S. B., Taylor, C. B., Malspeis. S., Rosner, B., Rockett, H. R., ... Colditz, G. A. (2003). Relation between dieting and weight change among preadolescents and adolescents. *Pediatrics, 112*, 900-906.

Fisher, K. (2011). *Uncover the pros and cons of fad dieting*. Available at: http://www.insidershealth.com/article/uncover_the_pros_and_cons_of_fad_dieting/3491

Fitness. (2014). *Uncover the pros and cons of fad dieting*. Available at: http://www.fitness.com/articles/242/uncover_the_pros_and_cons_of_fad_dieting.php

Food and Nutrition Board, Institute of Medicine. (2005). *Dietary Reference Intakes for Energy, Carbohydrate, Fiber, Fat, Fatty Acids, Cholesterol, Protein and Amino Acids. The National Academies Press*. Available at: http://www.nap.edu/catalog/10490/dietary-reference-intakes-for-energy-carbohydrate-fiber-fat-fatty-acids-cholesterol-protein-and-amino-acids-macronutrients

Food Standards Agency. (2010). *Your guide to the eatwell plate-helping you eat a healthier diet*. Available at: http://collections.europarchive.org/tna/20100927130941/http://food.gov.uk/multimedia/pdfs/publication/eat wellplateguide0310.pdf

Foster, G. D., Wyatt, H. R., Hill, J. O., McGuckin, B. G., Brill C., Mohammed, B. S., ... Klein S. (2003). A randomized trial of a low-carbohydrate diet for obesity. *The New England Journal of Medicine, 348*(21), 2082-2090. http://dx.doi.org/10.1056/NEJMoa022207

Foxcroft, L. (2011). *Calories & Corsets: A History of Dieting Over 2,000 Years* (1st ed). London: Profile Books.

Freeman, J. M., Kossoff, E. H., & Hartman, A. L. (2007). The Ketogenic Diet: One Decade Later. *Padiatrics, 119*, 535-543. http://dx.doi.org/10.1542/peds.2006-2447

Gardner, C. D., Kiazand, A., Alhassan, S., Kim, S., Stafford, R. S., Balise, R. R., ... King, A. C. (2007). Comparison of the Atkins, Zone, Ornish, and LEARN diets for change in weight and related risk factors among overweight premenopausal women: the A TO Z Weight Loss Study: a randomized trial. *JAMA, 297*(9), 969-977. http://dx.doi.org/10.1001/jama.297.9.969

Geissler, C. A., & Powers, H. J. (2005). *Human Nutrition* (11th ed). Philadelphia: Elsevier.

Giovannucci, E., Rimm, E. B., Stampfer. M. J., Colditz. G. A., Ascherio, A., & Willett, W. C. (1994). Intake of fat, meat, and fibre in relation to risk of colon cancer in men. *Cancer Research, 54*, 2390-2397.

Harris, J. L., Bargh, J. A., & Brownell, K. D. (2009). Priming effects of television food advertising on eating behavior. *Health Psychology, 28*(4), 404-413. http://dx.doi.org/10.1037/a0014399

He, K., Hu, F. B., Colditz, G. A., Manson, J. E., Willett, W. C., & Liu, S. (2004). Changes in intake of fruits and vegetables in relation to risk of obesity and weight gain among middle-aged women. *International Journal of Obesity, 28*, 1569-1574. http://dx.doi.org/10.1038/sj.ijo.0802795

Health and Social Care Information Centre. (2013). *Statistics on obesity, physical activity and diet.* Available at: http://www.hscic.gov.uk/catalogue/PUB10364

Health Research Funding. (2014). *Pros and Cons of Fad Diets.*
Available at: http://healthresearchfunding.org/pros-cons-fad-diets/

Howarth, C., Saltzman, E., & Roberts, B. (2001). Dietary Fiber and Weight Regulation, *Nutrition Reviews, 59*(5), 129-139. http://dx.doi.org/10.1111/j.1753-4887.2001.tb07001.x

Hughes, K. (2012). Calories & Corsets: A History of Dieting Over 2,000 Years by Louise Foxcroft – review. *The Guardian.*
Available at: http://www.theguardian.com/books/2012/jan/13/calories-and-corsets-louise-foxcroft-review

Human Diseases and Conditions. (2014). *Eating Disorders.*
Available at: http://www.humanillnesses.com/original/E-Ga/Eating-Disorders.html#ixzz3FZz0vdzs

Jebb, S., & Goldberg, G. (1998). Efficacy of very low-energy diets and meal replacements in the treatment of obesity. *Journal of Human Nutrition and Dietetics, 11*(3), 219-226.
http://dx.doi.org/10.1046/j.1365-277X.1998.00101.x

Ipatenco, S. (2014). *The disadvantages of the Grapefruit diet.*
Available at: http://www.livestrong.com/article/427630-the-disadvantages-of-the-grapefruit-diet

Kreipe, R. E., Strauss J., Hodgman, C. H., & Ryan R. M. (1989). Menstrual cycle abnormalities and subclinical eating disorders: A preliminary report. *Psychosomatic Medicine; 51*, 81-86.

Lang, T., & Rayner, G. (2007). Overcoming policy cacophony on obesity: an ecological public health framework for policymakers. *Obesity Reviews, 8*, 165-181. http://dx.doi.org/10.1111/j.1467-789X.2007.00338.x

Lanza, E., Schatzkin, A., Daston, C., Corle, D., Freedman, L., Ballard-Barbash, R., ... the PPT Study Group. (2001). Implementation of a 4-y, high-fiber, high-fruit-and-vegetable, low-fat dietary intervention: results of dietary changes in the Polyp Prevention Trial. *American Society for Clinical Nutrition, 74*(3), 387-401.

Mahon K. L., & Escott-Stump, S. (1996). *Krause's Food, Nutrition, and Diet Therapy* (9th ed). Philadelphia: W.B. Saunders Company.

Martin, A., Cherubini, A., Andres-Lacueva, C., Paniagua, M., & Joseph, J. (2002). Effects of fruits and vegetables on levels of vitamins E and C in the brain and their association with cognitive performance. *The Journal of Nutrition, Health and Aging, 6*(6), 392-404.

McClernon, F. J., Yancy, W. S., Eberstein, J. A., Atkins, R. C., & Westman, E. C. (2007). The Effects of a Low-Carbohydrate Ketogenic Diet and a Low-Fat Diet on Mood, Hunger, and Other Self-Reported Symptoms. *Obesity, 15*(1), 182-187. http://dx.doi.org/10.1038/oby.2007.516

McNight, C. (2013). *Health Risks of Yo-Yo Dieting.*
    Available at: http://www.livestrong.com/article/353915-health-risks-of-yoyo-dieting

Mind, Exercise, Nutrition… Do it!. (2015). *What we do.*
    Available at: http://www.mendcentral.org/aboutus/whatwedo

Medicines and Healthcare products Regulatory Agency. (2014). *Medicines and Healthcare products Regulatory Agency.* Available at: http://www.mhra.gov.uk/index.htm#page=DynamicListMedicines

Mirror Mirror Eating Disorders. (2014). *Society and Eating Disorders.*
    Available at: http://www.mirror-mirror.org/society.htm

National Health Services. (2013). *Latest obesity stats for England are alarming - Health News - NHS Choices.* Available at:
    http://www.nhs.uk/news/2013/02February/Pages/Latest-obesity-stats-for-England-are-alarming-reading.asp
    x

National Health Services. (2014a). *Obesity-Causes.*
    Available at: http://www.nhs.uk/Conditions/Obesity/Pages/Causes.aspx

National Health Services. (2014b). *Very low calorie diets.*
    Available at: http://www.nhs.uk/Livewell/loseweight/Pages/very-low-calorie-diets.aspx

NHS Choices. (2014). *Top diets review for 2014 - Live Well - NHS Choices.*
    Available at: http://www.nhs.uk/Livewell/loseweight/Pages/top-10-most-popular-diets-review.aspx

National Institutes of Health. (1998). *Clinical Guidelines On the Identification, Evaluation, and Treatment of Overweight and Obesity in Adults.* Available at: http://www.ncbi.nlm.nih.gov/books/NBK2003/pdf/TOC.pdf

New, S. A., Robins, S. P., Campbell, M. K., Martin, J. C., Garton, M. J., Bolton-Smith, C., ... Reid, D. M. (2000). Dietary influences on bone mass and bone metabolism: further evidence of a positive link between fruit and vegetable consumption and bone health. *American Journal of Clinical Nutrition, 71*, 142-151.

Pereira, M. A., & Ludwig, D. S. (2001). Dietary fiber and body-weight regulation: Observations and mechanisms, *Pediatric Clinics of North America, 48*(4), 969-980. http://dx.doi.org/10.1016/S0031-3955(05)70351-5

Polivy, J. (1996). Psychological consequences of food restriction, *Journal of the American Dietetic Association, 96*(6), 589-592. http://dx.doi.org/10.1016/S0002-8223(96)00161-7

Pritikin, N. (1981). *The pritikin permanent weight loss manual.* New York, NY: Grosset and Dunlap.

Reddy, S. T., Wang, C. Y., Sakhaee, K., Brinkley, L., & Pack, C. Y. C. (2002). Effect of low-carbohydrate high-protein diets on acid-base balance, stone-forming propensity, and calcium metabolism. *American Journal of Kidney Diseases, 40*(2), 265-274. http://dx.doi.org/10.1053/ajkd.2002.34504

Refsum, H., Ueland, P. M., Nygard, O. & Vollset, S. E. (1998). Homocystein and cardiovascular disease. *Annual Review of Medicine, 49*, 31-62. http://dx.doi.org/10.1146/annurev.med.49.1.31

Riccardi, G., Giacco, R., & Rivellese, A. A. (2004). Dietary fat, insulin sensitivity and the metabolic syndrome. *Clinical Nutrition, 23*(4), 447-456. http://dx.doi.org/10.1016/j.clnu.2004.02.006

Rotchford, L. (2013). Diets through history: The good, the bad and the scary. *CNN.* Available at: http://edition.cnn.com/2013/02/08/health/diets-through-history

Scientific Advisory Committee on Nutrition. (2015). *Carbohydrates and Health.* London: TSO.

Sears, B., & Lawren, B. (1995). *The Zone-a dietary road map.* New York, NY: Harper Collins.

Shulman, G. I. (2000). Cellular mechanisms of insulin resistance. *Journal of Clinical Investigation, 106*(2), 171-176. http://dx.doi.org/10.1172/JCI10583

Stern, D. (2015). *The disadvantages of Slim Fast.*
    Available at: http://www.livestrong.com/article/67235-disadvantages-slim-fast/

Stice, E., Cameron, R. P., Killen, J. D., Hayward, C., & Taylor, C. B. (1999). Naturalistic weight-reduction efforts prospectively predict growth in relative weight and onset of obesity among female adolescents. *Journal of Consulting & Clinical Psychology, 67*, 967-974. http://dx.doi.org/10.1037/0022-006X.67.6.967

Strecher, V. J., Seijts, G. H., Kok, G. J., Latham, G. P., Glasgow, R., DeVellis, B., ... Bulger, D. W. (1995). Goal setting as a strategy for health behavior change. *Health Education & Behavior, 22*, 190-200. http://dx.doi.org/10.1177/109019819502200207

Swinburn, B. A., Caterson, I., Seidell, J. C., & James, W. P. T. (2004). Diet, nutrition and the prevention of excess weight gain and obesity. *Public Health Nutrition, 7*(1A), 123-146. http://dx.doi.org/10.1079/PHN2003585

Thomas, S. L., Hyde, J., Karunaratne, A., Kausman, R., & Komesaroff, P. A. (2008). They all work...when you stick to them: A qualitative investigation of dieting, weight loss, and physical exercise, in obese individuals. *Nutrition Journal, 7*(34), 1-7. http://dx.doi.org/10.1186/1475-2891-7-34

Threapleton, D. E., Greenwood, D. C., Evans, C. E. L., Cleghorn, C. L., Nykjaer, C., Woodhead, C., ... Burley, V. J. (2013). Dietary fibre intake and risk of cardiovascular disease: systematic review and meta-analysis. *BMJ, 347*, f6879. http://dx.doi.org/10.1136/bmj.f6879

Tohill, B. C. (2005). *Dietary intake of fruits and vegetables and management of body weight.* Available at: http://www.who.int/dietphysicalactivity/publications/f%26v_weight_management.pdf

Truby, H., Baic, S., deLooy, A., Fox, K. R., Livingstone, M. B. E., Logan, C. M., ... Millward, D. J. (2006). Randomised control trial of four commercial weight loss programmes in the UK: initial findings from the BBC "diet trials". *BMJ, 332*(7555), 1309-1314. http://dx.doi.org/10.1136/bmj.38833.411204.80

Van den Berg, P., Neumark-Sztainer, D. R., Hannan, P. J., & Haines, J. (2007). Is dieting advice from magazines helpful or harmful? Five-year associations with weight-control behaviors and psychological outcomes in adolescents. *Pediatrics, 119*(1), e30-e37. http://dx.doi.org/10.1542/peds.2006-0978

Visscher, T. L. S., & Seidell, J. C. (2001). The Public Health Impact of Obesity. *Annual Review of Public Health, 22*, 355-375. http://dx.doi.org/10.1146/annurev.publhealth.22.1.355

Volek, J., Sharman, M., Gomez, A., Judelson, D., Rubin, M., Watson, G., ... Kraemer, W. (2004). Comparison of energy-restricted very low-carbohydrate and low-fat diets on weight loss and body composition in overweight men and women. *Nutrition & Metabolism (London), 1*(1), 13. http://dx.doi.org/10.1186/1743-7075-1-13

Weight-control Information Network. (2014). *Overweight and Obesity Statistics.* Available at: http://win.niddk.nih.gov/statistics/index.htm

World Health Organisation. (2015). *Obesity and overweight.* Available at: http://www.who.int/mediacentre/factsheets/fs311/en/

World Health Organisation/Food & Agriculture Organisation Expert Consultation. (2003). Joint Report: Diet, Nutrition and the Prevention of Chronic Diseases. *WHO Technical Report Series, 916.*

Westman, E. C., Feinman, R. D., Mavropoulos, J. C., Vernon, M. C., Volek, J. S., Wortman, J. A., ... Phinney, S. D. (2007). Low-carbohydrate nutrition and metabolism. *The American journal of Clinical Nutrition, 86*(2), 276-284.

Westman, E. C., Yancy, W. S., Edman, J. S., Tomlin, K. F. & Perkins, C. P. (2002). Effect of 6-Month Adherence to a Very Low Carbohydrate Diet Program. *The American Journal of Medicine, 113*, 30-36. http://dx.doi.org/10.1016/S0002-9343(02)01129-4

Wood, R. J., Volek, J. S., Liu, Y., Shachter, N. S., Contois, J. H., & Fernandez, M. L. (2006). Carbohydrate Restriction Alters Lipoprotein Metabolism by Modifying VLDL, LDL, and HDL Subfraction Distribution and Size in Overweight Men. *The Journal of Nutrition, 136*, 384-389.

Xu, H., Huang, X., Riserus, U., Krishnamurthy, V. M., Cederholm, T., Arnlov, J., ... Carrero, J. J. (2014). Dietary Fiber, Kidney Function, Inflammation, and Mortality Risk. *Clinical Journal of the Americal Society of Nephrology, 9*(12), 2104-2110. http://dx.doi.org/10.2215/CJN.02260314

6

Ellagic Acid May Improve Mechanical and Barrier Properties in Films of Starch

J. M. Tirado-Gallegos[1], D. R. Sepulveda-Ahumada[1], P. B. Zamudio-Flores[1], M. L. Rodríguez-Marin[2], F. Hernández-Centeno[3], V. Espinosa-Solis[4] & R. Salgado-Delgado[5]

[1] Fisiología y Tecnología de alimentos de la Zona Templada, Centro de Investigación en Alimentación y Desarrollo, A.C.-Unidad Cuauhtémoc. Cuauhtémoc, Chihuahua, México

[2] Conacyt Research Fellow at: Universidad Autónoma del Estado de Hidalgo, Ciudad Universitaria, Centro de Investigaciones Químicas, Carretera Pachuca-Tulancingo Km 4.5, 42183 Mineral de la Reforma, Hidalgo, México

[3] Departamento de Ciencia y Tecnología de Alimentos, Universidad Autónoma Agraria Antonio Narro, Buenavista, Saltillo, Coahuila, México

[4] Universidad Autónoma de San Luis Potosí. Coordinación Académica Región Huasteca Sur de la UASLP, km 5, carretera Tamazunchale-San Martin, 79960, Tamazunchale, S.L.P. México

[5] Departamento de Posgrado en Ingeniería Química y Bioquímica, Instituto Tecnológico de Zacatepec, Zacatepec, Morelos, México

Correspondence: P. B. Zamudio-Flores, Fisiología y Tecnología de alimentos de la Zona Templada, Centro de Investigación en Alimentación y Desarrollo, A.C.-Unidad Cuauhtémoc. Avenida Río Conchos s/n, Parque Industrial, Apartado postal 781, C.P. 31570. E-mail: pzamudio@ciad.mx

Abstract

Packaging increases the shelf life of food and facilitates its handling, transportation and marketing. The main packaging materials are plastics derived from petroleum, but their accumulation has given rise to environmental problems. An alternative is the use of biodegradable materials. In this regard, starch is an excellent choice because it is an abundant and renewable source with film-forming properties. However, the films obtained from starch have some limitations with respect to their mechanical and barrier properties. Several strategies have been developed in order to improve these limitations, ranging from the addition of lipids to the modification of the polymer structure. The aim of this review was propose the use of ellagic acid as a cross-linking agent that may improves the mechanical and barrier properties in films based on exists reports that phenolic compounds interact with starch-based materials decreasing their rate of retrogradation. Furthermore, ellagic acid is a powerful natural antioxidant, which would allow the production of active packaging with antioxidant properties, in addition to the improvement of the mechanical and barrier properties of starch films. In this concern more studies such as Fourier transform infrared spectroscopy, X-ray diffraction, differential scanning calorimetry and thermogravimetric analysis are necessary to verify the structural changes and interactions between starch and ellagic acid. We expect extensive use of it in the future of packaging materials.

Keywords: ellagic acid, starch, cross-linking, phenolic compounds, antioxidant

1. Introduction

Food packaging plays a key role in the conservation, distribution and marketing of food products. Packaging protects the product from mechanical, physical, chemical and microbiological damage (Falguera, Quintero, Jiménez, Muñoz, & Ibarz, 2011). Plastics are chemically synthesized polymers that are widely used in food packaging, as their production is relatively simple and inexpensive (Ghanbarzadeh, Almasi, & Entezami, 2011); however, its use has caused environmental problems (Marsh & Bugusu, 2007). An alternative is the use of edible films and coatings, which contribute to the reduction of environmental pollution. These materials can be obtained from renewable and biodegradable sources, such as polysaccharides, proteins, lipids, resins and their mixture thereof (Campos, Gerschenson, & Flores, 2011; Ribeiro, Vicente, Teixeira, & Miranda, 2007). Starch is one of the most abundant polysaccharides in nature. It has a great variety of botanical sources, and it is relatively inexpensive to isolate (Bertuzzi, Armada, & Gottifredi, 2007; Jiménez, Fabra, Talens, & Chiralt, 2012). Generally, starch films have poor mechanical properties and can be highly hygroscopic (Chiumarelli & Hubinger, 2012; Ghanbarzadeh et al., 2011). The addition of substances that generate intermolecular bonds, improving the integrity of the film, is among the strategies to improve the properties of starch films (Olivato, Grossmann, Bilck, & Yamashita, 2012). Ellagic acid, besides being a powerful antioxidant, can interact with polysaccharides as cross-linking agent, retaining its antioxidant properties (Kim et al., 2009). Generally, the cross-linking of starch involves the formation of esters between the carboxyl groups of the cross-linking agent (generally a carboxylic acid) and the OH of the glucose in starch (Kim et al., 2009; Reddy & Yang, 2010). Ellagic acid it is a polyphenolic acid without carboxyl groups ; however, it is the product of the hydrolysis of ellagitannins (glucose esters and phenolic compounds) (Ascacio-Valdés, Aguilera-Carbó, Rodríguez-Herrera, & Aguilar-González, 2013), whereby it should be possible to make this acid work as a cross-linking agent in the presence of oxidized starches, which are characterized by the presence of carboxyl groups, to obtain biodegradable films with suitable properties for use as food packaging or coating. This article describes the findings that support our hypothesis.

2. Mechanical Properties and Water Vapor Permeability, a Limitation of Starch Films

The use of plastic materials, mainly derived from petroleum, shows an upward trend. About 150 million tons of plastic are produced each year throughout the world, which generates large amounts of waste that pollute the environment (Shit & Shah, 2014). Among the various applications of petroleum polymers is their use for food packaging, which is why in recent years there has been an increasing interest in developing packaging from biodegradable polymers (Mali, Grossmann, Garcia, Martino, & Zaritzky, 2002). In this sense, starch is a polymer with great potential for the manufacture of this type of material, since it is abundant, biodegradable and cheap (Mali et al., 2002; Wilhelm, Sierakowski, Souza, & Wypych, 2003). Starch is a reserve polysaccharide that is found in plant materials in the form of semi-crystalline granules which, depending on the botanical source, vary in shape, size, structure and chemical composition, which affects their functional properties and is composed of two glucose polymer molecules linked by a glycosidic bond; one is a linear molecule known as amylose, and the other a branched molecule known as amylopectin (Campos et al., 2011; Jiménez et al., 2012; Smith, 2001; Tharanathan & Saroja, 2001; Wilhelm et al., 2003). Although this polysaccharide has shown great potential as a film-forming material and has been extensively studied as such, its films are not suitable for commercial use, mainly due to their poor mechanical properties and high affinity for water (Mali et al., 2002; Schmidt, Porto, Laurindo, & Menegalli, 2013). The mechanical and barrier properties of the films prepared from polymers are determined by the structure of the polymer , its nature, the presence of polar and non-polar groups in the polymer chain, the glass transition temperature (Tg) and the degree of cross-linking (Gajdoš, Galić, Kurtanjek, & Ciković, 2000; Mrkić, Galić, Ivanković, Hamin, & Ciković, 2006). Films made only with starch are brittle; however, the addition of plasticizers such as polyols, mainly low molecular weight compounds, decreases the intermolecular attraction between adjacent chains in the amorphous region (Donhowe & Fennema, 1993), which in turn increases the flexibility and elongation of the films while decreasing their tensile strength (Jiménez et al., 2012). Moreover, it has been found that plasticizers increase the hydrophilicity of films, thereby increasing their permeability to water vapor, oxygen and other gases (Jiménez et al., 2012; Mali et al., 2002; Mali, Grossmann, & Yamashita, 2010). The high presence of OH- groups of the starch and some plasticizers makes films highly sensitive to contact with water or air at a high relative humidity, which produces an increased permeability to water vapor (Schmidt et al., 2013). According to Forssell, Mikkilä, Moates, and Parker (1997), the glass transition temperature (Tg) should be considered as the most important parameter of the mechanical properties of amorphous and semi-crystalline materials; thus, the process of recrystallization of these materials should be controlled. Materials such as films made from thermoplastic starch (gelatinized starch plus plasticizer) are able to recrystallize. This molecular rearrangement is accelerated when these materials are stored above their Tg (Mali, Grossmann, García, Martino, & Zaritzky, 2006; van Soest & Vliegenthart, 1997; Y. Zhang, Rempel, & Liu, 2014). The recrystallization of starch films, and the consequent modification of their mechanical properties and permeability, is a consequence of the retrogradation process, which is characteristic of starch-based systems (Campos et al., 2011). Retrogradation occurs after gelatinization, and it is the result of a molecular arrangement that consists in the formation of hydrogen bonds between the oxygen of the carbon atom 6 and the OH- of carbon 2 of the glucose residues from the molecules of amylose, and the OH- group of carbon 2 and the OH- of carbon 6 of the glucose residues from the short chains of the amylopectin molecule, respectively (Figure 1) (Y. Zhang et al., 2014). Under proper conditions, the molecular arrangement can be of the crystalline type (Buléon, Colonna, Planchot, & Ball, 1998). Mali et al. (2006) characterized the thermal, mechanical and barrier properties of films obtained from different starches (corn, yam and cassava) under controlled storage conditions (90 days, 64% relative humidity, 20 °C). These authors using differential scanning calorimetry noted that the type of starch did not affect the Tg of the films, and that the variations in their mechanical properties and water vapor permeability were small. The authors noted also that starch-based films must have a Tg similar to the temperature at which they will be stored in order to reduce their crystallization (Mali et al., 2006).

The poor mechanical properties and high water vapor permeability of the starch films are closely related to the recrystallization or retrogradation of starch. Retrogradation, and the consequent increase in crystallinity, can be observed as the time of storage of films increases. Increasing the retrogradation kinetics, i.e., decreasing the rate of retrogradation in starch films, is one of the outstanding challenges in the field of packaging materials.

Recently, there has been an increasing interest in obtaining biodegradable active packaging; this has led to the use of natural extracts with antioxidant and antimicrobial properties. These extracts could also function as cross-linking agents. Cross-linking is a phenomenon that improves the stability and the mechanical properties of hydrocolloid-based polymeric matrices (Hager, Vallons, & Arendt, 2012; Silva-Weiss, Bifani, Ihl, Sobral, & Gómez-Guillén, 2014), and is characterized by the formation of intermolecular bonds between the polymer chains (Hager et al., 2012). It has been observed that the mechanical properties of films of cross-linked starch are significantly better than those of non-cross-linked starch (Ramaraj, 2007; Reddy & Yang, 2010). Many substances with antioxidant capacity are characterized by being phenolic compounds; ellagic acid, a phenolic acid that can form complexes with some proteins and polysaccharides, is one of the natural antioxidants (Kim et al., 2009).

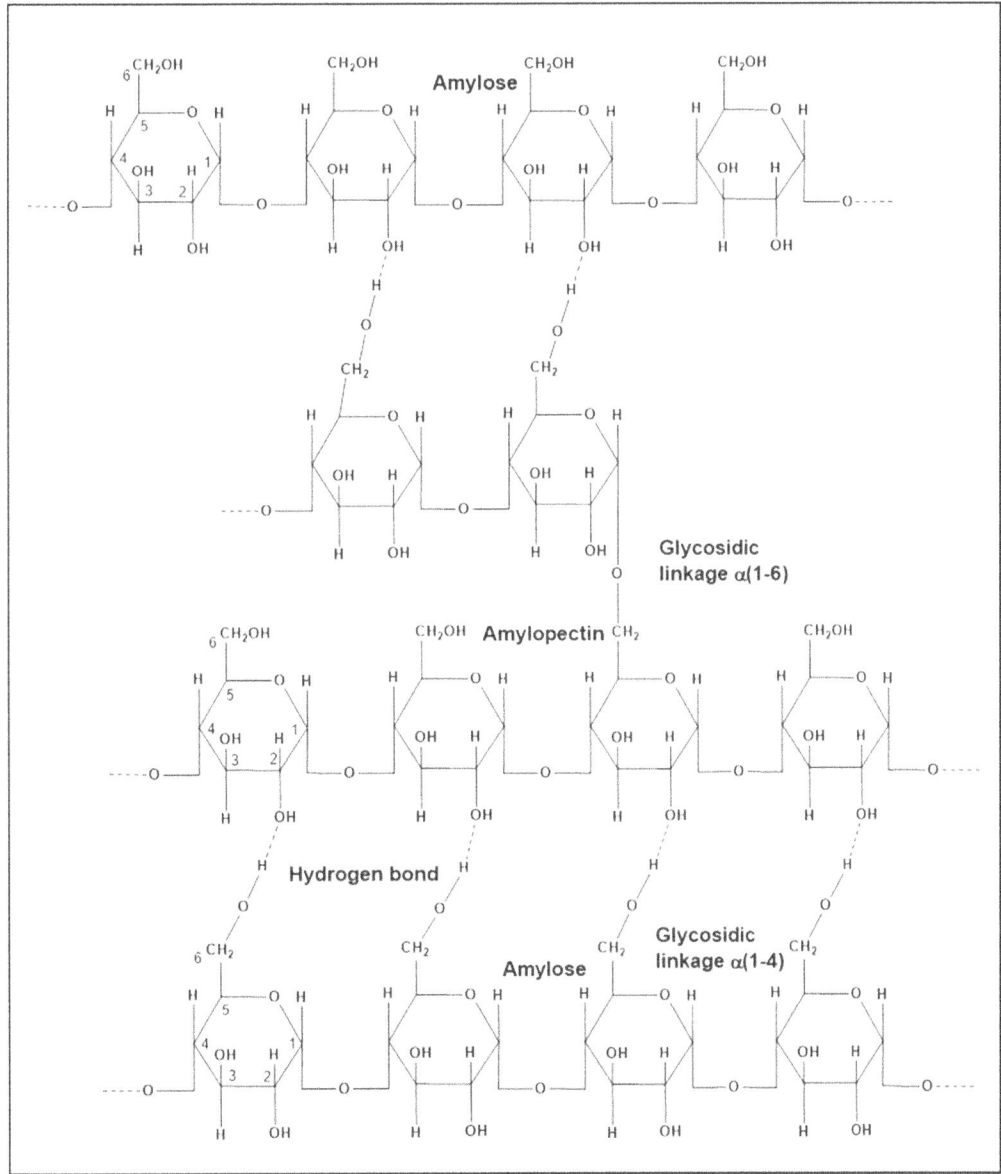

Figure 1. Formation of hydrogen bonds during starch retrogradation

3. Polyphenolic Compounds as Cross-linking Strategy in Biodegradable Films

Oxidation is one of the most common mechanisms among food degradation process, which cause a reduction in shelf life (Miller & Krochta, 1997). Oxidation alters the nutritional quality of foods, which may decrease and potentially lead to the formation of compounds with toxic effects. Oxidation causes the production of undesirable odors and affects other quality attributes such as color and texture (Finley & Given Jr, 1986). Recently, consumers have shown some concern regarding the use of synthetic chemicals in food such as aspartame, monosodium glutamate and high fructose corn syrup, preferring foods with natural antioxidants, which, in addition to being safe, have potential benefits to human health (Song, Bae, & Park, 2013). Ellagic acid is a phenolic acid derived (a dimer) from gallic acid (Komorsky-Lovrić & Novak, 2011). Ellagic acid is present in many fruits, especially in fruits such as red raspberries, pomegranates, nuts and in grape seeds (Kim et al., 2009; Priyadarsini, Khopde, Kumar, & Mohan, 2002). It can be found in free form or as a result of plant metabolism, or it can be found as ellagitannins, which are its precursors (Ascacio-Valdés et al., 2011; Ascacio - Valdés et al., 2014). Ellagitannins are ellagic acid esters and a polyol, usually glucose; upon contact with strong bases or acids, they undergo hydrolysis and release hexahydroadiphenine acid which forms ellagic acid by spontaneous lactonization (Figure 2).

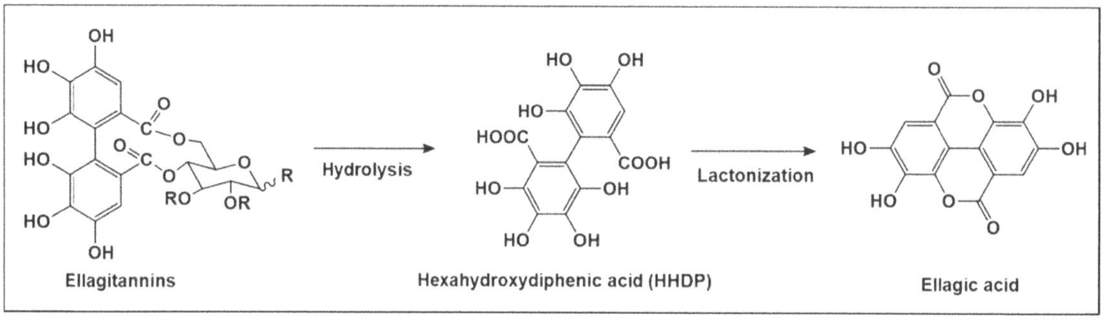

Figure 2. Hydrolysis of ellagitannins to ellagic acid

Ellagitannins, like other plant tannins, are found in vacuoles of intact cells. When plants are attacked by microorganisms (bacteria, fungi and viruses), they release phenolic compounds to prevent their tissues from becoming infected (Sepúlveda, Ascacio, Rodríguez-Herrera, Aguilera-Carbó, & Aguilar, 2011). Due to its presence in foods, and its antioxidant and antimicrobial properties, ellagic acid can be used as food additive (Komorsky-Lovrić & Novak, 2011); however in the EU countries ellagic acid has no status of the food additive. Ellagitannins can be found in some sub-products such as the shell of the pomegranate (*Punica granatum L.*) (Panichayupakaranant, Tewtrakul, & Yuenyongsawad, 2010). Ascacio-Valdés et al. (2013) analyzed the content of ellagic acid in the shell of ripe pomegranates; they reported a concentration of ellagic acid of 83.2 and 121.7 mg/g plant obtained by hydrolysis with methanolic hydrochloric acid and by hydrolysis with H_2SO_4 2 N, respectively. Compared with ellagic acid, the extracts from pomegranate shells have a similar antibacterial, anti-allergenic and anti-inflammatory activity (Panichayupakaranant et al., 2010).

At the moment, there are no reports of the effect of ellagic acid on the properties of biodegradable films regarding their use as packaging; however, there are studies with extracts or acids with others polyphenols similars to ellagic acid. For example, Kim et al. (2009) developed films of autoclaved chitosan with ellagic acid at different concentrations (0, 0.05, 0.1, 0.5 and 1%) as drug-eluting systems for the treatment of tumors in rats. Fourier transform infrared spectroscopy (FTIR) revealed that the autoclave process dissolved the amino groups of chitosan and the OH- of ellagic acid and the weak hydrogen bonds also affect changes in the FTIR spectrum. Siripatrawan and Harte (2010) obtained chitosan films that contained aqueous extracts of green tea as an antioxidant; the authors reported that the permeability coefficient decreased from 0.256 ± 0.023 to 0.087 ± 0.012 g mm m-2 d-1 kPa-1, while the density ranged from 1.21 ± 0.03 to 1.67 ± 0.03 g cm-3. These changes were attributed to the interactions between the functional groups of chitosan and the polyphenolic compounds of green tea extracts, as evidenced by FTIR spectra. Rivero, García, and Pinotti (2010) evaluated the cross-linking ability of tannic acid and its effect on chitosan-based biodegradable films. Adding tannic acid increased tensile strength up to 29% without affecting flexibility and decreased permeability to water vapor by 24% compared to films without acid; this was attributed by the authors to the effect of the cross-linking of the polymer matrix with tannic acid, which promoted a more rigid and compact structure, decreasing the elongation percentage.

Cross-linking is a strategy that improves the mechanical properties (Hager et al., 2012; Qiu, Hu, & Peng, 2013;

Reddy & Yang, 2010) and reduces the permeability to water vapor (Ghanbarzadeh et al., 2011; Olsson, Hedenqvist, Johansson, & Järnström, 2013) in biodegradable films. During cross-linking, the cross-linking agent promotes the formation of inter- and intra-molecular bonds in starch, generating films with a stronger structure and more uniform and compact surfaces (Qiu et al., 2013).

So far we have seen that phenolic compounds such as ellagic acid can be used in the production of films with antioxidant properties. It has been reported that ellagic acid can interact with polymers such as chitosan by autoclaving, which indicates the great thermal stability of this antioxidant, which has a melting point of 350 °C (Ascacio-Valdés et al., 2013). While it is true that there are no studies on starch films with ellagic acid, it would be interesting to know the effect of phenolic compounds on the physical and chemical properties of starch-based material.

4. Phenolic Compounds Interfere with Starch Retrogradation

It has been observed that the addition of some polyphenols purified from green and black tea can reduce the phenomenon of starch retrogradation. Wu, Chen, Li, and Li (2009) found that rice starch containing 10%, 14% or 20% of polyphenols showed no retrogradation in studies with DSC after 10 days of storage. The authors attributed this behavior to the interaction of the hydroxyl groups of tea polyphenols with the OH- groups of the starch molecules, which resulted in the formation of hydrogen bonds between polyphenols and starch. In a similar study, Xiao et al. (2011) reported that green tea polyphenols reduced rice starch retrogradation regardless of the content of amylose in rice. The results of the works cited above provide evidence of interactions between phenolic compounds and starches.

Moreover, Perazzo et al. (2014) developed biodegradable films based on cassava starch containing green tea extracts and palm oil extracts rich in carotenoids. They found that the mechanical properties and water vapor permeability of the films with extracts improved compared to controls (films with no extracts).

Zhu, Cai, Sun, and Corke (2008) evaluated the effect of 25 phenolic compounds on the paste- and texture-forming properties of wheat starch. The authors attributed the changes in the paste-forming profiles to possible interactions between the functional groups of phenolic compounds (methoxyl and hydroxyl) and amylose and amylopectin through hydrogen bonding and long-ranged interactions, such as van der Waals forces . With respect to the texture properties, phenolic acids reduced hardness and adhesion in greater proportion than flavonoids and other phenolic compounds. The possible interactions between amylose/amylopectin and phenolic compounds through hydrogen bonds and van der Waals forces could have affected the rearrangement of amylose and the retrogradation properties of amylopectin. The variation in the hardness of the gels was attributed to the different structures of the phenolic compounds, which can result in different degrees of interaction.

To date, it is believed that phenolic compounds interact with starch through hydrogen bonds. These interactions between phenolic compounds and starch can reduce the retrogradation process. Moreover, as mentioned above, ellagic acid can be found in nature interacting with glucose to form esters. Esters are the result of cross-linking reactions of starches; their formation requires the presence of carboxyls (carboxylic acids) and OH- groups. The presence of carboxyl groups is needed to attempt the possible cross-linking of ellagic acid with starch (Figure 3); one way to accomplish this is by previously oxidizing starch.

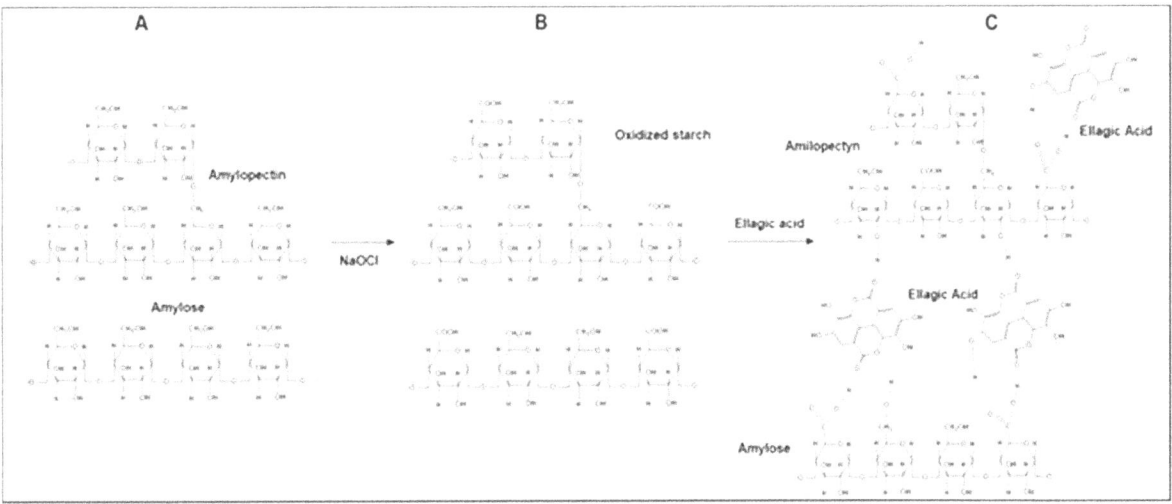

Figure 3. Cross-linking of starch with ellagic acid. (A) Native starch, (B) Oxidized starch and (C) Possible interaction of the cross-linking of oxidized starch with ellagic acid

5. Oxidized Starches: Inclusion of Carboxyl Groups in the Structure

In their native state, starches may have certain limitations for specific applications, while their modification increases their range of possible applications (Simsek, Ovando-Martínez, Whitney, & Bello-Pérez, 2012; Zamudio-Flores et al., 2015). The modification of starches is done to improve or change properties such as the paste-forming profile, viscosity, gelling properties, the stability of viscosity at different pH and shear stress values, retrogradation trends, surface properties, ionic character, among others (Abbas, Khalil, & Hussin, 2010; Ayoub & Rizvi, 2009; Kaur, Ariffin, Bhat, & Karim, 2012). The chemical modification of starch involves derivatization of its molecules through etherification, esterification, cationization, cross-linking, and oxidation. The oxidation of starch is a modification that has been practiced since the early nineteenth century (Steve, Qiang, & Sherry, 2005). Its use in food has increased because oxidized starches have high stability, low viscosity and binding properties (Sánchez-Rivera, García-Suárez, Velázquez del Valle, Gutierrez-Meraz, & Bello-Pérez, 2005). The oxidized starches used in the food industry are mainly modified with sodium hypochlorite (Sánchez-Rivera et al., 2005). During the oxidation of starch with sodium hypochlorite, the hydroxyl groups of the glucose residues are oxidized to carbonyl groups (C=O) and, finally, to carboxyl groups (COOH); the presence of these groups is used as an indicator of the extent of the oxidation process (Y. J. Wang & L. Wang, 2003). In order to evaluate the effect of the concentration of the oxidizing agent (sodium hypochlorite, NaClO), Y.-J. Wang and L. Wang (2003) oxidized corn starch with NaOCl (0.25 to 3%). The authors observed that the presence of carbonyl and carboxyl groups increased with the increase in the concentration of sodium hypochlorite in the reaction. Both amylose and amylopectin were degraded during oxidation, but amylose was more susceptible to oxidation. Similar results were reported by Sánchez-Rivera et al. (2005), who oxidized banana starches with different concentrations of sodium hypochlorite. Kuakpetoon and Y.-J. Wang (2001) investigated the effect of the botanical source of starch (corn, potato and rice) on the properties of starch oxidized by NaOCl. They analyzed the patterns of X-ray diffraction of the oxidized starches and found that the botanical source of starch had no effect on crystallinity, suggesting that oxidation takes place in the amorphous region of starch, which varies according to the type of starch. Kuakpetoon and Y. J. Wang (2006) evaluated the effect of amylose on the oxidation level of corn starches with different content of amylose. The authors used NaOCl (0.8, 2 and 5%) as oxidizing agent and found the highest concentrations of carboxyl and carbonyl groups in waxy starch, in which the concentration of NaOCl was higher (2 to 5%); they also reported that amylose was more sensitive to depolymerization, but that a higher content of it (70% amylose) hindered the formation of carboxyl groups. The authors explained the above by saying that it is probable that starches with up to 70% amylose do not allow the access of hypochlorite and water to their amorphous region, which would be necessary to carry out the oxidation of the OH- groups. As mentioned above, the modification of native starch changes its properties; thus, it is expected that the films obtained from modified starches also have different properties. Zamudio-Flores, Vargas-Torres, Pérez-González, Bosquez-Molina, and Bello-Pérez (2006) prepared films based on banana starch oxidized with three levels of NaOCl (0.5, 1.0 and 1.5%). Tensile strength increased with the oxidation level of starch, which probably was due to a greater oxidation. There were interactions between the carboxyl groups and the OH- of the glucose residues of starch through hydrogen bonds, which could have resulted in a greater integrity of the structure of the films. Moreover, S. D. Zhang, Zhang, Wang, and Wang (2009) evaluated the effect of the carbonyl groups on the properties of films based on oxidized corn starch plasticized with glycerol; the authors reported that the mechanical and thermal properties of the films improved with the increase of carbonyl groups, which in turn increased starch interactions (hydrogen bonds). Table 1 shows the concentration of carbonyl and carboxyl groups in starches oxidized with two concentrations of sodium hypochlorite. As can be seen, the carboxyl groups were present in greater concentration because they are the primary oxidation products. Similarly, it can also be seen that the presence of these groups depends on the type of starch.

Oxidation can also be performed with peroxides. Sangseethong, Termvejsayanon, and Sriroth (2010) oxidized commercial cassava starch (Manihot esculenta) using sodium hypochlorite and hydrogen peroxide at the same concentration (3% of oxidizing agent). They observed that the concentration of carboxyl groups concentration was higher when using sodium hypochlorite because the conversion of carbonyl to carboxyl groups was slower when peroxide was used.

Table 1. Content of carboxyl and carbonyl groups in oxidized starches

Starch	0.8 % NaOCl		2.0 % NaOCl		Reference
	Carbonyl (%)	Carboxyl (%)	Carbonyl (%)	Carboxyl (%)	
Potato	0.030	0.021	0.070	0.039	Kuakpetoon and Y.-J. Wang (2001)
Corn	0.030	0.050	0.060	0.140	
Rice	0.030	0.070	0.060	0.240	
Common corn			0.044	0.061	Y.-J. Wang and L. Wang (2003)
Waxy corn			0.048	0.061	
Banana			0.048	0.760	Sánchez-Rivera et al. (2005)
Common corn	0.020	0.030	0.070	0.180	Kuakpetoon and Y.-J. Wang (2006)
Waxy corn	0.020	0.030	0.050	0.160	
Amylose corn 50%	0.030	0.030	0.050	0.140	
Amylose corn 70%	0.050	0.040	0.060	0.080	

6. Conclusions

This article proposes the use of ellagic acid as cross-linking agent; ellagic acid is a powerful antioxidant that is found in some plant tissues in the form of ellagitannins, which are esters of ellagic acid, and some polyol such as the anhydroglucose molecule. Ellagic acid is derived from the hydrolysis of a glucose ester and polyphenols presents in plants, which leads us to believe that it may participate in cross-linking reactions in starches, either reacting with the hydroxyl groups of glucose residues or with the carboxyl groups of oxidized starches, generating esters. This could improve the mechanical and barrier properties of starch-based films because cross-linking promotes strong interactions at the molecular level that strengthen the structure of the films and can reduce the interaction with water, providing bioactive films with high antioxidant capacity.

Further experiments should be carried out adding ellagic acid to the formulation of films based on native and oxidized starches in order to measure the effect of the presence of carboxyl groups in the interaction between acid and starch and its effect on the physicochemical and structural properties of the films. Ellagic acid has been used to autoclave chitosan, indicating its high stability at elevated temperatures. It would be interesting to subject native and oxidized starch to an autoclave process with ellagic acid and to elaborate films based on the obtained starches, to assess the effect of this process on the physicochemical properties of such films. FTIR studies should performed on the obtained films in order to verify the structural changes in the starch resulting from its possible interaction with ellagic acid, while X-ray diffraction studies should be done to observe any changes in the amorphous/crystalline arrangement of starch. It would also be interesting to evaluate the mechanical and barrier properties of these films to determine how these properties are modified by the inclusion of ellagic acid in the polymer matrix. Clearly, these studies could be performed simultaneously with thermal analysis by differential scanning calorimetry (DSC) and thermogravimetric analysis (TGA) to evaluate the stability and thermal degradation of the polymer matrix. Finally, in order to characterize their functioning as active films, the antioxidant properties of these films should be evaluated.

Conflict of Interest

The authors declare that there are no conflicts of interest

Acknowledgments

The first author (JMTG) thanks to CONACYT for the doctoral scholarship. The authors declare that no conflict of interest exists related with this publication.

References

Abbas, K. A., Khalil, S. K., & Hussin, A. S. M. (2010). Modified starches and their usages in selected food products: a review study. *Journal of Agricultural Science, 2*(2), 90-100.

Ascacio-Valdés, J. A., Aguilera-Carbó, A., Rodríguez-Herrera, R., & Aguilar-González, C. (2013). Análisis de ácido elágico en algunas plantas del semidesierto Mexicano. *Revista Mexicana de Ciencias Farmacéuticas,*

44(2), 36-40.

Ascacio-Valdés, J. A., Buenrostro-Figueroa, J. J., Aguilera-Carbo, A., Prado-Barragán, A., Rodríguez-Herrera, R., & Aguilar, C. N. (2011). Ellagitannins: biosynthesis, biodegradation and biological properties. *Journal of Medicinal Plants Research, 5*(19), 4696-4703.

Ascacio-Valdés, J. A., Buenrostro, J. J., De la Cruz, R., Sepúlveda, L., Aguilera, A. F., Prado, A., . Aguilar, C. N. (2014). Fungal biodegradation of pomegranate ellagitannins. *Journal of Basic Microbiology, 54*(1), 28-34. http://dx.doi.org/10.1002/jobm.201200278

Ayoub, A. S., & Rizvi, S. S. H. (2009). An overview on the technology of cross-linking of starch for nonfood applications. *Journal of Plastic Film and Sheeting, 25*(1), 25-45. http://dx.doi.org/10.1177/8756087909336493

Bertuzzi, M. A., Armada, M., & Gottifredi, J. C. (2007). Physicochemical characterization of starch based films. *Journal of Food Engineering, 82*(1), 17-25. http://dx.doi.org/10.1016/j.jfoodeng.2006.12.016

Buléon, A., Colonna, P., Planchot, V., & Ball, S. (1998). Starch granules: structure and biosynthesis. *International Journal of Biological Macromolecules, 23*(2), 85-112. http://dx.doi.org/10.1016/S0141-8130(98)00040-3

Campos, C. A., Gerschenson, L. N., & Flores, S. K. (2011). Development of ddible films and coatings with antimicrobial activity. *Food and Bioprocess Technology, 4*(6), 849-875. http://dx.doi.org/10.1007/s11947-010-0434-1

Chiumarelli, M., & Hubinger, M. D. (2012). Stability, solubility, mechanical and barrier properties of cassava starch – Carnauba wax edible coatings to preserve fresh-cut apples. *Food Hydrocolloids, 28*(1), 59-67. http://dx.doi.org/10.1016/j.foodhyd.2011.12.006

Donhowe, I. G., & Fennema, O. (1993). The effects of plasticizers on crystallinity, permeability, and mechanical properties of methylcellulose films. *Journal of Food Processing and Preservation, 17*(4), 247-257. http://dx.doi.org/10.1111/j.1745-4549.1993.tb00729.x

Falguera, V., Quintero, J. P., Jiménez, A., Muñoz, J. A., & Ibarz, A. (2011). Edible films and coatings: Structures, active functions and trends in their use. *Trends in Food Science & Technology, 22*(6), 292-303. http://dx.doi.org/10.1016/j.tifs.2011.02.004

Finley, J. W., & Given Jr, P. (1986). Technological necessity of antioxidants in the food industry. *Food and Chemical Toxicology, 24*(10–11), 999-1006. http://dx.doi.org/10.1016/0278-6915(86)90280-2

Forssell, P. M., Mikkilä, J. M., Moates, G. K., & Parker, R. (1997). Phase and glass transition behaviour of concentrated barley starch-glycerol-water mixtures, a model for thermoplastic starch. *Carbohydrate Polymers, 34*(4), 275-282. http://dx.doi.org/10.1016/S0144-8617(97)00133-1

Gajdoš, J., Galić, K., Kurtanjek, Ž., & Ciković, N. (2000). Gas permeability and DSC characteristics of polymers used in food packaging. *Polymer Testing, 20*(1), 49-57. http://dx.doi.org/10.1016/S0142-9418(99)00078-1

Ghanbarzadeh, B., Almasi, H., & Entezami, A. A. (2011). Improving the barrier and mechanical properties of corn starch-based edible films: Effect of citric acid and carboxymethyl cellulose. *Industrial Crops and Products, 33*(1), 229-235. http://dx.doi.org/10.1016/j.indcrop.2010.10.016

Hager, A.-S., Vallons, K. J. R., & Arendt, E. K. (2012). Influence of Gallic Acid and Tannic Acid on the Mechanical and Barrier Properties of Wheat Gluten Films. *Journal of Agricultural and Food Chemistry, 60*(24), 6157-6163. http://dx.doi.org/10.1021/jf300983m

Jiménez, A., Fabra, M., Talens, P., & Chiralt, A. (2012). Edible and biodegradable starch films: a review. *Food and Bioprocess Technology, 5*(6), 2058-2076. http://dx.doi.org/10.1007/s11947-012-0835-4

Kaur, B., Ariffin, F., Bhat, R., & Karim, A. A. (2012). Progress in starch modification in the last decade. *Food Hydrocolloids, 26*(2), 398-404. http://dx.doi.org/10.1016/j.foodhyd.2011.02.016

Kim, S., Liu, Y., Gaber, M. W., Bumgardner, J. D., Haggard, W. O., & Yang, Y. (2009). Development of chitosan–ellagic acid films as a local drug delivery system to induce apoptotic death of human melanoma cells. *Journal of Biomedical Materials Research Part B: Applied Biomaterials, 90*(1), 145-155.

Komorsky-Lovrić, Š., & Novak, I. (2011). Determination of ellagic acid in strawberries, raspberries and blackberries by square-wave voltammetry. *International Journal of Electrochemical Science, 6*(10), 4638-4647.

Kuakpetoon, D., & Wang, Y. J. (2001). Characterization of different starches oxidized by hypochlorite.

Starch/Stärke,   *53*(5),   211-218.
http://dx.doi.org/10.1002/1521-379X(200105)53:5<211::AID-STAR211>3.0.CO;2-M

Kuakpetoon, D., & Wang, Y. J. (2006). Structural characteristics and physicochemical properties of oxidized corn starches varying in amylose content. *Carbohydrate Research, 341*(11), 1896-1915. http://dx.doi.org/10.1016/j.carres.2006.04.013

Mali, S., Grossmann, M. V. E., Garcia, M. A., Martino, M. N., & Zaritzky, N. E. (2002). Microstructural characterization of yam starch films. *Carbohydrate Polymers, 50*(4), 379-386. http://dx.doi.org/10.1016/S0144-8617(02)00058-9

Mali, S., Grossmann, M. V. E., García, M. A., Martino, M. N., & Zaritzky, N. E. (2006). Effects of controlled storage on thermal, mechanical and barrier properties of plasticized films from different starch sources. *Journal of Food Engineering, 75*(4), 453-460. http://dx.doi.org/10.1016/j.jfoodeng.2005.04.031

Mali, S., Grossmann, M. V. E., & Yamashita, F. (2010). Filmes de amido: produção, propriedades e potencial de utilização. *Semina: Ciências Agrárias, 31*(1), 137-156.

Marsh, K., & Bugusu, B. (2007). Food packaging-roles, materials, and environmental issues. *Journal of Food Science, 72*(3), R39-R55. http://dx.doi.org/10.1111/j.1750-3841.2007.00301.x

Miller, K. S., & Krochta, J. M. (1997). Oxygen and aroma barrier properties of edible films: a review. *Trends Food Science Technology, 8*, 228-237. http://dx.doi.org/10.1016/S0924-2244(97)01051-0

Mrkić, S., Galić, K., Ivanković, M., Hamin, S., & Ciković, N. (2006). Gas transport and thermal characterization of mono- and di-polyethylene films used for food packaging. *Journal of Applied Polymer Science, 99*(4), 1590-1599. http://dx.doi.org/10.1002/app.22513

Olivato, J. B., Grossmann, M. V. E., Bilck, A. P., & Yamashita, F. (2012). Effect of organic acids as additives on the performance of thermoplastic starch/polyester blown films. *Carbohydrate Polymers, 90*(1), 159-164. http://dx.doi.org/10.1016/j.carbpol.2012.05.009

Olsson, E., Hedenqvist, M. S., Johansson, C., & Järnström, L. (2013). Influence of citric acid and curing on moisture sorption, diffusion and permeability of starch films. *Carbohydrate Polymers, 94*(2), 765-772. http://dx.doi.org/10.1016/j.carbpol.2013.02.006

Panichayupakaranant, P., Tewtrakul, S., & Yuenyongsawad, S. (2010). Antibacterial, anti-inflammatory and anti-allergic activities of standardised pomegranate rind extract. *Food Chemistry, 123*(2), 400-403. http://dx.doi.org/10.1016/j.foodchem.2010.04.054

Perazzo, K. K. N. C. L., de Vasconcelos Conceição, A. C., dos Santos, J. C. P., de Jesus Assis, D., Souza, C. O., & Druzian, J. I. (2014). Properties and antioxidant action of actives cassava starch films incorporated with green tea and palm oil extracts. *PloS one, 9*(9), e105199. http://dx.doi.org/10.1371/journal.pone.0105199

Priyadarsini, K. I., Khopde, S. M., Kumar, S. S., & Mohan, H. (2002). Free radical studies of ellagic acid, a natural phenolic antioxidant. *Journal of Agricultural and Food Chemistry, 50*(7), 2200-2206. http://dx.doi.org/10.1021/jf011275g

Qiu, L., Hu, F., & Peng, Y. (2013). Structural and mechanical characteristics of film using modified corn starch by the same two chemical processes used in different sequences. *Carbohydrate Polymers, 91*(2), 590-596. http://dx.doi.org/10.1016/j.carbpol.2012.08.072

Ramaraj, B. (2007). Crosslinked poly(vinyl alcohol) and starch composite films. II. Physicomechanical, thermal properties and swelling studies. *Journal of Applied Polymer Science, 103*(2), 909-916. http://doi:10.1002/app.25237

Reddy, N., & Yang, Y. (2010). Citric acid cross-linking of starch films. *Food Chemistry, 118*(3), 702-711. http://dx.doi.org/10.1016/j.foodchem.2009.05.050

Ribeiro, C., Vicente, A. A., Teixeira, J. A., & Miranda, C. (2007). Optimization of edible coating composition to retard strawberry fruit senescence. *Postharvest Biology and Technology, 44*(1), 63-70. http://dx.doi.org/10.1016/j.postharvbio.2006.11.015

Rivero, S., García, M. A., & Pinotti, A. (2010). Crosslinking capacity of tannic acid in plasticized chitosan films. *Carbohydrate Polymers, 82*(2), 270-276. http://dx.doi.org/10.1016/j.carbpol.2010.04.048

Sánchez-Rivera, M. M., García-Suárez, F. J. L., Velázquez del Valle, M., Gutierrez-Meraz, F., & Bello-Pérez, L. A. (2005). Partial characterization of banana starches oxidized by different levels of sodium hypochlorite.

Carbohydrate Polymers, 62(1), 50-56. http://dx.doi.org/10.1016/j.carbpol.2005.07.005

Sangseethong, K., Termvejsayanon, N., & Sriroth, K. (2010). Characterization of physicochemical properties of hypochlorite- and peroxide-oxidized cassava starches. *Carbohydrate Polymers, 82*(2), 446-453. http://dx.doi.org/10.1016/j.carbpol.2010.05.003

Schmidt, V. C. R., Porto, L. M., Laurindo, J. B., & Menegalli, F. C. (2013). Water vapor barrier and mechanical properties of starch films containing stearic acid. *Industrial Crops and Products, 41*(0), 227-234. http://dx.doi.org/10.1016/j.indcrop.2012.04.038

Sepúlveda, L., Ascacio, A., Rodríguez-Herrera, R., Aguilera-Carbó, A., & Aguilar, C. N. (2011). Ellagic acid: biological properties and biotechnological development for production processes. *African Journal of Biotechnology, 10*(22), 4518-4523.

Shit, S. C., & Shah, P. M. (2014). Edible Polymers: Challenges and Opportunities. *Journal of Polymers, 2014*, 13. http://dx.doi.org/10.1155/2014/427259

Silva-Weiss, A., Bifani, V., Ihl, M., Sobral, P. J. A., & Gómez-Guillén, M. C. (2014). Polyphenol-rich extract from murta leaves on rheological properties of film-forming solutions based on different hydrocolloid blends. *Journal of Food Engineering, 140*(0), 28-38. http://dx.doi.org/10.1016/j.jfoodeng.2014.04.010

Simsek, S., Ovando-Martínez, M., Whitney, K., & Bello-Pérez, L. A. (2012). Effect of acetylation, oxidation and annealing on physicochemical properties of bean starch. *Food Chemistry, 134*, 1796-1803. http://dx.doi.org/10.1016/j.foodchem.2012.03.078

Siripatrawan, U., & Harte, B. R. (2010). Physical properties and antioxidant activity of an active film from chitosan incorporated with green tea extract. *Food Hydrocolloids, 24*(8), 770-775. http://dx.doi.org/10.1016/j.foodhyd.2010.04.003

Smith, A. M. (2001). The Biosynthesis of Starch Granules. *Biomacromolecules, 2*(2), 335-341. http://dx.doi.org/10.1021/bm000133c

Song, H. S., Bae, J. K., & Park, I. (2013). Effect of heating on DPPH radical scavenging activity of meat substitute. *Preventive Nutrition and Food Science, 18*(1), 80–84. http://dx.doi.org/10.3746/pnf.2013.18.1.080

Steve, W. C., Qiang, L., & Sherry, X. X. (2005). Starch modifications and applications. In W. C. Steve (Ed.), *Food carbohydrates: Chemistry, physical properties, and applications* (pp. 357-405). Boca Raton, FL: CRC Press.

Tharanathan, R. N., & Saroja, N. (2001). Hydrocolloid-based packaging films-alternate to synthetic plastics. *Journal of Scientific and Industrial Research, 60*(7), 547-559.

van Soest, J. J. G., & Vliegenthart, J. F. G. (1997). Crystallinity in starch plastics: consequences for material properties. *Trends in Biotechnology, 15*(6), 208-213. http://dx.doi.org/10.1016/S0167-7799(97)01021-4

Wang, Y.-J., & Wang, L. (2003). Physicochemical properties of common and waxy corn starches oxidized by different levels of sodium hypochlorite. *Carbohydrate Polymers, 52*(3), 207-217. http://dx.doi.org/10.1016/S0144-8617(02)003041

Wilhelm, H. M., Sierakowski, M. R., Souza, G. P., & Wypych, F. (2003). Starch films reinforced with mineral clay. *Carbohydrate Polymers, 52*(2), 101-110. http://dx.doi.org/10.1016/S0144-8617(02)00239-4

Wu, Y., Chen, Z., Li, X., & Li, M. (2009). Effect of tea polyphenols on the retrogradation of rice starch. *Food Research International, 42*(2), 221-225. http://dx.doi.org/10.1016/j.foodres.2008.11.001

Xiao, H., Lin, Q., Liu, G.-Q., Wu, Y., Tian, W., Wu, W., & Fu, X. (2011). Effect of green tea polyphenols on the gelatinization and retrogradation of rice starches with different amylose contents. *Journal of Medicinal Plants Research, 5*(17), 4298-4303.

Zamudio-Flores, P. B., Vargas-Torres, A., Pérez-González, J., Bosquez-Molina, E., & Bello-Pérez, L. A. (2006). Films prepared with oxidized banana starch: mechanical and barrier properties. *Starch/Stärke, 58*(6), 274-282. http://dx.doi.org/10.1002/star.200500474

Zamudio-Flores, P. B., Ochoa-Reyw, E., Ornelas-Paz, J. de J., Tirado-Gallegos, J. M., Bello-Pérez, L. A., Rubio-Ríos, A., & Cárdenas-Felix, R. G. (2015). Caracterización fisicoquímica, mecánica y estructural de películas de almidones oxidados de avenay plátano adicionadas con betalaínas. *Agrociencia, 49*, 483-498.

Zhang, S. D., Zhang, Y. R., Wang, X. L., & Wang, Y. Z. (2009). High carbonyl content oxidized starch prepared by hydrogen peroxide and its thermoplastic application. *Starch/Stärke, 61*(11), 646-655. http://dx.doi.org/10.1002/star.200900130

Zhang, Y., Rempel, C., & Liu, Q. (2014). Thermoplastic starch processing and characteristics-a review. *Critical Reviews in Food Science and Nutrition, 54*(10), 1353-1370. http://dx.doi.org/10.1080/10408398.2011.636156

Zhu, F., Cai, Y.-Z., Sun, M., & Corke, H. (2008). Effect of phenolic compounds on the pasting and textural properties of wheat starch. *Starch/Stärke, 60*(11), 609-616. http://dx.doi.org/10.1002/star.200800024

7

Minimally Processed Jackfruit: Opportunity for the Foodservice Industry

Rafidah A. Ramli[1], Azila Azmi[1], Noorsa R. Johari[1] & Syuhirdy M. Noor[1]

[1]Faculty of Hotel & Tourism Management, Universiti Teknologi MARA Pulau Pinang, Malaysia

Correspondence: Rafidah Aida Ramli, Faculty of Hotel & Tourism Management, Universiti Teknologi MARA Pulau Pinang, 13500 Pulau Pinang, Malaysia. E-mail: wanfidah@ppinang.uitm.edu.my

Abstract

Jackfruit is found in abundance in Malaysia. However the use of jackfruit as fresh dessert or in fruit platter has not been fully utilized in the foodservice industry due to the short shelf life of fresh jackfruit. The aim of this paper is to compare two simple calcium-infusion methods in ripe jackfruit pulps. Calcium is infused into the jackfruit pulps through immersion and vacuum-infusion techniques. The addition of calcium into the jackfruit enhanced the firmness and textural integrity of the fruit, thus prolonging the shelf life to 10 days. The minimally processed jackfruit exhibit fresh-like quality and longer shelf stability. It provides a variation in the fruits offered in foodservice establishment.

Keywords: jackfruit, calcium, minimal processing, shelf life, texture

1. Introduction

The consumption of fresh fruits and vegetables are growing as consumers become more aware about its health benefit (Allende, Tomás-Barberán, & Gil, 2006; Bennik, Vorstman, Smid, & Gorris, 1998). However, the current condition where people eat more meals outside their homes have increase the demand for convenient foods that retain as much fresh characteristics as possible (Ahvenainen, 1996; Corbo, Campaniello, Speranza, Bevilacqua, & Sinigaglia, 2015). Jackfruit, a local fruit of Malaysia is rich in nutrients. It contains carbohydrate and is high in fibre, minerals, carotenoid and thiamine (Jagadeesh et al, 2007; Willet, 2002). It has a distinctive aroma with sweet crisp flesh that can be eaten raw or cooked in a variety of sweet or savoury dishes. However, marketing of jackfruit has been hindered by the size of the fruit and difficulty in handling. As the whole fruit usually weigh up to more than 45kg, currently the fruits are marketed whole or separated fruitlets packed in polyethylene bags.

As the edible portion only makes up 30-35% of the fruits, the jackfruit is highly potential for minimal process. Pre-cut fruits lose some of its textural integrity due to the release of enzymes that disintegrate the cell wall. This presents an opportunity for the foodservice industry to cater to this increasing demand.

Jackfruit tree is an evergreen tree, easily grown in Asia (Jagadeesh et al., 2007). Every part of the tree can be used, either as food (Ismail & Kaur, 2013), medicine (Jagtap & Bapat, 2010) or building materials (Haq, 2006). The Malaysian Department of Agriculture has identified jackfruit as one of the Entry Point Projects for premium fruits in the Malaysian Economic Transformation Programme for export to the Middle East and Europe

Minimally processed jackfruit has a short shelf life as it remains in respiratory condition and physiologically active (Saxena, Bawa, & Raju, 2009). Various methods to prolong the keeping-quality of jackfruit have been proposed (Jayaprahash, 2013; Saxena, Bawa & Raju, 2012; Saxena, Saxena, Raju & Bawa, 2013). The market for fresh cut fruits has grown rapidly in recent years. Ohlsson (2002) suggested that minimal processing is replacing more traditional preservation methods as minimally processed products are better in term of sensorial quality and nutritional aspects. The foodservice industries are also focusing on getting more pre-prepared ingredients to minimize handling and to reduce operating cost (Ahvenainen, 1996). The International Fresh-cut Produce Association (IFPA) defines fresh-cut products as fruit or vegetables that have been trimmed and/or peeled and/or cut into 100% usable product that is bagged or pre-packaged to offer consumers high nutrition, convenience, and flavour while still maintaining its freshness (Lamikanra, 2002). The treatment done on minimally processed produce resulted tissue wounding, which cause deleterious effect on the produce (Silva, Bastos, Wurlitzer, Barros and Mangan, 2002). The minimally processed fruits are generally evaluated based on colour, texture, physicochemical characteristics and sensory attributes (Rico, Martín-Diana, Barat, & Barry-Ryan, 2007). In minimally processed products, the greatest hurdles is the changes in the physical attributes of the fruits (Soliva-Fortuny & Martín-Belloso, 2003; Toivonen & Brummell, 2008). Some of the methods currently employed in the fresh cut industry include modified atmosphere packaging (Saxena, Bawa, & Srinivas Raju, 2008), mild heat treatment (Son & Usol, 2001), edible coating (Dávila-Aviña et al., 2011) and calcium infusion (Gras, Vidal, Betoret, Chiralt & Fito, 2003).

As calcium has been associated in the maintenance of texture in various fruits, its inclusion in the bulbs is expected to retain the crispness of the jackfruit. This paper will look at two different methods of calcium infusion into the bulbs of ripe jackfruit. The methods evaluated were immersion and vacuum-infusion in 1% calcium solution. This paper would emphasise on the textural characteristics of jackfruits bulbs stored at 8 ° C.

2. Method

2.1 Jackfruit

Jackfruits *Artocarpus heterophyllus* var CJ1 were obtained from the Crystal Fruit Farm in Gopeng, Perak. The fruits were harvested 123 days after anthesis and ripen at room temperature between 3 - 5 days. The fruits were ripe when its skin becomes soft to slight pressure and emits a sweet aroma. Chemical characteristics of ripe jackfruit is discussed by Ong, Nazimah, Osman, Quek, Voon, Mat Hashim, Chew and Kong (2006) The ripe fruits were cut and the fruitlets removed and deseeded. The control and treated samples were kept at 8° C before further tests was conducted.

2.2 Calcium-infusion

Two infusion methods were used to introduce calcium into the jackfruit. These were immersion in the calcium solution at 8 ° C and vacuum-infusion for 15 minutes. These methods were evaluated as it can be easily done in any foodservice establishment. The calcium solution was prepared by dissolving calcium chloride at 1% concentration in distilled water. For the immersion method, the deseeded fruitlets were arranged in a shallow plastic trays and the solution was added until all the fruitlets were immersed completely. The trays were covered and placed in the chiller at 8 ° C between 18 - 24 hours. The solution was discarded and the fruitlets were drained.

The vacuum-infusion method requires a vacuum chamber that can accommodate a tray and connected to a vacuum pump. In this study, the tray containing deseeded jackfruit fruitlets were arranged in a tray filled with calcium chloride solution and placed in a laboratory size desiccator. The desiccator was attached to a vacuum pump and vacuum was applied for 15 minutes until no bubbles were observed in the solution. Jackfruits from both treatments were placed in polystyrene trays, cling-wrapped and kept in the chiller.

2.3 Physicochemical Characteristics

Several tests were conducted to evaluate the effects of calcium infusion in fresh cut jackfruit. The tests conducted include texture analysis, ascorbic acid content, and colour.

2.4 Texture

Texture analysis was done using Texture Analyzer (Stable Micro System version 1.05, UK). Samples were separated from the seeds and cut length-wise to 1 cm width strips. A total of 10 replicates were done for firmness test.

2.5 Ascorbic Acid Content

The ascorbic acid content of the ripe jackfruit pulps was determined by the AOAC method (AOAC Official Method Number 967.21, 1995). Analysis was done in triplicate.

2.5 Colour Evaluation

The colour of the jackfruit pulps were assessed using the Minolta Colorimeter Spectrophotometer CM 3500 (Minolta Co. Ltd., Japan). Jackfruit samples were cut into 2 cm x 2 cm squares. Values of L and b* were recorded. L value indicates the lightness of the sample, with 100 being very light and 0 black. A positive b* denotes the sample to be yellow, while a negative b* means blue. Data is an average of six different samples.

2.6 Statistical Analysis

Analyses of variance on data in this study were conducted using the SPSS 11.0 Window software. The one way analysis of variance (ANOVA) was done to determine if there was any significant difference and Games-Howell test was performed with level of significant of $p \leq 0.05$.

3. Results

3.1 Physicochemical Characteristics

Texture is a very important aspect in fresh cut fruits acceptance. According Saxena et. al. (2013), texture is one of the determining factors in the acceptability of fresh cut jackfruit. The firmness of jackfruit bulbs after 14 days of storage is shown in Table 1.

All the samples showed a decrease in firmness as storage days progressed. During immersion, the diffusion of immersion liquid into the ripe jackfruit matrix caused an increase in the turgor pressure. The increased turgidity and the effect of Ca^{2+} in the ripe jackfruit pulps were correlated by the apparent firmness at day 0. However, the firmness of the immersed ripe jackfruit pulps decreased after day 3 of storage, followed by increased texture data recorded on day 10 and 14.

Table 1. Changes in firmness of jackfruit during storage

Storage Day	Firmness (N)		
	Untreated Jackfruit	Immersed for 18 hours	Vacuumed-infused for 15 minutes
0	0.38	0.43	0.54
3	0.21	0.32	0.47
7	0.15	0.34	0.52
10	0.15	0.43	0.55
14	0.13	0.32	0.52

ANOVA and Games-Howell test revealed a significant difference between firmness of the pulps due to treatment at $P < 0.05$.

Table 2. ANOVA effect of treatment on the firmness of the jackfruit pulps

ANOVA					
	Sum of Squares	df	Mean Square	F	Sig.
Between Groups	.250	2	.125	25.314	.000
Within Groups	.059	12	.005		
Total	.309	14			

Table 3. Games-Howell test for effect of treatment on firmness of jackfruit pulps

(I) 1	(j) 1	Mean difference (i-j)	Std. Error	Sig.	95% confidence interval	
					Lower bound	Upper bound
Control	Immersed	-.16400*	.05263	.045	-.3236	-.0044
	Vacuumed	-.31600*	.04802	.003	-.4757	-.1563
Immersed	Control	.16400*	.05263	.045	.0044	.3236
	Vacuumed	-.15200*	.02905	.004	-.2405	-.0635
Vacuumed	Control	.31600*	.04802	.003	.1563	.4757
	Immersed	.15200*	.02905	.004	.0635	.2405

*. The mean difference is significant at the 0.05 level.

The increase in texture data observed in immersed samples on day 10 and 14 were not due to increase in firmness of the pulps but rather due to stringiness of the pulps. On both days (day 10 and 14), the blade was not able to execute a clean cut on the pulps. The stringiness was brought about by the disintegration of the cell walls which caused the pulps to be fibrous. Therefore, the infusion of calcium into the ripe jackfruit pulps matrix through immersion method failed to maintain the firmness of the pulps during storage. The immersed ripe jackfruit lost its firmness by day 3

In vacuum-infused jackfruit, the firmness of the jackfruit pulps was maintained until day 14 of storage. According to Collins and Wiley (1967), vacuum treatment removes intercellular gases from the tissues, facilitating the penetration of calcium into the tissues.

In both immersed and vacuum-infused ripe jackfruit pulps, infusions were done at room temperature and 8 ° C respectively. Therefore, the cell wall structures were undamaged. Lamikanra and Watson (2004) found at low temperature penetration of calcium into fruit cells were improved. Calcium-fortification of plant materials had been shown to modify the structural and mechanical properties of the products (Gras et al., 2003) through the formation of calcium pectate gels that strengthen the cell wall structures (Toivonen & Brummell, 2008). However, in this study, jackfruit immersed in calcium solution failed to maintain the structural integrity as proved by the declining firmness. The gases in the tissues of the immersed jackfruit may have expanded and rupture the tissues during storage, causing loss of firmness (Collins and Wiley, 1967).

Table 4 showed the ascorbic acid content of the jackfruit during. The untreated samples had a much higher ascorbic acid content compared to treated samples.

Table 4. Ascorbic acid content of control and treated jackfruit during storage

Storage Day	Ascorbic Acid (mg/100g sample)		
	Untreated Jackfruit	Immersed for 18 hours	Vacuumed-infused for 15 minutes
0	3.7	1.9	2.6
3	1.7	1.2	2.4
7	1.5	1.2	2.4
10	0.6	0.9	0.9
14	0.6	0.6	0.7

Table 5. ANOVA effect of treatment on the ascorbic acid content of the jackfruit pulps

ANOVA					
	Sum of Squares	df	Mean Square	F	Sig.
Between Groups	1.089	2	.545	.609	.560
Within Groups	10.740	12	.895		
Total	11.829	14			

Table 6. Games-Howell test for the effect of treatment on the ascorbic acid content of the jackfruit pulps

(I) Treatment	(J) Treatment	Mean Difference (I-J)	Std. Error	Sig.	95% Confidence Interval	
					Lower Bound	Upper Bound
Control	Immersed	.4600	.6066	.742	-1.495	2.415
	Vacuumed	-.1800	.7003	.964	-2.222	1.862
Immersed	Control	-.4600	.6066	.742	-2.415	1.495
	Vacuumed	-.6400	.4643	.408	-2.061	.781
Vacuumed	Control	.1800	.7003	.964	-1.862	2.222
	Immersed	.6400	.4643	.408	-.781	2.061

ANOVA and Games-Howell tests revealed there are no significance differences for ascorbic acid content of the jackfruit pulps due to treatment at $P < 0.05$. As ascorbic acid is water-soluble immersing the cut jackfruit pulps into the calcium solution may have caused the ascorbic acid to leach out. However, during storage, all the samples showed a decrease in the ascorbic acid content, recording a value of 0.6 mg ascorbic acid for both untreated and immersed jackfruits; and 0.7 mg per 100 gram samples for vacuumed jackfruit.

The ascorbic acid is sensitive to oxidation and a decline in its activity can be expected during storage (Ali, Masud, Abbasi, Mahmood, & Hussain, 2013). The wound caused by cutting the jackfruit released enzymes which brought about the reduction of the ascorbic acid content in the untreated jackfruit pulps (Bett et al., 2001).

The higher ascorbic acid content at day 0 in vacuumed-treated jackfruit may be attributed to the minimal amount of time in solution compared to immersion-treated jackfruit. However, the rate of ascorbic acid loss is higher in untreated jackfruit. Calcium had been proven to be effective in retaining ascorbic acid (Ali et al., 2013)

Table 7. Colour data of stored jackfruit

	Control after 14 days	Immersed after 14 days	Vacuumed after 14 days
ΔL	-5.34 ± 0.04	-6.04 ± 0.03	-9.38 ± 0.12
Δb^*	-10.07 ± 0.08	-5.59 ± 0.02	-3.82 ± 0.08

n = 6

Colour is among the first attributes evaluated by the consumers (Toivonen & Brummell, 2008). Changes in the lightness coefficient, L* and b* values in the Hunter colour chart of the samples is tabulated in Table 7. The values reported are in reference to control sample at day 0. All the samples (untreated and treated with calcium) recorded a decrease in the L* and b* values. Negative L-values indicate darker colour as compared to the fresh sample after 14 days of storage at 8 °C. All the samples were also less yellow than the untreated jackfruit.

The changes in colour implied the jackfruit bulbs may have undergone browning reaction. The same was observed in dried jackfruit bulbs (Saxena, Maity, Raju, & Bawa, 2012), with decreasing L* and b* values. Saxena et al (2012) reported similar browning and degradation of carotenoid in air dried jackfruit. Thus, calcium treatment does not prevent browning in jackfruit pulps.

4. Conclusion

Consumer awareness about the health benefits of consuming more fruits have resulted an increase in fresh cut fruit demand. In this experiment, two calcium-infusion methods were evaluated to assess the changes in firmness, ascorbic acid content and colour of minimally processed jackfruit pulps. The result showed that infusions of calcium into the tissue of jackfruit pulps were able to retain the textural integrity of the fruit up to at least 3 days by immersion method and up to 10 days by vacuum infusion method. The ascorbic acid contents of jackfruits treated with calcium was comparable to untreated samples. A slight browning was detected in all the samples after 14 days of storage. Thus, in a foodservice application, where jackfruit can be minimally processed for the convenience of the consumer; introduction of calcium by vacuum infusion to retain its textural characteristics is achievable.

References

Ahvenainen, R. (1996). New approaches in improving the shelf life of minimally processed fruit and vegetables. *Food Science & Technology, 71*(June), 179-187. http://doi.org/10.1016/0924-2244(96)10022-4

Ali, S., Masud, T., Abbasi, K. S., Mahmood, T., & Hussain, I. (2013). Influence of CaCl2 on biochemical composition, antioxidant and enzymatic 2 activity of apricot at ambient storage. *Pakistan Journal of Nutrition, 12*(5), 476-483. Retrieved from
http://www.scopus.com/inward/record.url?eid=2-s2.0-84882741909&partnerID=40&md5=a5f1456f3d93e0823beaa7b853ad2cf2

Allende, A., Tomás-Barberán, F. a., & Gil, M. I. (2006). Minimal processing for healthy traditional foods. *Trends in Food Science and Technology, 17*(9), 513-519. http://doi.org/10.1016/j.tifs.2006.04.005

Bennik, M., Vorstman, W., Smid, E., & Gorris, L. (1998). The influence of oxygen and carbon dioxide on the growth of prevalent Enterobacteriaceae andPseudomonasspecies isolated from fresh and controlled-atmosphere-stored vegetables. *Food Microbiology, 15*(5), 459-469.
http://doi.org/10.1006/fmic.1998.0187

Corbo, M., Campaniello, D., Speranza, B., Bevilacqua, A., & Sinigaglia, M. (2015). Non-Conventional Tools to Preserve and Prolong the Quality of Minimally-Processed Fruits and Vegetables. *Coatings, 5*(4), 931-961. http://doi.org/10.3390/coatings5040931

Dávila-Aviña, J. E. D. J., Villa-Rodríguez, J., Cruz-Valenzuela, R., Rodríguez-Armenta, M., Espino-Díaz, M., Ayala-Zavala, J. F., ... González-Aguilar, G. (2011). Effect of edible coatings, storage time and maturity stage on overall quality of tomato fruits. *American Journal of Agricultural and Biological Sciences, 6*(1), 162-171. http://doi.org/10.3844/ajabssp.2011.162.171

Gras, M. L., Vidal, D., Betoret, N., Chiralt, a., & Fito, P. (2003). Calcium fortification of vegetables by vacuum impregnation: Interactions with cellular matrix. *Journal of Food Engineering, 56,* 279-284. http://doi.org/10.1016/S0260-8774(02)00269-8

Ismail, N., & Kaur, B. (2013). Consumer Preference for Jackfruit Varieties in Malaysia. *Journal of Agribusiness*

Marketing, 6, 37-51.

Jagadeesh, S. L., Reddy, B. S., Swamy, G. S. K., Gorbal, K., Hegde, L., & Raghavan, G. S. V. (2007). Chemical composition of jackfruit (Artocarpus heterophyllus Lam.) selections of Western Ghats of India. *Food Chemistry, 102*(1), 361-365. http://doi.org/10.1016/j.foodchem.2006.05.027

Jagtap, U. B., & Bapat, V. a. (2010). Artocarpus: A review of its traditional uses, phytochemistry and pharmacology. *Journal of Ethnopharmacology, 129*(2), 142-166. http://doi.org/10.1016/j.jep.2010.03.031

Jayaprahash C, L. J. (2013). Development and Evaluation of Shelf Stable Retort Pouch Processed Readyto- Eat Tender Jackfruit (Artocarpus heterophyllus) Curry. *Journal of Food Processing & Technology, 4*(10). http://doi.org/10.4172/2157-7110.1000274

Ong, B. T., Nazimah, S. A., Osman, A., Quek, S. Y., Voon, Y. Y., Hashim, D. M., Chew, P. M., & Kong, Y. W. (2006). Chemical and flavour changes in jackfruit (Artocarpus heterophyllus Lam.) cultivar J3 during ripening. *Postharvest Biology and Technology, 40*(3), 279-286. http://doi.org/10.1016/j.postharvbio.2006.01.015

Rico, D., Martín-Diana, a. B., Barat, J. M., & Barry-Ryan, C. (2007). Extending and measuring the quality of fresh-cut fruit and vegetables: a review. *Trends in Food Science & Technology, 18*(7), 373-386. http://doi.org/10.1016/j.tifs.2007.03.011

Saxena, A., Bawa, A. S., & Srinivas, Raju, P. (2008). Use of modified atmosphere packaging to extend shelf-life of minimally processed jackfruit (Artocarpus heterophyllus L.) bulbs. *Journal of Food Engineering, 87*(4), 455-466. http://doi.org/10.1016/j.jfoodeng.2007.12.020

Saxena, A., Bawa, a. S., & Raju, P. S. (2009). Optimization of a multitarget preservation technique for jackfruit (Artocarpus heterophyllus L.) bulbs. *Journal of Food Engineering, 91*(1), 18-28. http://doi.org/10.1016/j.jfoodeng.2008.08.006

Saxena, A., Bawa, a. S., & Raju, P. S. (2012). Effect of Minimal Processing on Quality of Jackfruit (Artocarpus heterophyllus L.) Bulbs Using Response Surface Methodology. *Food and Bioprocess Technology, 5*(1), 348-358. http://doi.org/10.1007/s11947-009-0276-x

Saxena, A., Maity, T., Raju, P. S., & Bawa, A. S. (2012). Degradation Kinetics of Colour and Total Carotenoids in Jackfruit (Artocarpus heterophyllus) Bulb Slices During Hot Air Drying. *Food and Bioprocess Technology, 5*(2), 672-679. http://doi.org/10.1007/s11947-010-0409-2

Saxena, A., Saxena, T. M., Raju, P. S., & Bawa, A. S. (2013). Effect of Controlled Atmosphere Storage and Chitosan Coating on Quality of Fresh-Cut Jackfruit Bulbs. *Food and Bioprocess Technology, 6*(8), 2182-2189. http://doi.org/10.1007/s11947-011-0761-x

Soliva-Fortuny, R. C., & Martín-Belloso, O. (2003). New advances in extending the shelf-life of fresh-cut fruits: A review. *Trends in Food Science and Technology, 14*(9), 341-353. http://doi.org/10.1016/S0924-2244(03)00054-2

Son, A. T., & Usol, L. (2001). C : Food Chemistry and Toxicology Use of Mild Heat Pre-treatment for Quality Retention of Fresh-cut Cantaloupe Melon, *70*(1).

Toivonen, P. M. A., & Brummell, D. A. (2008). Biochemical bases of appearance and texture changes in fresh-cut fruit and vegetables. *Postharvest Biology and Technology, 48*(1), 1-14. http://doi.org/10.1016/j.postharvbio.2007.09.004

Vii, P. (2002). Minimal Processing, 217-234.

8

Drying Kinetics of Microwave-Dried Vegetable Amaranth (*Amaranthus dubius*) Leaves

Saheeda Mujaffar[1] & Alex Lee Loy[1]

[1]Food Science and Technology Unit, Department of Chemical Engineering, The University of the West Indies, St. Augustine, Trinidad and Tobago, West Indies

Correspondence: Saheeda Mujaffar, Food Science and Technology Unit, Department of Chemical Engineering, The University of the West Indies, St. Augustine, Trinidad and Tobago, West Indies. E-mail: saheeda.mujaffar@sta.uwi.edu

Abstract

The effect of microwave power level (200, 500, 700 and 1000W) on the drying behaviour of amaranth (*Amaranthus dubius*) leaves was investigated. Higher microwave power levels effected faster drying and there was an increase in drying rate constant (k) as microwave power level increased from 200 to 1000W and an increase in diffusivity (D_{eff}) values from 3.04 x 10^{-10} to 2.82 x 10^{-9} m^2/s. Leaves dried at 1000W power level however showed noticeable scorching after 540s of drying. Drying at the lower microwave power levels occurred in the constant and falling rate period, while at the higher power levels drying occurred in the falling rate period after an initial warm-up phase. Amaranth leaves could be dried at 700W power from an initial moisture content of 6.00 g H$_2$O/g DM (85.7% wb) to 0.08 g H$_2$O/g DM (7.6% wb) in 11.5 min. Overall, of the twenty-two thin layer models applied to the *MR* data, the Alibas model gave the best fit in terms of both the root mean square error (RMSE) and the chi-square statistic (χ^2).

Keywords: amaranth; microwave drying; Fick's Law; thin-layer modelling

1. Introduction

Vegetable amaranth (*Amaranthus sp.*) grows easily in the Caribbean and the edible portion (succulent leaves and young shoots) is widely used throughout the region as a cooked leafy vegetable. It is commonly called 'spinach' and 'chorai bhaji' in Trinidad and 'calalloo' in Jamaica. Vegetable amaranth (*Amaranthus sp.*) can also be used as a leafy green in salads and recipe preparations, often interchangeably with spinach, and is gaining popularity internationally as a less expensive option to kale. Amaranth leaves reportedly contain a high proportion of high quality protein, vitamins A and C, calcium, iron and fiber (Borneo & Aguirre, 2008, Rodriguez, Perez, Romel, & Dufour, 2011; Alegbejo, 2013; Andini, Yoshida, & Ohsawa, 2013).

The postharvest life of amaranth greens is relatively short because of rapid wilting of the tender foliage. The leaves and soft leaf stalks are used within a few hours of purchase at produce markets or stored in polyethylene bags in refrigerated storage (4°C) for use within one week. Like spinach leaves, drying of amaranth presents an opportunity to minimize wastage, preserve the leaves for future use, and add value by converting the leaves to a form that can be easily added to various food preparations. Dried amaranth flakes can potentially be used in various food preparations such as soups, sauces, dips and smoothies. Until recently, drying of vegetable amaranth leaves as a method to extend shelf-life was not widely investigated. Research works on dried amaranth have focused primarily on quality and nutrient content of the dried material (Fathima, Begum, & Rajalakshmi, 2001; Rodriguez et al., 2011; Aletor & Abiodun, 2013; Peter, Elizabeth, Judith, & Hudson, 2014) and one study looked at the use of the ground, dried amaranth leaves in green pasta as a substitute for spinach (Borneo & Aguirre, 2008). Published works on drying of vegetable amaranth are limited. Akonor & Amankwah (2012) investigated the thin-layer drying kinetics of solar-dried *Amaranthus hybridus* leaves. The leaves were dried to a moisture content 8.5% (wb) over a period of two drying days. They reported a short sample heating phase and a short constant rate period followed by a falling rate period. An effective diffusivity (D_{eff}) value of 1.95 x 10^{-9} m^2/s was reported and the Page model was found to best describe the moisture ratio data. The open sun drying of *Amaranthus* sp. was compared with drying in a greenhouse-type solar dryer (P. Singh, S. Singh, B.R. Singh, J. Singh and S.K. Singh, 2014). Leaves were subjected to three pre-treatments, namely magnesium chloride (0.1%), sodium bicarbonate (0.1%), and potassium metabisulfite (2%). Drying curves were presented and moisture ratio values were tabulated. Pre-treated samples took less time to dry for both sun and solar-dried samples. Drying rates were higher in the greenhouse solar dryer. A short constant rate period was observed, based on the drying curves and they reported that the greenhouse dried samples were superior in quality.

There is a growing interest in microwave drying of leafy greens and herbs as a viable alternative to the conventional hot air drying method (Soysal, 2004; Ozbek & Dadali, 2007; Alibas Ozkan, B. Akbudak & N. Akbudak, 2007; Sharada, 2013). The main advantages of this drying method compared with hot air drying are a reduction in drying time, lower impact on product quality and less energy use (Zhang, Tang, Mujumdar & Wang, 2006). Microwaves are a form of electromagnetic radiation that effect heating of foods by a process of dielectric heating (Decareau, 1985; Zhang et al., 2006). Based on previous work, the three common factors that would appear to have an impact on the drying rate and quality of microwave-dried leafy vegetables are microwave power level, loading density (sample size), and pre-treatments. With respect to microwave drying of amaranth in particular, documentation is limited but suggests the potential for this method of drying (Fathima et al., 2001; Rajeswari, Bharati, Naik & Naganur, 2013). Fathima et al. (2001) investigated the quality of microwave drying of selected greens, including hot water-blanched amaranth leaves (*Amaranth* sp.). The maximum power output of the microwave oven and the loading density or sample size used were not disclosed. They reported that leaves could be dried in 12 min to a moisture value of 5% (wb). They noted that reconstituted amaranth leaves had acceptability scores similar to fresh leaves. Rajeswari et al. (2013) compared the quality and shelf life of pre-treated *Amaranthus tricolor* leaves (sample size not given) dried for 2h 37min at 60°C in a cabinet drier and 3min 31s in a microwave oven at 900W power. The moisture content of microwave dried control samples averaged 3.12%. Following drying, dehydrated leaves were packed in polypropylene pouches and stored in airtight aluminium containers for six months under ambient conditions. No works were found on the thin layer modelling of drying data of amaranth leaves.

The objective of this study was therefore to investigate the microwave drying of amaranth (*Amaranthus dubius*). Given the lack of information on the application of this drying method to this raw material, the work will follow the typical well-established approach to air drying kinetics (generation of drying curves, rate curves, effective diffusion coefficients and thin-layer modelling), as well as works done on microwave drying of spinach leaves and similar leafy vegetables and herbs. The objective of this study was to investigate the effect of microwave output power (200-1000W) on the drying behaviour of amaranth leaves, and model the drying data using twenty-two thin layer models.

2. Method

2.1 Raw Material Handling and Preparation

Amaranth shoots (*Amaranthus dubius*) was purchased at local market and stored at 4°C in a refrigerator until use. Prior to drying, the leaves with petioles left intact were separated from the main hardy stems (Dadali et al., 2008) using a pair of scissors. Leaves that were damaged or wilted were discarded and a total weight of 2kg of leaves was used. The weight, length and width of leaves averaged 2g, 11cm and 6cm, respectively.

2.2 Microwave Drying

Drying was carried out using a 34L oven capacity Amana MCS10TS Menumaster Commercial digital microwave oven (Accelerated Cooking Products (ACP), Cedar Rapids, IA, USA) with the following technical features: 3.5kV, 1000W, 120V, 60 Hertz. Time and microwave output power adjustments were done via a digital control panel located at the front of the oven. A sample size of 20g of leaves was selected as the amount that could be spread in a thin layer onto the sample plate. Preliminary experiments revealed that blanching resulted in deterioration of leaf quality during drying when compared with untreated leaves. Leaves were spread in single layer onto a ceramic plate and the total weight of the plate with the leaves was recorded before placing into the oven. Leaves were first dried at full power (1000W) and at 30s intervals the plate with the sample taken out, quickly weighed and returned to the oven. Drying was continued until there was no further change in weight or until burning of leaves was observed. At the end of drying, leaves were allowed to cool, packaged in re-sealable plastic storage bags and stored in laboratory desiccators until analysis. The drying procedure was repeated at power levels 700W, 500W and 200W and drying runs at each power level were repeated five (5) times.

2.3 Analytical Methods

Sample weights (g) during microwave drying were recorded (0.01±0.005g) using an Explorer Pro Balance, Model EP2102C (Ohaus Corporation, NJ, USA). Moisture content of the fresh and dried samples was done using a Halogen Moisture Analyzer HB43-S (Mettler Toledo-AG, Zurich, Switzerland) set at 105°C. Water activity (a_w)

was measured using an Aqua Lab CX-2 1021 water activity meter (Decagon Devices Inco., Pullman, WA, USA). Surface colour of leaves was measured using a CR-410 Choma Meter (Konica Minolta Sensing Americas, Inc., NJ, USA). Hunter values (*L*, *a*, *b*) were recorded in four samples. The maximum for "*L*" value is 100 (white) and the minimum is zero (black). Positive "*a*" value is red, negative "*a*" is green, while positive "*b*" value is yellow and negative "*b*" is blue (Hunterlab, 2008). *Hue angle (°), Chroma* and *Total colour difference (ΔE)* between fresh and dried leaves were calculated as given in Equations 1 through 3.

$$Hue = Arc \tan (b/a) \tag{1}$$

$$Chroma = (a^2 + b^2)^{1/2} \tag{2}$$

$$\Delta E = [(L_0 - L)^2 + (a_0 - a)^2 + (b_0 - b)^2]^{1/2} \tag{3}$$

Energy consumption for microwave drying was obtained using Equation 4 (Hebbar, Vishwanathan & Ramesh, 2004).

$$E_t = P.t \tag{4}$$

2.4 Drying Data Analysis

Drying data was analysed using the standard approach to drying studies, that is, the generation of drying curves, drying rate curves and moisture ratio (*MR*) data. Moisture content (g H_2O)/g DM) of the leaves during the drying process was back-calculated based on the final moisture content value at the end of drying (Mujaffar & Sankat, 2005; 2014). Effective moisture diffusivity (D_{eff}) values were calculated based on Fick's Second Law of Diffusion and through the calculation of the Moisture Ratio (*MR*) based on the first term of the solution by Crank (1975) in Equation 5. The drying rate constant (*k*) was determined from a plot of ln *MR* versus time (*t*) based on Equation 6 and the rate constants (*k*) determined from the slopes were then used to calculate the diffusivity values (Equation 7), using 1 mm as the leaf thickness (Simha & Gugalia, 2013).

$$MR = (M-M_e) / (M_o-M_e) = (8/\pi^2) \exp [-\pi^2 D_{eff} t/ 4X^2] \tag{5}$$

$$MR = (M-M_e) / (M_o-M_e) = Ae^{-kt} \tag{6}$$

$$k = \pi^2 D_{eff}/4X^2 \tag{7}$$

For this study, a total of twenty-two (22) empirical and semi-empirical thin layer models (Alibas, 2014; W. Silva, C. Silva, Gama and Gomes, 2014) were applied to the *MR* data (Table 1). As is now a common practice in thin-layer drying studies, the performance (fit) of the models is assessed through the use of the coefficient of determination (R^2), root mean square error (*RMSE*) and the chi-square statistic (χ^2). Model fit was done using Curve Expert Professional software (Version 2.3), Version 2.3.0 (Hyams, 2016).

Table 1. Thin Layer Drying Models

Model name	Equation
Newton	$MR = exp (-kt)$
Page	$MR = exp(-kt^n)$
Modified Page	$MR = exp(-kt)^n$
Henderson and Pabis	$MR = a \, exp (-kt)$
Modified Henderson and Pabis	$MR = a \, exp (-kt) + b \, exp (-gt) + c \, exp (-ht)$
Logarithmic	$MR = a \, exp (-kt) + c$
Two-Term	$MR = a \, exp (-k_0 t) + b \, exp (-k_1 t)$
Two-Term Exponential	$MR = a \, exp (-k t) + (1-a) \, exp (-k \, a \, t)$
Wang & Singh	$MR = 1+at+bt^2$
Verma	$MR = a \, exp(-kt)+(1-a) \, exp(-gt)$
Hii	$MR = a \, exp(-kt^n) + c \, exp(-gt^n)$
Midilli	$MR = a \, exp (-kt^n) + b \, t$
Peleg	$MR = 1 - (x/(a+bx))$
Weibull distribution	$MR = a - b \, exp (-kt^n)$
Diffusion approach	$MR = a \, exp(-kt)+(1-a) \, exp(-kbt)$
Aghbashlo et al.	$MR = -k_1t / (1 + k_2t)$
Logistic	$MR = a_0 / ((1 + a \, exp (kt))$
Jena and Das	$MR = a \, exp (-kt + b \, t^{1/2}) + c$
Demir et al.	$MR = a \, exp (-kt^n) + c$
Simplified Fick's Diffusion (SFFD) equation	$MR = a \, exp (-c \, (t/L^2))$
Modified Page Equation-II	$MR = exp (-k \, (t/L^2))^n$
Alibas	$MR = a \, exp (-kt^n + b \, t) + g$

2.5 Statistical Analysis

Further Regression Analysis and ANOVA were carried out using GENSTAT statistical software (VSN International, Hampstead, UK) and Post hoc analysis carried out using "Rapid publication-ready MS-Word tables for one-way ANOVA" (Assaad, Hou, Zhou, Carroll & Wu, 2015).

3. Results and Discussion

3.1 Colour and General Observations

Generally, as the leaves dried, they underwent changes such as a colour change to a lighter green, the development of a grassy, hay-like odour, and some yellowing. As drying proceeded, moisture was seen on the surface of the leaves followed by evaporation of this water as the leaves appeared to become more brittle in texture. The higher the power level, the faster the development of these changes. Leaves dried at 1000W were very brittle and smelled "cooked" after 4 min of drying and beyond 9 min were burnt. Leaves dried at 200W power level appeared soggy for most of the drying process. With the exception of leaves dried at 200W microwave power level, fully dried leaves could be easily crushed into flakes by hand. From general observations, leaves dried at 700W power appeared to have the most appealing appearance and texture. These leaves remained intact as whole leaves after drying, but could be easily crushed to flakes or blended to a powder. Zhang et al. (2006) noted cautioned that uneven heating and textural changes can occur during microwave drying alone and success of microwave heating with high moisture contents often depends on uniformity of heating. Excessive heating along edges and corners may also lead to overheating and possible scorching and development of off-flavours (Nijhuis et al., 1998).

Table 2 gives the colour attributes of fresh and dried leaves. There were significant differences ($p \leq 0.05$) between fresh and dried leaves with regard to all colour attributes values. Overall, there was an increase in "L" values in dried leaves at all power levels, indicating lightening of the leaves. With respect to "a" values, leaves dried at the lower power levels (200 and 500W) were similar in value indicating that the change in greenness was not significant. Leaves dried at the higher power levels (700 and 1000W) showed higher negative "a" values, indicating that the green colour intensified. Overall, there was an increase in "b" values in dried leaves at all power levels, indicating yellowing of the leaves during microwave drying.

Table 2. Colour attributes of fresh and dried amaranth leaves.

Attribute	Fresh	Dried 200W	Dried 500W	Dried 700W	Dried 1000W
L	27.89 ± 0.439^c	34.64 ± 0.336^a	31.56 ± 1.06^b	33.65 ± 1.07^{ab}	31.39 ± 0.559^b
a	-5.852 ± 0.424^a	-5.532 ± 0.355^a	-5.852 ± 0.169^a	-8.51 ± 0.427^b	-7.725 ± 0.33^b
b	6.928 ± 0.331^c	12.09 ± 0.305^{ab}	11.14 ± 0.291^b	12.94 ± 0.631^a	11.1 ± 0.389^b
Chroma	9.071 ± 0.524^c	13.3 ± 0.398^b	12.59 ± 0.304^b	15.49 ± 0.761^a	13.52 ± 0.508^b
Hue	-49.92 ± 0.74^a	-65.48 ± 1.07^d	-62.28 ± 0.668^c	-56.69 ± 0.0871^b	-55.19 ± 0.212^b
ΔE	Reference	8.55 ± 0.35^a	5.83 ± 0.55^b	8.77 ± 1.26^a	5.78 ± 0.72^b

Values are means ± SEM, n = 4 per treatment group.

Means in a row without a common superscript letter differ ($P<0.05$) as analyzed by one-way ANOVA and the LSD test.

The change in "L" values are an indicator for measuring browning and the change in "b" values may be due to decomposition of chlorophyll and carotenoid pigments, non-enzymatic Maillard browning and formation of brown pigments (Dadali, Demirhan, & Ozbek, 2007b). Di Cesare, Forni, Viscardi and Nani (2003) reported higher chlorophyll levels in microwave-dried basil leaves compared with air-dried and freeze-dried samples, concluding that the green colour of the leaves was better preserved during microwave-drying. With respect to colour change during microwave-drying of spinach, Alibas Ozkan et al. (2007) observed a change in "L", "a" and "b" values, reported that the least changes were observed at a power level of 750W as power increased from 90 to 1000W. As shown in this study, Shaw, Meda, Tabil and Opoku (2007) found that microwave-dried coriander leaves had higher "L^*" values compared with fresh leaves, indicating lightening of the leaves. Unlike what was found in this study, there was a decrease in "a^*" and "b^*" values. Shaw et al. (2007) also noted that when compared with fresh leaves, the microwave-dried leaves had lower colour change index values compared with oven-dried samples. Dadali et al. (2007b) investigated the colour change kinetics of spinach undergoing microwave drying and reported that the values of "L" and "b" decreased, while values of "a" and total colour change (ΔE) increased as leaves were dried. They reported that the changes in colour attributes depended on the ratio of microwave output power to sample amount.

3.2 Drying Curves

Yield (%) calculated using weight data averaged 16.6% for leaves dried at 200W and between 15.5 and 14.5% for those dried at 500 to 1000W. Rajeswari et al. (2013) reported an average yield of 13.5% for amaranth leaves dried at 900W power for 3.5 min. As given in Table 3, as the power level decreased from 1000 to 700W, the time taken to attain equilibrium moisture content increased approximately 2-fold from 7.3 to 11.5 min. Further decreasing the power level to 500W again effected a 2-fold increase in time, however, further decreasing from 500 to 200W power resulted in a 7-fold increase in drying time to 135.7 min. The reduction in drying time with increased power level is documented for other leafy materials. Alibas Ozkan et al. (2007) also found a marked decline in drying time of spinach leaves, with the drying time obtained using 90W power being 13.81 times longer than the process using 1000W power. Dadali, Demirhan, and Ozbek (2007a) reported that drying time of spinach leaves was reduced by 80.55% when a microwave power of 900W was used instead of 180W. Sharada (2013) found the drying time of spinach leaves to be 2.86 times longer at 90W power level compared with 350W. Similar results were also documented for parsley, mint and basil leaves (Soysal, 2004; Ozbek & Dadali, 2007; Dermirham & Ozbek; 2011).

Table 3. Moisture content and water activity values of microwave-dried amaranth

	Fresh	Dried 200W	Dried 500W	Dried 700W	Dried 1000W
MC (g H$_2$O)/g DM	6.00 ± 0.19^a	0.14 ± 0.01^b	0.04 ± 0.05^b	0.08 ± 0.004^b	0.10 ± 0.01^b
a$_w$	0.976 ± 0.004^a	0.565 ± 0.029^b	0.407 ± 0.016^c	0.299 ± 0.059^d	0.434 ± 0.031^c
Time to Eqm (min)	NA	135.7 ± 3.3^a	19.8 ± 0.7^b	11.5 ± 0.3^c	7.3 ± 0.1^d

Values are means ± SEM, n = 5 per treatment group.

Means in a row without a common superscript letter differ ($P<0.05$) as analyzed by one-way ANOVA and the LSD test.

Initial moisture content and water activity (a$_w$) values of fresh amaranth leaves averaged 6.00 g H$_2$O/g DM (85.7% wb) and 0.976, respectively. A plot of moisture content (g H$_2$O/g DM) versus time for leaves dried at the various microwave power levels are given in Figure 1.

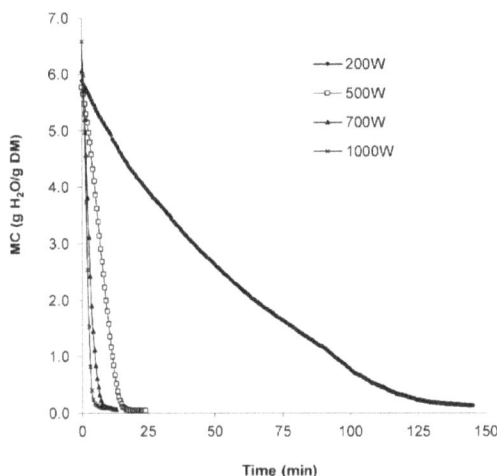

Figure 1. Effect of microwave power level on moisture content of amaranth leaves

Moisture content was significantly affected by drying time and microwave power level, as well as a time-power interaction ($p \leq 0.001$). Increasing the power level from 200 to 500W resulted in an increase in moisture reduction, however the greatest effect was observed as the power level increased from 200 to 500W. After 17 min of drying, the moisture values of samples dried at 200, 500, 700 and 1000W averaged 4.45, 0.07, 0.08 and 0.10 g H$_2$O/g DM, respectively.

The equilibrium moisture content and water activity values of microwave-dried samples are also given in Table 3. Moisture values of dried samples were found to be significantly different from fresh leaves ($p \leq 0.001$) but there was no effect of microwave power. Water activity values of dried samples were found to be significantly different from fresh leaves ($p \leq 0.001$), and while there was a small treatment effect ($p \leq 0.05$) there was no discernible trend. Equilibrium moisture content values for dried samples at 1000W to 200W microwave power levels ranged from 0.04 to 0.14 g H$_2$O/g DM (4.1% to 12.3% wb). Moisture values of leaves dried at 700W power averaged 0.08 g H$_2$O/g DM (7-8% wb).

The initial and equilibrium moisture content values and the effect of increasing microwave power level as seen in this study are in agreement with their results of studies done on the microwave drying and air of amaranth as well as spinach. Fathima et al. (2001) reported a moisture content of 5% (wb) for hot water-blanched amaranth leaves (*Amaranth* sp.) dried at 100% power (microwave wattage not given) for 720 s. Rajeswari et al. (2013) compared the quality and shelf life of *Amranthus tricolor* leaves dried for 2h 37min at 60°C in a cabinet drier and 3 min 31 s in a microwave oven at 900W power. The moisture content of microwave dried control samples averaged 3.12% (wb), while those of oven dried samples averaged 3.25% (wb). Singh et al. (2014) found the initial moisture content of amaranth leaves to average 6.55 to 7.27 g H_2O/g DM (87-88% wb) and microwave-dried untreated leaves to range from 0.05 to 0.06 g H_2O/g DM (5-6% wb).

In the present study the moisture content values of microwave-dried leaves after 11.5 min at 700W power output are similar to those reported by Akonor and Amankwah (2012) for the solar drying of *Amaranthus hybridus* leaves. The initial moisture content of the leaves averaged 85.8% (wb) and leaves were solar-dried for two days to a final moisture content of 8.5% (wb). Singh et al. (2014) reported that a reduction in moisture content in amaranth leaves from 87.4 to 5.1% (wb) after 450 min of drying in a greenhouse-type solar dryer. Research work done on the microwave drying of spinach leaves also revealed that microwave power level had an important effect on moisture reduction, with higher moisture reduction at higher power levels (Alibas Ozkan et al., 2007; Dadali et al., 2007a; Sharada, 2013). For the microwave-drying of basil leaves, Seyedabadi (2015) noted that increasing microwave power increases the thermal gradient in the samples and thus increases evaporation rate, which results in faster drying. The drying time of basil leaves could be reduced by 82% by the application of 900W power instead of 90W power.

The drying energy consumption for 20g samples of amaranth leaves are given in Figure 2. Energy consumption decreased as microwave power increased, with the most noticeable decline of 452 to 165 W.h occurring when microwave power increased from 200 to 500W. At power levels from 500 to 1000W, energy consumption values were not statistically different. Alibas Ozkan et al. (2007) also reported higher energy consumption values for leaves dried at lower power levels of 90 to 160W microwave power levels, with values at 350 to 1000W power levels being similar to each other.

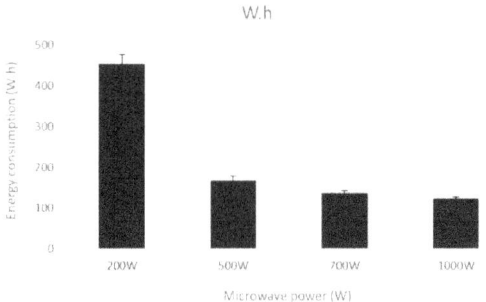

Figure 2. The drying energy consumption of amaranth leaves at different power levels.

3.3 Drying Rate Curves

Plots of drying rate (rate of change in moisture content) as a function of time and average moisture content are shown in Figure 3a and b, respectively. As shown in Figure 3a, the higher the microwave power level the higher the drying rate during the first 5 min of drying. Drying rates at 1000 and 700W power levels increased rapidly during the first 90s of drying, then decreased with time. For leaves dried at 500W power, drying rate in increased gradually during the first 5 min of drying, remained steady for 5 min, then declined. For leaves dried at 200W power, rate was steady for the duration of the drying period. Drying rates after 1.75 min averaged 0.10, 0.30, 1.52 and 2.40 g H_2O/g DM/min for leaves dried at 200, 500, 700 and 1000W, respectively.

Similar results were seen for microwave drying of spinach (Alibas Ozkan et al., 2007) who reported that the maximum moisture was removed between the first 120 to 150 s of drying at 1000W microwave power, adding that about 60% of the drying process was completed by that time. At the lowest microwave power level of 90W, maximum drying rate of spinach leaves occurred after 270s of drying, with 7.91% of the drying process being completed at that time.

As shown in Figure 3b, drying at 200W power level occurred mainly in the constant rate period. After an initial warm up period, drying at 500W power level occurred in the constant rate period for the moisture content values between 4.5 to 1.0 g H_2O/g DM, following which there was a falling rate period. For leaves dried at 700 and 1000W power levels, there was an increase in drying rate at the start of drying between moisture levels of 6.5 to

4.5 g H_2O/g DM. At moisture values lower than 3.0 to 3.5 g H_2O/g DM, drying occurred in the falling rate period. Similar changes in drying rate with time and with increasing microwave power levels were reported for spinach, mint and parsley leaves (Soysal, 2004; Alibas Ozkan et al., 2007; Ozbek & Dadali, 2007; Sharada, 2013). According to those authors, high moisture values of leaves at the initial stages of drying results in higher absorption of microwave power and higher drying rates due to higher moisture diffusion. Decrease in moisture as drying progresses results in a decrease in absorption of microwaves, a drop in temperature, and therefore a fall in drying rate.

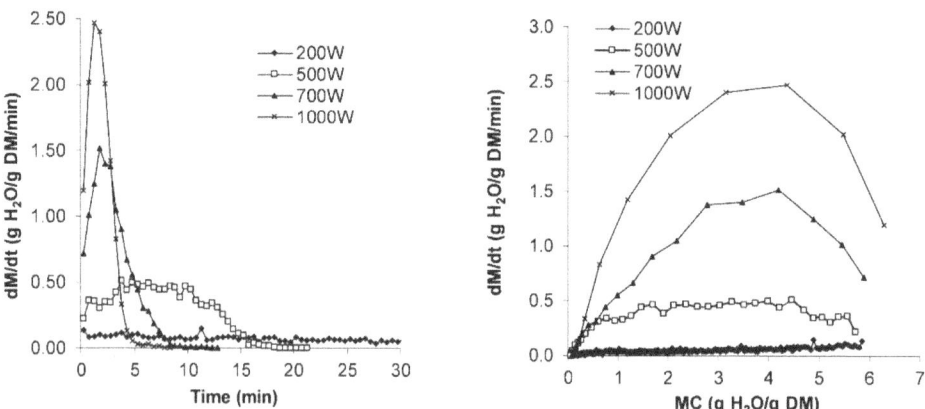

Figure 3. Drying rate curves of amaranth leaves at various microwave power levels

Constant rate drying has also been reported by some authors for the microwave drying of leafy materials. Soysal (2004) also found that constant rate drying dominated the drying time of parsley leaves dried at 360-900W power levels, with the critical moisture content averaging 1.75 g H_2O/g DM. For grape leaves (Alibas, 2014), constant rates of drying occurred at higher moisture values, with the critical moisture content (start of the falling rate period) averaging 2.5 g H_2O/g DM, similar to what was found in this study. Ozbek and Dadali (2007) also observed a short warm-up phase at higher microwave power levels for the microwave-drying of mint leaves. Drying during the constant rate period is not unexpected during microwave drying, as the internal pressure gradient generated by the microwave field effectively pumps water to the surface, wetting the surface and the rate of drying is dependent upon the evaporation of this free moisture from the surface. Other authors have reported drying during the falling rate only for the drying of spinach and celery leaves (Alibas Ozkan et al., 2007; Demirham & Ozbek, 2011). Demirham and Ozbek (2011) added that the critical moisture content was equal to the initial moisture content of celery leaves which meant that the microwave drying process at 180 to 900W power levels was entirely controlled by mass transfer resistance.

3.4 Drying Rate Constants (K) and Diffusion Coefficients (D_{eff})

The drying rate constants (k) for the initial minutes of microwave drying were determined from plots of the ln free moisture (ln MR) as a function of drying time. Drying rate constants (Table 4) were significantly different at the different power levels ($p \leq 0.001$). There was a 4.5-fold increase in drying rate constant (k) as microwave power level increased from 200 to 1000W.

Table 4. Drying rate constants (k) and diffusion coefficients (D_{eff}) for microwave-dried amaranth leaves

Microwave power level (W)	k (1/s)	[a]D_{eff} (m^2/s)	R^2
200	0.0003 ± 0.000^a	3.04×10^{-10}	0.9781
500	0.0012 ± 0.0001^a	1.22×10^{-10}	0.9882
700	0.0125 ± 0.0006^b	1.27×10^{-9}	0.9943
1000	0.0278 ± 0.0008^c	2.82×10^{-9}	0.9981

Values are means ± SEM, n = 5 per treatment group.

Means in a column without a common superscript letter differ ($P<0.05$) as analyzed by one-way ANOVA and the LSD test.

[a]D_{eff} = k $(4X^2/\pi^2)$ where X = half thickness 0.05cm

With respect to drying rate constants reported in the literature, Dadali et al. (2007a) determined that for *Spinacia oleracea* L. the drying rate constant determined using the Page model increased from 0.0436 to 0.3455 /min as microwave output power increased from 180 to 900W while Alibas Ozkan et al. (2007) dried 50g samples and found the drying constants to increase from 0.0107 to 0.1283 /min as microwave output power increased from 90 to 1000W. This trend is in agreement with that seen in this study. Diffusivity (D_{eff}) values for amaranth leaves calculated from Equation 7 increased from 3.04 x 10^{-10} to 2.82 x 10^{-9} m^2/s. The low moisture diffusivity values of leaves dried at 200W power level is expected, since these leaves looked moist for most of the drying process.

Simha and Gugalia (2013) found that the effective moisture diffusivity values (D_{eff}) for *Spinacia oleracea* leaves dried at 180 and 300W microwave power levels ranged between 6.085 to 9.736 x 10^{-9} m^2/s. Microwave-drying studies on *Spinacia oleracea* L. by Dadali et al. (2007a) and Alibas Ozkan et al. (2007) did not report on diffusivity values. With respect to D_{eff} values for other microwave-dried leafy greens and herbs, Demirhan and Ozbek (2011) reported moisture diffusivity values for microwave-dried celery leaves of 0.343 x 10^{-10} to 1.714 x 10^{-10} m^2/s as microwave power increased from 180 to 900W using 25g celery leaf samples. Alibas (2014) reported *D*-values for microwave-dried celery leaves ranging from 1.595 x 10^{-10} to 6.377 x 10^{-12} m^2/s at power levels of 90 to 1000W. An increase with microwave power level has been attributed to the increased activity of water molecules at higher power levels.

3.5 Thin Layer Models

A plot of Moisture Ratio (*MR*) versus time (*s*) calculated based on Equation 6 is given in Figure 4. Moisture ratio values calculated for the falling rate period were significantly affected by drying time, microwave output power and a time-power interaction ($p \leq 0.001$). A trend similar to the drying curves was observed, where increasing the power level from 200W to 500W effected a noticeable decline in moisture ratio. The moisture ratio values of leaves dried at 500W power were similar to those dried at 700W power. Further increasing the power level to 1000 W resulted in a slight decrease in moisture ratio values.

While no previous works were found on the moisture ratio curves of microwave-dried amaranth leaves, a greater decline in *MR* values with increased microwave power level has been reported for other leafy materials including spinach, mint, celery and grape leaves, (Alibas Ozkan et al., 2007; Dadali et al., 2007a; Ozbek & Dadali, 2007; Demirham & Ozbek, 2011; Alibas, 2014). Dadali et al. (2007a) reported when microwave output power decreased from 900 to 180W the time for the moisture ratio of spinach leaves to reach zero increased from 3.8 to 18 minutes, respectively.

With respect to drying of amaranth leaves by conventional drying methods, Akonor and Amankwah (2012) presented *MR* curves for the drying of *Amaranthus hybridus* leaves during solar cabinet drying. They noticed a constant rate period preceding the falling rate period as evidenced by an "elbowing" of the moisture ratio curves at the start of the drying process.

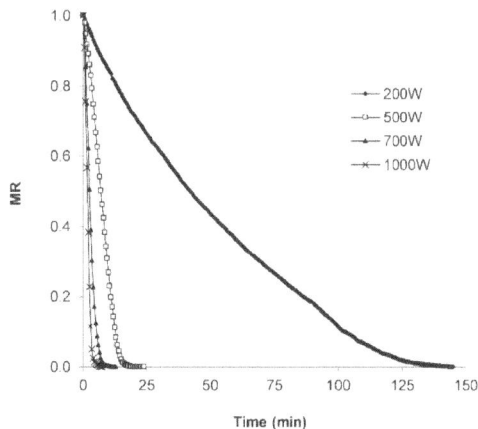

Figure 4. Moisture ratio curves for amaranth leaves at different microwave power levels

Many current drying studies also focus on the application of traditionally used thin layer models (empirical, semi-empirical and theoretical) to describe the moisture ratio (*MR*) data. The performance (fit) of the models is assessed through the use of the coefficient of determination (R^2), root mean square error (*RMSE*) and the chi-square statistic (χ^2) as is the common practice in thin-layer drying studies. Of the twenty-two thin layer models applied to the *MR* data, the coefficients for the four models which best fit the data are given in Table 5. Beyond these four models, the other models showed regression coefficients of less than 0.900 and some models simply failed.

Table 5. Thin layer models and constants in order of best fit for amaranth leaves at different microwave power levels

Power Level	Model	Model Constants							R^2	RMSE	χ^2
		k	n	a	b	k_1	g	k_2			
200W	Alibas	0.9089	1.0054	1.0000	0.9246		-0.0140		0.9962	0.021197	0.000654
	Aghbashlo et al.					0.0060		-0.0022	0.9864	0.040117	0.001839
	Wang & Singh			-0.0061					0.9785	0.050482	0.002912
	Peleg			112.50	0.6654				0.9423	0.082731	0.007822
500W	Alibas	0.2425	1.0109	0.9867	0.2515		-0.0079		0.9988	0.011230	0.000155
	Aghbashlo et al.					0.0039		-0.0013	0.9922	0.028278	0.000864
	Wang & Singh			-0.0037					0.9784	0.047107	0.002397
	Peleg			166.38	0.7081				0.9517	0.070417	0.005355
700W	Aghbashlo et al.					0.0013	0.0000	-0.0007	0.9940	0.026533	0.000734
	Alibas	0.9862	1.0004	1.1599	0.9870		-0.1120		0.9863	0.040257	0.001805
	Wang & Singh			-0.0016	0.0000				0.9805	0.047931	0.002395
	Peleg			562.19	0.5067				0.9572	0.071050	0.005263
1000W	Alibas	1.0350	1.0000	1.2535	1.0348		-0.2678		0.9987	0.010268	0.000107
	Peleg			3648.64	0.5420				0.9984	0.011654	0.000137
	Wang & Singh			-0.0002					0.9965	0.016924	0.000288
	Aghbashlo et al.					0.0002		-0.0001	0.9959	0.018409	0.000341

For leaves dried at 200, 700 and 1000W microwave power levels, the Alibas model (Alibas, 2012) gave the best fit in terms of both the root mean square error (RMSE) and the chi-square statistic (χ^2). At the 500W power level, the Alibas model was second best and a close second to the Aghbashlo et al. model. According to the author (Alibas, 2012), the Alibas model derived from the Midilli model. The constants (k and n) as well as the coefficients (a, b and c) depend on the operating conditions, which for microwave drying, includes the weight of the material and the microwave power level. Figure 5 gives the comparison of the predicted MR values with the experimental values for all treatments based upon the Alibas model, and for the Aghbashlo et al. model for leaves dried at 500W power level.

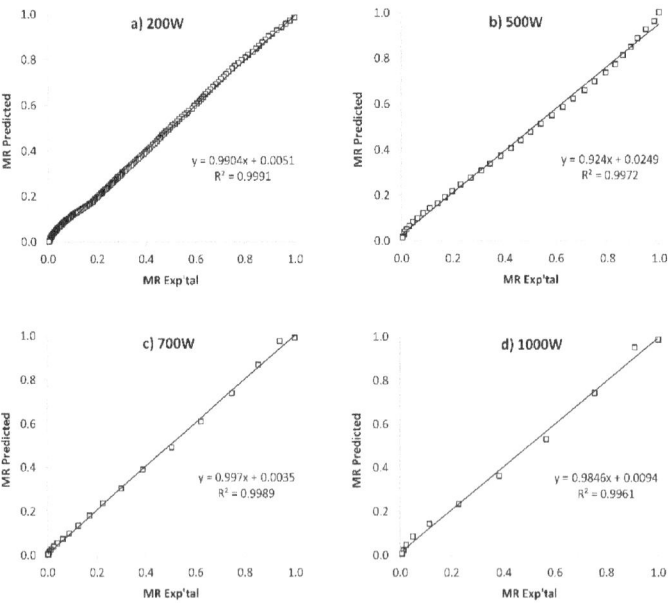

Figure 5. Comparison of predicted versus experimental moisture ratio values for amaranth leaves dried at different microwave power levels. Slope of straight line equations and high R^2 values give an indication of good agreement of values

No works were found on the modelling of microwave drying of amaranth leaves, however, Akonor and Amankwah (2012) modelled the data for the drying of *Amaranthus hybridus* leaves during solar cabinet drying using five well-known semi-empirical drying models. Based on the lowest RMSE and χ^2 values, the Page model best fit the MR data.

With regard to microwave drying of spinach leaves, Dadali et al. (2007a) reported when microwave output power decreased from 900 to 180W the time for the moisture ratio of spinach leaves to reach zero increased from

3.8 to 18 minutes, respectively. They tested the fit of eight thin layer drying models, and observed that the Page model was generally the most appropriate to describe the *MR* data. Alibas Ozkan et al. (2007) also presented plots of *MR* versus time for microwave-dried *Spinacia oleracea* L. leaves dried at different power levels, reporting that the Page model adequately described the *MR* data. Sharada (2013) found that out of four thin layer models (Newton, Henderson and Pabis, Midilli and Page) the Midilli model best described the microwave drying behaviour of spinach (*Spinacia oleracea*) leaves. Simha and Gugalia (2013) also found the Midilli model to best describe the drying data for microwave-dried *Spinacia oleracea* leaves.

With respect to the microwave drying of other leafy materials, Soysal (2004) gave *MR* plots for microwave-dried parsley leaves and modelled the data using the Page model. Ozbek and Dadali (2007) presented similar *MR* plots for microwave-dried mint leaves, applying ten drying models and reporting the Midilli model as best describing the drying data. The Midilli model was also recommended for the modelling of *MR* data for microwave-dried celery leaves (Demirham & Ozbek, 2011). Seyedabadi (2015) found the logarithmic model to best describe drying data for basil leaves dried at 90 to 900W microwave power levels.

4. Conclusions

From the results of this study, microwave drying appears to be a feasible drying method for the rapid drying of amaranth leaves. Microwave power level has a significant impact on the drying rates and quality of dried samples. An increase in power level resulted in more rapid drying, with the risk of scorching increasing at 1000W power. Drying at 200W power level was the least favourable drying treatment in terms of drying rate and overall appearance. The optimum drying power level based on drying rates and on quality and appearance of the leaves was found to be 700W for a maximum of 11.5 min for 20g samples. These leaves remained intact as whole leaves but could be easily crushed to flakes or blended to a powder. Drying at this microwave power level occurred in the falling rate period at moisture values below 4.5 g H_2O/g DM, following an initial warm-up period. The drying data was successfully analysed through the use of a drying rate constant (k) and moisture diffusivity (D_{eff}) and the Alibas model successfully applied to the *MR* data.

Nomenclature

A	Drying constant
a_w	Water activity
D_{eff}	Diffusion coefficient (m^2/s)
DM	Dry matter (g)
E_t	Energy consumption (W.h)
FW	Fresh weight (g)
k	Drying rate constant (1/h)
L_0, a_0, b_0	Colour attributes at time = 0
L, a, b	Colour attributes at time = t
M	Moisture content (g H_2O/g DM) at time = t
M_o	Initial Moisture Content (g H_2O/g DM)
M_e	Equilibrium Moisture Content (g H_2O/g DM)
MR	Moisture Ratio
P	Microwave power
R^2	Coefficient of determination
RMSE	Root Mean Square Error
X	Half-thickness of sample (m)
$k, k_0, k_1, a, b, c, g, n$	Model constants
t	Time (min)
wb	Wet basis (g H_2O/100g FW)
χ^2	Chi-Square

References

Akonor, P., & Amankwah, E. (2012). Thin layer drying kinetics of solar-dried *Amaranthus hybridus* and *Xanthosoma sagittifolium* Leaves. *Journal of Food Technology, 10*(3), 92-96. http://doi.org/10.4172/2157-7110.1000174

Alegbejo, J. O. (2013). nutritional value and utilization of amaranthus (*Amaranthus* spp.) - A Review. *Bayero Journal of Pure and Applied Sciences, 6*(1), 136-143. http://dx.doi.org/10.4314/bajopas.v6i1.27

Aletor, O., & Abiodun, A. R. (2013). Assessing the effects of drying on the functional properties and protein solubility of some edible tropical leafy vegetables, *Research Journal of Chemical Sciences, 3*(2), 20-26.

Alibas, I. (2014). Microwave, air and combined microwave-air drying of grape leaves (*Vitis vinifera* L.) and the determination of some quality parameters. *International Journal of Food Engineering,* 10(1), 69-88. http://10.1515/ijfe-2012-0037

Alibas, I. (2012). Microwave drying of grapevine (*Vitis vinifera* L.) leaves and the determination of some quality parameters. *Journal of Agricultural Sciences, 18,* 43-53.

Alibas Ozkan, I., Akbudak, B., & Akbudak, N. (2007). Microwave drying characteristics of spinach. *Journal of Food Engineering, 78*(2), 577-583. http://doi.org/10.1016/j.jfoodeng.2005.10.026

Andini, R., Yoshida, S., & Ohsawa, R. (2013). Variation in protein content and amino acids in the leaves of grain, vegetable and weedy types of amaranths. *Agronomy, 3*(2), 391-403. http://doi.org/10.3390/agronomy3020391

Assaad, H. I., Hou, Y., Zhou, L., Carroll, R. J., & Wu, G. (2015). Rapid publication-ready MS-Word tables for two-way ANOVA. *SpringerPlus, 4*(1), 33. http://doi.org/10.1186/s40064-015-0795-z

Borneo, R., & Aguirre, A. (2008). Chemical composition, cooking quality, and consumer acceptance of pasta made with dried amaranth leaves flour. *LWT-Food Science and Technology, 41*(10), 1748-1751. http://doi.org/10.1016/j.lwt.2008.02.011

Crank, J. (1975). *The mathematics of diffusion* (2nd ed.). UK: Oxford University Press.

Dadali, G., Demirhan, E., & Ozbek, B. (2007a). Microwave heat treatment of spinach: drying kinetics and effective moisture diffusivity. *Drying Technology, 25*(10), 1703-1712. http://doi.org/10.1080/07373930701590954

Dadali, G., Demirhan, E., & Ozbek, B. (2007b). Color change kinetics of spinach undergoing microwave drying. *Drying Technology, 25*(10), 1713-1723.

Dadali, G., Demirhan, E., & Ozbek, B. (2008). Effect of drying conditions on rehydration kinetics of microwave dried spinach. *Food and Bioproducts Processing, 86*(4), 235-241. http://doi.org/10.1016/j.fbp.2008.01.006

Decareau, R. V. (1985). *Microwaves in the food processing industry*. Orlando: Academic Press Inc.

Demirhan, E., & Ozbek, B. (2011). Thin-layer drying characteristics and modeling of celery leaves undergoing microwave treatment. *Chemical Engineering Communications, 198*(7), 957-975. http://doi.org/10.1080/00986445.2011.545298

Di Cesare, L. F., Forni, E., Viscardi, D., & Nani, R. C. (2003). Changes in the chemical composition of basil caused by different drying procedures. *Journal of Agricultural and Food Chemistry,* 51, 3575-3581. http://10.1021/jf021080o

Fathima, A., Begum, K., & Rajalakshmi, D. (2001). Microwave drying of selected greens and their sensory characteristics. *Plant Foods for Human Nutrition, 56*(4), 303-311. http://doi.org/10.1023/A:1011858604571

Hebbar, H.U., Vishwanathan, K. & Ramesh, M. (2004). Development of combined infrared and hot air dryer for vegetables. Journal of Food Engineering 65(4):557-563. http://doi.org/10.1016/j.jfoodeng.2004.02.020

HunterLab (2008). *Hunter L, a, b Color Scale*. Application Note 8(9). Virginia, USA: Hunter Associates Laboratory Inc.

Hyams, D. G. (2016). CurveExpert Professional 2.3. http://www.curveexpert.net.

Mujaffar, S., & Sankat, C. K. (2005). The air drying behaviour of shark fillets. *Canadian Biosystems Engineering / Le Genie Des Biosystems Au Canada, 47,* 11-21.

Mujaffar, S., & Sankat, C. K. (2014). Modelling the drying behaviour of unsalted and salted catfish (*Arius* sp.) slabs. *Journal of Food Processing and Preservation, 39 (6),* 1385-1398. DOI: 10.1111/jfpp.12357

Nijhuis, H. H, Torringa, H. M., Muresan, S. Yuksel, D. Leguijt, C., & Kloek, W. (1998). Approaches to improving the quality of dried fruit and vegetables. *Trends in Food Science and Technology, 9,* 13-20.

Ozbek, B., & Dadali, G. (2007). Thin-layer drying characteristics and modelling of mint leaves undergoing microwave treatment. *Journal of Food Engineering, 83*(4), 541-549. http://doi.org/10.1016/j.jfoodeng.2007.04.004

Peter, C., Elizabeth, K., Judith, K., & Hudson, N. (2014). Retention of β-carotene, iron and zinc in solar dried amaranth leaves in Kajiado County, Kenya., *International Journal of Sciences: Basic and Applied* Research (IJSBAR), *13*(2), 329-338.

Rajeswari, R., Bharati, P., Naik, K. R., & Naganur, S. (2013). Dehydration of amaranthus leaves and its quality evaluation. *Karnataka Journal of Agricultural Sciences, 26*(2), 276-280.

Rodriguez, P., Perez, E., Romel, G., & Dufour, D. (2011). Characterization of the proteins fractions extracted from leaves of *Amaranthus dubius* (*Amaranthus* spp.). *African Journal of Food Science, 5*(7), 417-424

Seyedabadi, E. (2015). Drying kinetics modelling of basil in microwave dryer. *Agricultural Communications, 3*(4), 37-44.

Sharada, S. (2013). Microwave drying of *Spinacia oleracea*, IJRET: *International Journal of Research in Engineering and Technology, 2*, 481-486.

Shaw, M., Meda, V., Tabil, J., & Opoku, A. (2007). Drying and color characteristics of coriander foliage using convective thin -layer and microwave drying. *Journal of Microwave Power & Electromagnetic Energy, 41*(2), 56-65.

Silva, W. P., Silva, C. M. D. P. S., Gama, F. J. a., & Gomes, J. P. (2014). Mathematical models to describe thin-layer drying and to determine drying rate of whole bananas. *Journal of the Saudi Society of Agricultural Sciences, 13*(1), 67-74. http://doi.org/10.1016/j.jssas.2013.01.003

Simha, P., & Gugalia, A. (2013). Thin layer drying kinetics and modelling of *Spinacia oleracea* leaves. *International Journal of Applied Engineering Research, 8*(9), 1053-1066.

Singh, P., Singh, S., Singh, B. R., Singh, J., & Singh, S. K. (2014). The drying characteristics of amaranth leaves under greenhouse type solar dryer and open sun. *Green Journal of Agricultural Sciences, 6*(2014). http://doi.org/10.15580/GJAS.2014.6.040314174

Soysal, Y. (2004). Microwave drying characteristics of parsley. *Biosystems Engineering, 89*(2), 167-173. http://doi.org/10.1016/j.biosystemseng.2004.07.008

VSN International Limited (Rothamsted Experimental Station). (2011). GenStat Release 10.3 Discovery Edition.

Zhang, M., Tang, J., Mujumdar, A. S., & Wang, S. (2006). Trends in microwave-related drying of fruits and vegetables. *Trends in Food Science and Technology, 17*(10), 524-534. http://doi.org/10.1016/j.tifs.2006.04.011

9

Evaluation of Children's Lunch Box Contents by Photograph and Their Relationship with Mothers' Concern

Tomoko Osera[1,2], Setsuko Tsutie[3], Misako Kobayashi[2] & Nobutaka Kurihara[1]

[1]Hygiene and Preventive Medicine, Graduate School of Life Science, Kobe Women's University, Japan

[2]Takakuradai Kindergarten attached to Kobe Women's University, Japan

[3]Clinical Nutrition Management, Graduate School of Life Science, Kobe Women's University, Japan

Correspondence: Nobutaka Kurihara, Graduate School of Life Science, Kobe Women's University, 2-1 Higashisuma-Aoyama, Suma, Kobe, Japan. E-mail: kurihara@suma.kobe-wu.ac.jp

Abstract

Japanese kindergarten children usually bring lunch prepared by mothers. The contents may be influenced by mothers' food concerns. We investigated the relationship between mothers' concerns and children's lunch box contents and preferences. Lunch boxes of 209 children were digitally photographed for 4 days at a private kindergarten in Japan. The amounts of rice, main dishes, vegetables and fruits in the lunch boxes were estimated by measuring the area occupied by each in the photograph; a questionnaire, including questions on mothers' concerns and children's preferences, was completed by mothers. Vegetable amounts in the lunch boxes were significantly related to mother's concerns for their children's lunch. Compared with estimated vegetable amounts below 11%, the amounts above 11% indicated that the number of foods disliked by children was lower, and mothers reported a higher rate of mindfulness towards vegetables and lower rate towards frozen food and believed that they prepared a balanced lunch. Thus, vegetable amounts in children's lunch boxes, estimated using photographs, may predict mothers' food concerns and children's balanced/unbalanced diets.

Keywords: Children's Lunch Box, Photograph, Vegetables, Mothers' concern

1. Introduction

Mothers' attitude toward children acquiring healthy food habits may have an influence on children's consumption. The aim of this study is to investigate the relationship between mothers' concern for lunch box and children's lunch box contents. In the study, we investigate to mother's concern for children pay attention to their lunch box. In Japan, children at nursery facilities usually eat school lunches, while those at kindergartens usually bring packed lunches from home. There are no standards for ensuring that children bring healthy packed lunches in the UK (Rees, Richards & Gregory, 2008, Evans et al., 2010a) where approximately half of the school pupils get packed lunch from home (Smithers et al., 2000) similar trends have been observed in Japan, and it seems that foods cooked by mothers are included in children's packed lunches.

Mothers must be careful regarding the contents of the packed lunches. However, one study (Rees, Richards & Gregory, 2008) reported that in the UK, very few packed lunches contained vegetables, and fruit intake was particularly low for those having a school meal. Moreover, Evans et al. (2010a) reported that few packed lunches met the school meal standards. In the US, 41% of the elementary school children brought lunch to school on any given day, of which 45% brought snacks (Hubbard et al., 2013). In Australia, almost all children had some form of 'junk food' in their packed lunches, with a mean of 3.1 ± 0.1 servings (Sanigorski et al., 2005). This indicates that children's packed lunch contains few vegetables in these countries. Similarly, the packed lunches of Japanese children may also contain unbalanced diets, although little research has been conducted on this subject in the kindergarten children. Thus, the contents may be influenced by the mothers' food concern. In this study, we investigated the varieties of foods contained in the children's packed lunches by mothers' with various food concerns.

Early childhood is the most important period for establishing healthy eating habits and controlling children's preferences. Childhood is a sensitive period for development of food acceptance patterns (Nicklas et al., 2001, Ilingworth & Lister, 1964, Cashidan, 1994). Likes and dislikes are easily acquired during the early years of life

(Birch, 1979, Birch & Fisher, 1998); Vereeken and Maes (2010) reported that young children's dietary habits were associated with the mothers' nutritional knowledge and attitudes. Thus, the mother's role is important in establishing a child's eating habits and preferences. In this study, we compared the frequency of eating a school lunch with that of eating a packed lunch. We investigated the relationship between the frequency of bringing a packed lunch to the school, mothers' concerns for the foods in lunches packed by them and children's preferences.

It has been well established that the foods consumed, particularly vegetables, are very important for overall health. Consumption of vegetables prevents conditions, such as cardiovascular diseases, stroke, hypertension, diabetes, obesity, and certain types of cancer, leading to enhanced human longevity (Zhang et al., 2011, Apped et al., 1997, World Cancer Research Fund & American Institute for Cancer Research, 2007, Carter et al., 2010). Food intake in individuals is influenced by many factors, such as age, sex, eating habits, knowledge of nutrition, individual preferences, overall health and social status (Osera et al., 2016a, MacFarlane, Crawford & Worsley, 2010, Bauer et al., 2008, Hlimi et al., 2012). Among these, preferences are one of the primary determinants of food intake (Cook, Wardle & Gibson. 2003, Dovey et al., 2008). We hypothesized that the contents of children's packed lunch reflect mothers' concerns for children's diet.

In order to test this hypothesis, we conducted an investigation in Japan to determine the percentage of space in children's lunch box accounted for by the vegetables and the effect of food preference as well as the mothers' concerns on this percentage. In addition, the relationship between mothers' concerns and contents of children's lunch boxes was assessed.

2. Method

2.1 Participant (Subject) Characteristics

We photographed children's lunch boxes that were assigned an ID number using a digital camera on three occasions over the course of a school year at a private kindergarten in Hyogo Prefecture, Japan, from April 2013 to March 2014. Using photographs, we classified the contents of each packed lunch into five groups: 1) rice (e.g., staple foods with grains), 2) main dishes (e.g., meat, poultry, eggs and fish), 3) vegetables, 4) fruits and 5) sweets (e.g., jelly or Jell-O) (Figure 1). We also measured the percentage of space accounted for by each food group in 209 lunch boxes. The amount of food from each group was then estimated in the lunch boxes of 209 children by measuring the area in the photograph with an image-analysing program. We performed the estimations by measuring the area in the photograph using Image J, and calculated the average of the three occasions for each food group.

The mothers were adequately informed about the objectives and methods of this investigation, and they answered the questionnaire voluntarily, with the right to withdraw at any time during the study. Individual privacy was strictly protected throughout the investigation. Under these conditions, the mothers agreed to cooperate with the scientific investigation while their children were included in this study. The study was performed after receiving approval from the principals of the kindergarten and nursery facilities. The study was also approved by the president of the kindergarten and the Kobe Women's University people's ethics committee.

Figure 1. Children's packed lunch with ID number

2.2 Research Design

Data were analysed together with a questionnaire regarding mothers' concerns for children's lunch boxes and their eating habits at the end of the investigation. In March 2014, questionnaires were distributed to mothers of the children who attended a private kindergarten in Hyogo Prefecture, Japan. The questionnaire sheets were distributed to the mother of each child (i.e., a mother with two children in kindergarten received two sheets and answered the questions regarding each child). In total, questionnaires from 264 (79%) children were collected; of them, 209 were included in the study. Mothers of 209 corrected all study that submitted questionnaire and their children bring packed lunch all day.

2.3 Questionnaire

The questionnaire completed by the mothers included 50 questions and had three sections: 1) children's food habits (18 questions), 2) mothers' food habits (14 questions) and mothers' concerns for their children's lunch box contents (18 questions). We defined the mother's concern for her child's lunch box as 'preparation time', 'consideration of the nutritional balance for the child's packed lunch' and 'the frequency of using frozen food'. We were obvious this questionnaire's reliability. This questionnaire is a revised version of the one originally created by the Japan Sports Council that was used to determine the food habits of elementary school children and junior high school students (Japan Sports Council, 2010). We revised it to specifically determine the food habits of children aged between 3 and 5 years and their guardians. These improvisations proved to be adequate, and we checked the reliability of the revised version. Specifically, questions regarding mothers' concerns for children's packed lunches were added to the original version. For the calculated data presented in Table 1, the median percentage of vegetables was 11%. Banalization of data was performed to divide it into two groups as shown in Table 2 and Figure 2.

Figure 2. Examples of children's packed lunches estimated vegetable amounts

Photographs on the left side show packed lunches with vegetable amounts below 11%; a) and b).

Those on the right side show packed lunches with vegetable over 11%; c) and d).

2.4 Statistics and Data Analysis

We used a five-point or a four-point rating scale in our research, with the highest point indicating good food habits. The foods that the children disliked were chosen from a list of 55 foods selected from those available at regular school lunches and often disliked by children as shown in our previous study (Osera et al., 2012). SPSS version 21.0 J was used for all statistical analyses. In addition, all data were analysed using a Fisher's exact test and Mann–Whitney U test. Study consisted of only the questionnaire; we attempted to determine a relation between the contents of the lunch box and the answers to the questionnaire regarding food habits. Children's lunch box and questionnaire had the same ID (Figure 1).

3. Results

Using photographs, the contents of each lunchbox were classified into five groups: 1) rice, 2) main dishes, 3) various kinds of vegetables, 4) fruits and 5) sweets. The percentages of area for each of the five groups were 37.9% ± 9.2%, 37.3% ± 10.0%, 11.9% ± 7.8%, 8.9% ± 7.9% and 2.0% ± 3.7%, respectively (Table 1, Figure 3). The areas for vegetables, including broccoli and tomato, were combined and presented as a single area.

Table 1. Estimated amount of each group in lunch boxes

	Mean	±	SD	Max	Min
Rice	37.9	±	9.2	63	9
Main dishes	37.3	±	10	62	6
Vegetables	11.9	±	7.8	38	0
Fruit	8.9	±	7.9	32	0
Sweets	2.0	±	3.7	17	0

Note. N = 209

A total of three packed lunches per person.

Average of 209 children's lunch boxes.

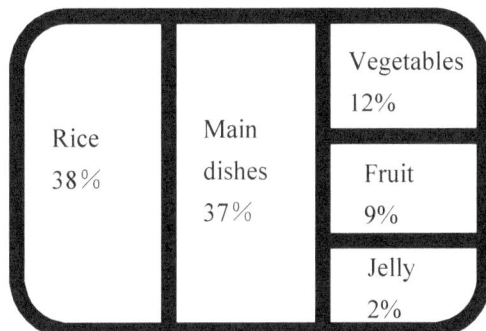

Figure 3. Average of children's lunch box contents

For the calculated data presented in Table 1, the median percentage of vegetables was 11%. Banalization of data was performed to divide it into two groups as shown in Table 2 and Figure 2. The number of foods disliked by children was smaller when the percentage of the 'vegetables' area was above 11% ($p < 0.01$, Table 2). A greater percentage of the guardians associated with this area were mindful of vegetables, while only a small percentage used frozen food ($p < 0.01$, each). In cases with the percentage of the 'vegetables' area below 11%, these results were even lower. Therefore, the percentages of 'vegetables' area above or below 11% were significantly related with mothers' consideration of a balanced lunch box ($p < 0.01$). In addition, watching TV and school lunch fun was significantly related to the 'vegetables' area ($p < 0.05$, each). Thus, the percentage of 'vegetables' area may be related to children's unbalanced or balanced diets and mothers' concern for food.

Table 2. Relationship between the side dish area, children's food habits and mothers' concerns for the children's packed lunch

	Side dishes area						
	Below 11%			Above 11%			
	N	Mean	±	SD	N	Mean	±	SD	P value
---	---	---	---	---	---	---	---	---	---
Children who disliked food[+]	72	8.06	±	8.09	75	4.69	±	4.11	0.006**
Watching TV during dinner[#]	90	2.69	±	1.5	89	3.17	±	1.49	0.025*
Children who enjoy school lunch[#]	90	4.10	±	0.97	89	4.40	±	0.78	0.031*
Guardians who think lunch box is balanced[#]	90	3.07	±	0.79	87	3.53	±	0.76	0.000***
Guardians who are mindful of vegetable amount in lunch box[#]	90	3.38	±	0.99	88	3.82	±	0.65	0.006**
Guardians who do not use frozen food in packed lunch[#]	90	2.62	±	1.1	88	2.97	±	1.08	0.038*

Note. When we calculated the data from Table 1, the percentage of the median values of vegetables was 11%. Banalization is the process of dividing the data into two groups presented above.

The amount of the 'vegetables' area below 11% vs. above 11% was assessed for significance using a Mann–Whitney U test.

* $p < 0.05$, ** $p < 0.01$, *** $p < 0.001$

This binary distinction was performed according to the median value.

+ Guardians selected the children's disliked foods from 55 food items.

A five-point scale; the highest point indicates good habits.

A total of 52.9% of the children had a lunch box that contained some vegetables. We investigated whether tomatoes and other types of vegetables were present in their lunch box (Table 3). We found that 49.5% of the lunch boxes contained at least one tomato. Therefore, the percentage of 'vegetables' area or the presence of tomatoes in children's lunch box was related to their well-balanced diets and mothers' concerns for food.

'Guardians who think that packed lunch is well-balance' was related to the number of different types of vegetables in lunch boxes. Moreover, we found a relationship between the number of different types of vegetables or the presence of tomatoes and 'nutritional balance' (p < 0.01 for both; Table 4). In addition, there was a relationship between the presence of tomatoes, 'nutritional balance' and 'the time taken by mothers to prepare the packed lunch' (Table 5). Furthermore, the presence of tomatoes in the lunch box was determined to be related to children's balanced diets and mothers' concerns for food.

Table 3. Relationship between the types of vegetables and presence of tomatoes

		Tomato			
		Presence		Absence	
		N	%	N	%
The types of	1	11	(27.5)	29	(72.5)
vegetables	2	22	(56.4)	17	(43.6)
in the lunch box	3	14	(63.6)	8	(36.4)
	4	6	(75.0)	2	(25.0)
	5	1	(100)	0	(0)
	6	1	(100)	0	(0)
	All	55	(49.5)	56	(50.5)

Note. N = 111

We randomly chose 124 from the 209 total lunch boxes.

The percentage of the lunch boxes which contained no vegetables was 10.

Table 4. Relationship between the types of vegetables and the mothers' concerns for children's packed lunch

The kinds of vegetables	0		1		2		3		Fisher's exact test
in the lunch box	N	%	N	%	N	%	N	%	
Guardians who think that packed lunch is well-balanced									
Always good	2	(66.7)	0	(0)	1	(33.3)	0	(0)	
Usually	2	(5.4)	7	(18.9)	15	(40.5)	13	(35.1)	
No concern	4	(8.2)	18	(36.7)	13	(26.5)	14	(28.6)	0.006[**]
Sometimes not good	1	(7.7)	8	(61.5)	3	(23.1)	1	(7.7)	
Never	2	(66.7)	0	(0)	1	(33.3)	0	(0)	
Mothers who consider nutritional balance to be the most important thing									
Very much	9	(12.3)	30	(41.1)	21	(28.8)	13	(17.8)	0.021[*]
Not at all	4	(7.8)	10	(19.6)	18	(35.3)	19	(37.3)	

Note. The types of vegetables were assessed using a Fisher's exact probability test; [*] p < 0.05, [**] p < 0.01

Table 5. Relationship between the presence of tomatoes and the mothers' concerns for children's packed lunch

	Tomato				Fisher's exact test
	Presence		Absence		
	N	%	N	%	
Mothers who consider nutritional balance to be the most important thing while preparing the packed lunch					
Very much	29	(56.9)	22	(43.1)	0.027*
Not at all	26	(35.6)	47	(64.4)	
Lunch box preparing time					
Over 45 min	2	(22.2)	7	(77.8)	
30 min~	13	(36.1)	23	(63.9)	
20 min~	22	(61.1)	14	(38.9)	0.013*
10 min~	6	(37.5)	10	(62.5)	
5 min~	0	(0)	4	(100)	
Less than 5 min	3	(100)	0	(0)	

Note. Tomatoes present in the lunch box as assessed by a Fisher's exact test; * $p < 0.05$

4. Discussion

In this study, we aimed to investigate the percentage of space in children's lunch boxes accounted for by the vegetables and the effect of food preference and mothers' concerns on this percentage as well as the relationship between mothers' concerns for children's lunch boxes and their contents.

The percentage of 'vegetables' area in the lunch boxes was 11.9% ± 7.8%. The majority of children's lunchbox capacity was either 360mL or 450mL. The percentage of 'vegetables' consumed was smaller than the increase in vegetable consumption as stipulated by Health Japan 21 (the second edition) (The Minister of Health, Labour and Welfare, 2013). This policy aims to ensure an intake of 350 g vegetables per day. Therefore, the percentage of vegetables in the lunches in this study was very far from the ideal proportion. More research is needed to determine the strategies for increasing this percentage.

In addition, when the percentage of the 'vegetables' area was above 11%, a larger percentage of mothers reported a mindfulnes towards vegetables ($p < 0.01$) and prepared a well-balanced packed lunch ($p < 0.01$); a smaller percentage of them used frozen food, and the number of foods disliked by their children was smaller ($p < 0.01$; Table 2) than when the area was below 11%. It has been reported that there is a strong correlation of the intake of fruits and vegetables with preferences and accessibility. The correlation observed between the children's and parents' intake of fruits and vegetables indicates that the parents' habits are a potential determinant for that of the children's (Bere & Kleep, 2004). It has also been reported that infants who received repeated dietary exposure to a particular food tend to prefer its flavour and, even, consume it more (Forestell & Mennella, 2007). Therefore, it is important to try to reduce the food types which children dislike during childhood by cooking well-balanced packed lunches, which includes not only fish and meat but also vegetables. By calculating the percentage of 'vegetables' area, mothers' food concerns and a balanced or unbalanced diet in children can be predicted.

When the percentage of 'vegetables' area was above 11%, the amount of time the children watched TV was smaller, and the number of children who considered school lunch to be enjoyable was greaterthan when the area was below 11% ($p < 0.05$ for both; Table 2). Thus, it may be an indirect relation and not a direct one. Kristiansen et al. (2013) suggested that a higher academic background of the parents was associated with a smaller amount of time spent watching TV; lower frequency of watching TV in the child's bedroom; greater amount of exercise; consumption of more fruits and vegetables, fewer sweets and less soft drinks and fast food and with more regular meals. Children's propensities to consume high-fat and high-sugar foods were positively associated with high-risk television behaviours (Lissner et al., 2012). Our previous study suggested that children's preferences were related to enjoying the school lunch (Osera et al., 2014). The percentage of 'vegetables' area may influence not only children's preferences, contents of their lunch boxes and mothers' concerns for these contents but also the amount of time spent watching TV and enjoying the school lunch. These two indirect additional items, i.e., the amount of time spent watching TV and enjoying the school lunch may be related to the children's lifestyle and preferences for particular food types. Thus, by calculating the percentage of 'vegetables' area in the lunch box, mothers' food concerns for children's packed lunch and children's preferences, as well as the amount of time the children spend watching TV and enjoying a school lunch can be predicted.

Table 3 shows that in the children's lunch box containing vegetables, half of the items were a combination of tomatoes and other vegetables. Almost all the children consumed one item of vegetables, and it was only a mini tomato, which is almost 10 g. Overall, the percentage of 'vegetables' area in the lunch boxes was smaller than the recommended Japanese dietary intake. Rogers investigated the primary school children's packed lunches in the UK. The total intake of fruit and vegetables for boys and girls was found to be 53.7 g and 66.6 g, respectively. Although the nutritional guidelines recommend that lunchtime meals should include one portion of both fruits and vegetables, the actual intake clearly fell far short of this, irrespective of meal type (Rogers et al., 2007).

Mothers' concerns for children's packed lunch were related to the vegetable items in the lunch boxes ($p < 0.01$; Table 4). These result shows that more types of vegetables in the lunch box indicate the mother's concern to be higher. Our previous study suggested that a mother's attitude toward her child's acquisition of healthy food habits has an effective influence on children's consumption of soybean products and that a mother's positive attitude toward soybean products may be influence her children's consumption of soybean products (Osera et al., 2016 b). The presence of tomatoes was related with a greater concern about the nutritional balance and a shorter cooking time (Table 5). Therefore, the presence of tomatoes in the children's lunch box may be referred a good index. Hubbard et al. (2014) suggested that those who design school wellness policies should take initiatives to work collaboratively with parents to improve the quality of foods brought from home. Moreover, Blissett (2011) suggested that the context of an authoritative parenting and feeding style is associated with better fruit and vegetable consumption during the childhood years. It is recommended that children's lunch box includes a tomato because it is easy to consume about 10 g of brightly coloured vegetables.

Our study is cross-sectional; thus, we could not clearly demonstrate a cause and effect relationship. Practical intervention aimed at parents must be implemented as a long-term solution (Cleghorn et al., 2009). The smart lunch box intervention, targeting both parents and children, has led to small improvements in the food and nutritional content of the children's packed lunch (Evans et al., 2010b). In order to do so, a more effective strategy that may be appropriate for use in an intervention is needed. In future, we will attempt to perform a study on how to improve the contents of children's lunch boxes. However, the findings from the present study suggest target areas for improvement of intervention strategies.

In conclusion, the percentage of consumption of vegetables in the studied Japanese children was far lower than the recommended consumption per day in Japan. The percentage of 'vegetables' area in the lunch box seems to be related to mothers' concerns for vegetables and to children's preferences. In addition, vegetable amounts in packed lunch may be related to mothers' concerns for vegetables and children's preferences.

Acknowledgments

We thank all the children, their guardians and the teachers at the kindergartens and nursery schools for their participation and cooperation with our questionnaire.

References

Apped, L. J., Moore, T. J., Obarzened, E., William, M. V., Laura, P. S., Frank. M. S., George, A. B., … David, W. H. (1997). A clinical trial of the effects of dietary patterns on blood pressure. The *New England Journal of Medicine, 336*(16), 1117-1124. http://dx.doi.org/10.1056/NEJM199704173361601

Bauer, K. W., Larson, N. I., Nelson, M. C., Story, M., & Neumark-Sztainer, D. (2008). Socio-environmental, personal and behavioral predictors of fast-food intake among adolescents. *Public Health Nutrition, 12*(10), 1767-1774. http://dx.doi.org/10.1017/S1368980008004394

Bere, E., & Kleep, K. (2004). Correlates of fruit and vegetable intake among Norwegian schoolchildren: parental and self-reports. *Public Health Nutrition, 7*(8), 991-998. http://dx.doi.org/10.1079/PHN2004619

Birch, L. L. (1979). Preschool children's food preferences and consumption patterns. *Journal of Nutrition Behavior, 11*(4), 189-192. http://dx.doi.org/10.1016/S0022-3182(79)80025-4

Birch, L. L., & Fisher, J. (1998). Development of eating behaviors among children and adolescents. *Pediatrics, 101*, 539-549.

Blissett, J. (2011). Relationships between parenting style, feeding style and feeding practices and fruit and vegetable consumption in early childhood. *Appetite, 57*, 826-831.

http://dx.doi.org/ 10.1016/j.appet.2011.05.318

Carter, P., Gray, L.J., Troughton, J., Kamlesh, K., & Melanie, J. D. (2010). Fruit and vegetable intake and incidence of type 2 diabetes mellitus: systematic review and meta-analysis. *BMJ, 341*, c4229. http://dx.doi.org/10.1136/bmj.c4229

Cashidan, E. A. (1994) A sensitive period for learning about food. *Human Nature, 5,* 279-291. http://dx.doi.org/10.1007/BF02692155

Cleghorn, C. L., Evans, C. E. L., Kitchen, M. S., & Cade, J. E. (2009). Details and acceptability of a nutrition intervention program designed to improve the contents of children's packed lunches. *Public Health Nutrition, 13*(8), 1254-1261. http://dx.doi.org/10.1017/S1368980009991509

Cook, L., Wardle, J., & Gibson, E. L. (2003). Relationship between parental report of food Neophobia and everyday food consumption in 2-6 year old children. *Appetite. 41,* 205-206.

Dovey, T. M., Staples, P. A., Gibson, E. L., & Halford, J. C. (2008). Food neophobias and 'picky/fussy' eating in children: A review. *Appetite, 50,* 181-193. http://dx.doi.org/10.1016/j.appet.2007.09.009

Evans, C. E. L., Greenwood, D. C., Thomas, J. D., & Cade, J. E. (2010a). A cross-sectional survey of children's packed lunches in the UK: food- and nutrient-based results. *Journal of Epidemiology & Community Health, 64,* 977-983. http://dx.doi.org/10.1136/jech.2008.085977

Evans, C. E. L., Greenwood, D. C., Thomas, J. D., Cleghorn, C. L., Kitchen, M. S., & Cade, J. E. (2010b). SMART lunch box intervention to improve the food and nutrient content of children's packed lunches: UK wide cluster randomised controlled trial. *Journal of Epidemiology & Community Health, 64,* 970-976. http://dx.doi.org/10.1136/jech.2008.085837

Forestell, C. A., & Mennella, J. A. (2007). Early determinants of fruit and vegetable acceptance. *Pediatrics, 120,* 1247-1254. http://dx.doi.org/10.1542/peds.2007-0858

Hlimi, T., Skinner, K., Hanning, R. M., Martin, I. D., & Tsuji, L. J. (2012). Traditional food consumption behavior and concern with environmental contaminants among Cree schoolchildren of the Mushkegowuk territory. *International Journal of Circumpolar Health, 71,* 17344. http://dx.doi.org/10.3402/ijch.v71i0.17344

Hubbard, K., Must, A., Eliasziw, M., Folta, S., & Goldberg, J. (2013). What elementary school children bring from Home to Eat at School. *J Acad Nutr Diet, 113*(9), A91.

Hubbard, K., Must, A., Eliasziw, M., Folta, S. C., & Goldberg, J. (2014). What's in Children's Backpacks: Foods Brought from Home. *Journal of the Academy of Nutrition and Diet, 114,* 1421-1431. http://dx.doi.org/10.1016/j.jand.2014.05.010

Ilingworth, R. S. & Lister, J. (1964). The critical or sensitive period, with special reference to certain feeding problems in infants and children. The *Journal of Pediatrics, 65,* 839-848.

http://dx.doi.org/10.1016/S0022-3476(64)80006-8

Japan Sports Council. The report of food habits of elementary school children and junior high school students. 2010. http://www.jpnsport.go.jp/anzen/anzen_school/tyosakekka/tabid/1490/Default.aspx (accessed 16 February 2016)

Kristiansen, H., Juliuson, P. B., Eide, G. E., Roelants, M., & Bjerknes, R. (2013). TV viewing and obesity among Norwegian children: the importance of parental education. *Acta Paediatrica, 102*(2), 199-205. http://dx.doi.org/10.1111/apa.12066

Lissner, L., Lanfer, A., Gwozdz, W., Olafsdottir, S., Eiben, G., Moreno, L A., ... Reisch, L. (2012) Television habits in relation to overweight, diet and taste preferences in European children: the IDEFICS study. *European Journal of Epidemiology, 27,* 705-715. http://dx.doi.org/10.1007/s10654-012-9718-2

MacFarlane, A., Crawford, D., & Worsley, A. (2010). Associations between parental concern for adolescent weight and the home food environment and dietary intake. *Journal Nutrition Education and Behavior, 42*(3), 152-160. http://dx.doi.org/10.1016/j.jneb.2008.11.004

Nicklas, T. A., Baranowski, T., Baranowski, J. C., Cullen, K., Rittenberry, L., & Olvera, N. (2001). Family and childcare provider influences of preschool children's fruit, juice, and vegetables consumption. *Nutrition Reviwew, 59*(7), 224-235.

Osera, T., Tsutie, S., Kobayashi, M., & Kurihara, N. (2012). Relationship of mother's food preferences and attitudes with children's preferences, *Food Nutrition Sciences, 3,* 1461-1466.

http://dx.doi.org/10.4236/fns.2012.310190

Osera, T., Tsutie, S., Kobayashi, M., & Kurihara, N. (2014). A Retrospective Study on the Relationship of Changes in Likes/Dislikes with Food Habits in 4- and 6-Year-Old children. *European Journal of Nutrition*

& Food Safety, 4(4), 604-613.

Osera, T., Tsutie, S., Kobayashi, M., & Kurihara, N. (2016a). Associations between Children's Food Preferences and Food Habits towards Healthy Eating in Japanese Children. *Journal of Child & Adolescent Behavior. 4*(3), 1000292. http://dx.doi.org/10.4172/2375-4494.1000292

Osera, T., Tsutie, S., Kobayashi, M., & Kurihara, N. (2016b). Using Soybean Products in School Lunch for Health Education may improve Children's Attitude and Guardians' Knowledge in Kindergarten. *Journal of Child & Adolescent Behavior, 4*(5), 1000310. http://dx.doi.org/10.4172/2375-4494.1000310

Rees, G. A., Richards, C. J., & Gregory, J. (2008). Food and nutrient intakes of primary school children: a comparison of school meals and packed lunches. *Journal of Human Nutrition and Dietetics, 21*, 420-427. http://dx.doi.org/10.1111/j.1365-277X.2008.00885.x

Rogers, I. S., Ness, A. R., Hebditch, K., Jones, L. R., & Emmett, P. M. (2007). Quality of food eaten in English primary schools: school dinners vs packed lunches. *European Journal of Clinical Nutrition, 61*, 856-864. http://dx.doi.org/10.1038/sj.ejcn.1602592

Sanigorski, A. M., Bell, A. C., Kremer, P. J., & Swinburn, B. A. (2005). Lunchbox contents of Australian school children: room for improvement. *European Journal of Clinical Nutrition,* 59, 1310-1316. http://dx.doi.org/10.1038/sj.ejcn.1602244

Smithers, G., Gregory, J., Bates, C. L., Prentice, A., Jackson, L. V., & Wenlock, R. (2000). The National Diet and Nutrition Survey: young people aged 4-18 years. *Nutrition Bulletin, 25*(2), 105-111. http://dx.doi.org/10.1046/j.1467-3010.2000.00027.x

The Minister of Health, Labour and Welfare Health Japan 21 (the second term) 2013. http://www.mhlw.go.jp/seisakunitsuite/bunya/kenkou_iryou/kenkou/kenkounippon21/en/kenkounippon21 (accessed 3 April 2016)

Vereecken, C., & Maes, L. (2010). Young children's dietary habits and associations with the mothers' nutritional knowledge and attitudes. *Appetite, 54*, 44-51. http://dx.doi.org/10.1016/j.appet.2009.09.005

World Cancer Research Fund & American Institute for Cancer Research. Food, Nutrition and Physical Activity and the Prevention of Cancer: A Global Perpsective. 2007.Washington, DC: WCRF/AICR. http://www.aicr.org/assets/docs/pdf/reports/Second_Expert_Report.pdf. (accessed 16 February 2016)

Zhang, X., Shu, X. O., Xian, Y. B., Yang, G., Li, H., Gao, J., Cai, H., Gao, Y. T., & Zheng, W. (2011). Cruciferous vegetable consumption is associated with a reduced risk of total and cardiovascular disease mortality. *American Journal of Clinical Nutrition, 94*, 240-242. http://dx.doi.org/10.3945/ajcn.110.009340

10

Evaluation of the Quality of Composite Maize-Wheat Chinchin Enriched with *Rhynchophorous phoenicis*

Ojinnaka, M. C.[1], Emeh, T. C.[2] & Okorie, S. U.[2]

[1] Department of Food Science & Technology, Michael Okpara University of Agriculture, Umudike, Abia State, Nigeria

[2] Department of Food Science & Technology, Imo State University, Owerri, Imo State, Nigeria

Correspondence: Ojinnaka, M. C., Department of Food Science & Technology, Michael Okpara University of Agriculture, Umudike, Abia State, Nigeria. E-mail: mcojinnaka@yahoo.co.uk

Abstract

The purpose of this research was to develop and evaluate a snack product (chin-chin) from composite maize-wheat flour blends enriched with edible palm weevil (*Rhyhnchophorus phoenicis*) paste. The maize-wheat chin-chin enriched with *R. phoenicis* were subjected to acceptability test using twenty member semi-trained panelist. The moisture, fat, protein and carbohydrate compositions of the snack samples had significant differences in their values. Sample 5M5R90W (containing 5% maize flour and *Rhyhnchophorus phoenicis* paste and 90% wheat flour) had the highest protein value of 19.05% while the least value 9.39% was obtained by sample 100M0R0W (100% maize flour alone). Sample 100M0R0W containing 100% maize flour also had the highest carbohydrate value of 75.24%. There was no significant difference in the ash and crude fiber contents of the chin-chin samples enriched with edible palm weevil paste. There were significant differences (P≤0.05) in the functional properties of maize-wheat composite flour blends. Their wettability values ranged from 46.67 – 200 while the swelling index, bulk density and oil absorption capacity showed no significant difference in their values. The result of the mineral analysis showed phosphorus, magnesium and sodium had significant differences in their values in the range of 317.55 – 376.75mg/100g; 5.60 -13.60mg/100g;59.0 – 70.3mg/100g, respectively. There were no significant differences in the sensory attributes of the chin-chin samples. The result showed that an acceptable chin-chin product can be processed with the inclusion of the larva of edible palm weevil with maize-wheat composite flour to enhance the nutritional quality of the product.

Keywords: maize, wheat, *Rhyhnchophorus phoenicis*, chin-chin

1. Introduction

Sustainable diets have been defined as diets with low environmental impacts which contribute to food and nutrition security and to healthy life for present and future generations. Sustainable diets are protective and respectful of biodiversity and ecosystems, culturally acceptable, accessible, affordable; nutritionally adequate, safe and healthy; while optimizing natural and human resources (FAO, 2012). Insects could be of great interest as a possible solution to food safety and environmental sustainability of food production due to their capability to serve as an important source of protein and other nutrients as well as have ecological and economic advantages (Belluco et al., 2013). Insects constitute quality food and feed, have high feed conversion ratios, and emit low levels of greenhouse gases (Huis, 2013). An upsurge of interest in the use of insects as sustainable diets should be encouraged as many of them are nutritionally, economically and ecologically important.

Entomophagy (human consumption of insects) has been practiced since mankind first made an appearance on planet earth. According to López and Shanley (2004), insects have played an important role in the history of human nutrition in Africa, Asia and Latin America. Hundreds of species are still eaten. Among the most importantorders of insects consumed in Nigeria are Coleoptera, Hymenoptera, Isoptera, Lepidoptera, Odonata, Orthoptera and they are highly priced (Fasoranti & Ajiboye, 1993). Notable examples ofthese are the palm weevil, *Rhynchophorus phoenicis*, termites, *Macrotermes nigeriense* (queen, king and reproductives), *Cirinaforda*, and variegated grasshopper, *Zonocerus variegatus* (Adedire & Aiyesanmi, 1999).

African insects are rich in protein and usually processed to tasty food products which are used as flavour intensifiers in soups and stews and also add protein to protein-poor diets. Ordinarily, insects are not used as emergency food sources during shortages, but are included as a planned part of the diet throughout the year or when seasonally available (Inyang & Iduh, 1996). Among the people living in south of the Sahara, the spectrum of hunger is endemic. This makes the insects unconventionally interesting to study because they remain under exploited and not recovered. However, the physical and chemical properties of their proteins in food systems during processing, storage, preparation and consumption is affected (Fennema, 1996).

People especially in areas where insects were not consumed for a long time prefer incorporating insects into the food in a way they are not visible, having accepted only the idea that the insects have a nutritional value. In practice, dried insects may be crushed or pulverized, and raw or boiled insects ground or mashed, making their insect form unrecognizable. They become masses of protein and lipids that can be mixed with other foodstuffs such as grain, ground meat and mashed potatoes to make a variety of dishes (Mitsuhashi, 2010). Some recipes of such dishes have been published (Borkovcová et al., 2009)

The larva of *Rhynchophorus phoeniciss*, a Coleoptera of Curculionidae family is used as traditional food in several countries. It is a delicious meal in many parts of Cameroon and other countries in Africa where it is found. The high cost of animal protein has directed interest towards several insects as potential sources of proteins for humans. Among the insects species, *R. phoénicis* larvae are considered the major sources of dietary lipids and proteins. They are consumed worldwide, especially in developing and under developed countries where consumption of animal protein may be limited because of economic, social, cultural or religious factors (Cerda et al., 1999). *Rhynchophorus* spp. are major pests of date palms, coconut palms, oil palms and sugarcane (Aldryhim & Al-Bukiri, 2003). *Rhynchophorus phoenicis* is rich in protein, inexpensive and underutilized by the industries, meanwhile it offers the same benefits as other meat products with less fat when deffated. They contain in this delipidated form over 80% of high quality protein with high content of essential amino acids (USDANAL, 2005) and can be useful in many food applications (Prinyawiwatkul et al., 1993). Though they are very destructive, their nutritional potentials have endeared them to man.

Chin-chin is a fried snack popular in West Africa. It is a sweet, hard, donut- like baked or fried dough of wheat flour, eggs and other customary baking items (Akubor, 2004). Chin-chin is easily one of the most favoured food item, a much relished African pastry which could serve as a dessert, snack and also as a popular street food. Chin-chin enjoys a very special place in the heart as well as stomachs of West African population (Mepba et al., 2007a). Chin-chin is one food item, that invites a great deal of flexibility in terms of the ingredient used and method of preparation involved; while some like to eat it hard and crunchy, others prefer a softer easier to chew version. This research was conducted to evaluate the nutrient composition of chin-chin from composite maize-wheat flour blends enriched with larva of edible palm weevil (*Rhynchophorus phoenicis)*.

2. Materials and Methods

2.1 Material Collection

Live larvae of edible palm weevil (*Rhynchophorus phoenicis)* were supplied from Orlu in Imo State. The method of Womeni et al. (2012) was used in processing *Rhynchophorus phoenicis* into paste for chin-chin production. The method described by Omueti et al. (2009) was used in processing yellow maize (*Zea mays*) into flour. The method of production of chin-chin described by Adegunwa et al. (2014) was used.

2.2 Nutrient Analysis of Chin-Chin Enriched with Rhynchophorus phoenicis

The chin-chin samples were analyzed for moisture, ash, protein and fat contents using the method described by AOAC (2000). The mineral components (phosphorus, magnesium, calcium, potassium and sodium) were analyzed using an Atomic Absorption Spectrophotometer (AAS, Model SP9, Pychicham UK).

2.3 Determination of Functional Properties of Composite Wheat-Maize Flour Blends

2.3.1 Bulk Density

Bulk density of flour samples were determined by weighing the sample (50g) into 100ml graduated cylinder, then tapping the bottom ten times against the palm of the hand and expressing the final volume as g/ml.

2.3.2 Wettability

The method of Onwuka (2005) was used. Into a 25ml graduated cylinder with a diameter of 1cm, 1g of sample was added. A finger was placed over the open end of the cylinder which was invested and clamped at a height of 10cm from the surface of a 600ml beaker containing 500ml of distilled water. The finger was removed and the rest material allowed to be dumped. The wettability is the time required for the sample to become completely wet.

2.3.3 Oil Absorption Capacity

Two grams (2g) of sample was mixed with 20ml of oil in a blender at high speed for 30sec. Samples were then allowed to stand at 30^0C for 30 minutes then centrifuged at 1,000rpm for 30 minutes. The volume of supernatant in a graduated cylinder was noted. Density of water was taken to be 1g/ml and that of oil determined to be 0.93g/ml.

2.3.4 Swelling Index Determination

Three grams (3g) portions (dry basis) of each flour were transferred into clean, dry graduated (50ml) cylinders. Flour samples were gently leveled into it and the volumes noted. Distilled water (30ml) was added to each sample; the cylinder was swirled and allowed to stand for 60 minutes while the change in volume(swelling) was recorded every 15 minutes. The swelling power of each flour sample was calculated as a multiple of the original volume as done by Ukpabi and Ndumele (1990).

2.4 Sensory Evaluation

The sensory attributes - colour, taste, texture, flavour, aroma, appearance and general acceptability were evaluated by twenty member semi-trained panelist using a 9- point hedonic scale with 1 representing the least score (dislike extremely) and 9 the highest score (like extremely). Analysis of Variance (ANOVA) was performed on the data gathered to determine differences, while the least significant test was used to detect significant differences among the means (Iwe, 2002, 2007).

3. Results and Discussion

3.1 Nutritional Composition

The results of the proximate composition of maize-wheat composite chin-chin enriched with palm weevil are presented in Table 2. There was significant difference ($P \leq 0.05$) in the moisture content of the chin-chin enriched with edible palm weevil paste. The moisture content ranged from 4.35 – 5.34% with sample 20M5R75W (20%maize flour: 5% *Rhynchophorus phoneicis* paste: 75% wheat flour) having the highest value of 5.34% while sample 5M5R90W (5% maize four: 5% *Rhynchophorus phoneicis* paste: 90% wheat flour) had the lowest value of 4.35%. There was reduction in moisture contents as the rate of addition of wheat flour increased. Adegunwa et al. (2014) reported moisture content values of 3.98 – 5.04 in composite millet-wheat chin-chin. Sanni et al. (2006) reported that the lower the moisture content of a product to be stored the better the shelf stability of such products. Low moisture ensures higher shelf stability of dried product. The values obtained for moisture in Table 2 were minimal and may not have adverse effect on the quality attributes of the product (Kure et al., 1998). There was no significant difference in the ash content of the samples. It was observed that as the level of wheat flour substitution increased the ash content also increased. The ash content values were in the range of 1.85 – 3.21%. The ash content indicates a rough estimation of the mineral content of product (Adegunwa et al., 2014). There was also no significant difference in the crude fibre content of the chin chin. The values were in the range of 1.11 – 1.25%. Falola et al. (2014) reported similar values (0.77 – 2.15%) when they produced chin-chin from modified cocoyam starch.

Table 1. Blend formulation of chin-chin samples enriched with edible palm weevil (*Rhynchophorus phoenicis*)

Samples	Maize flour	*Rhynchophorus phoenicis paste*	Wheat flour
20M5R75W	20	5	75
15M5R80W	15	5	80
10M5R85W	10	5	85
5M5R90W	5	5	90
100M0R0W	100	0	0

Table 2. Nutrient composition of maize-wheat composite chin-chin enriched with *Rhynchophorus phoenicis*

SAMPLES PARAMETERS	20M5R75W	15M5R80W	10M5R85W	5M5R90W	100M0R0W	LSD
Moisture content	5.34 [a]±0.16	4.97 [b] ±0.13	4.38 [c]±0.02	4.35[cd]±0.08	5.31[a]±0.02	0.18
Ash	2.18 [a]±0.02	2.69 [a]±0.03	2.80 [a]±0.06	3.21[a]±0.04	1.85 [a] ±0.01	-
Crude fibre	1.15 [a]±0.01	1.13 [a]±0.01	1.11[a]±0.01	1.10[a]±0.02	1.25 [a] ±0.03	-
Fat	9.93 [d]±1.05	10.28 [c]±0.31	11.39 [b] ±0.08	15.82[a]±0.06	6.97[e]±0.17	0.91
Protein content	11.84 [d]±0.10	14.29[c]±0.10	17.21 [b] ±0.10	19.05[a]±1.11	9.39[e]±0.10	0.93
Carbohydrate	69.56 [b] ±1.06	66.63[c]±0.33	63.10[d]±0.18	55.79[c]±0.05	75.24 [a] ±0.07	0.91

Means in the same row with the same superscript are not significantly different at P<0.05
20M5R75W (20% maize flour : 5% Rhynchophorus phoenicis paste: 75% wheat flour)
15M5R80W (15% maize flour : 5% Rhynchophorus phoenicis paste: 80% wheat flour)
10M5R85W(10% maize flour : 5% Rhynchophorus phoenicis paste: 85% wheat flour)
5M5R90W (5% maize flour : 5% Rhynchophorus phoenicis paste: 90% wheat flour)
100M0R0W (100% maize flour : 0% Rhynchophorus phoenicis paste: 0% wheat flour)

There was significant difference (P<0.05) in the fat content of the samples. The samples had fat content values of 6.97 – 15.82%. The high fat content values especially the 15.82% recorded in sample 5M5R90W (5% maize flour: 5% *Rhynchophorus phoneicis* paste :90% wheat flour) could be due to the edible palm weevil added to the samples and the margarine used in the sample preparations. Opara et al. (2012) reported a high lipid content of 54.20% for larva of *R. phoenicis*. Kiin-Kabari and Ogbonda (2010) reported fat contents of 19.6 and 16.3% in *R. phoenicis* enriched fillers used for pies and sandwich production. Fasasi (2009) reported that low fat content in a dry product will help in increasing the shelf life of the sample by decreasing the chances of rancidity and also contribute to low energy value of the food product while high fat content product will have high energy value and promotes lipid oxidation. Edible insects contain good quality fatty acid especially long chain omega-3 fatty acids such as alpha-linoleic acid, eicosapentaenoic acid (Yang et al., 2006). The reason for insects containing long-chain PUFAs and different fatty acid compositions is linked with the diet and enzymatic activity in the insects (Mlcek et al., 2014). Lipids are necessary in food because they increase palatability and retain the flavor of food (Aiyesanmi & Oguntokun, 1996). They also play a structural and physiological role.

The protein content of the chin-chin samples ranged from 9.39 - 19.05%. The lowest mean protein content was recorded for Sample 100M0R0W 9.39%, while the highest mean protein content was recorded for Sample 5M5R90W (5% maize four: 5% *Rhynchophorus phoneicis* paste: 90% wheat flour) 19.05%. The high protein content in Sample 5M5R90W 19.05% maybe as a result of high wheat flour used in that sample; while the significant variation in the protein value of the samples could be attributed to the different proportions used in the formulation. The crude protein content of edible palm weevil has been reported to be very high, ranging from 28.42% (Banjo et al. 2006) to 71.63% (Braide & Nwaoguikpe, 2011). Idolo (2010) reported protein content of 9.96% in *R. phoenicis* on wet basis while he put that on dry basis at 25.16%. He also observed that the protein content of wheat buns enriched with larvae of *R. phoenicis* increased progressively in proportion to the percentage of larvae added. Kinyuru et al. (2009) in their study on the process development of wheat buns enriched with edible termites *(Marcrotermes subhylanus)* reported protein content of 15.63% in sample with 5% termite substitution with wheat flour compared to 10.60% protein in the wheat buns without termite paste addition. The larvae of *Rhychophorus phoenicis*have been reported to be a rich source of digestible proteins able to make up for the dietary imbalance as they form real sources of food for man and other animals (Fasoranti, 1997). Kiin-Kabari and Ogbonda (2013) reported that fillers enriched with *R. phoenics* paste for pies and sandwich had higher protein contents of 16.4 and 12.4% compared to the common meat pie and sandwich fillers which had protein contents of 11.2 and 9.9% respectively. Insects, a traditional food in many parts of the world, are highly nutritious and especially rich in proteins and thus represent a potential food and protein source. The high protein content is an indication that the insects can be of value in man and animal ration and can eventually replace higher animal protein usually absent in the diet of rural dwellers in developing countries (Banjo et al., 2006). Ekpo and Onigbinde (2005) have reported high level of leucine, lysine and threonine in the insect larva. Lysine and threonine are limiting amino acids in wheat, rice, cassava and maize based diets prevalent in the developing world (Ozimek et al., 1985). The inclusion of the larva into these staples would enhance the nutritional quality in these diets.

There were significant differences in the carbohydrate content of the chin-chin samples enriched with edible palm weevil. The values were in the range of 55.79 – 75.24%. Similar values have been reported for chin-chin produced from composite millet-wheat (Adegunwa et al., 2014) and in the use of modified cocoyam starch in composite for chin-chin production (Falola et al., 2014). Ekop et al. (2010) reported carbohydrate content of 22.70 dry weight for *R. phoenicis* (palm weevil).

There were no significant differences in the calcium and potassium contents of the chin-chin enriched with edible palm weevil paste (Table 3). But there were significance differences in the phosphorus, magnesium and sodium compositions of the samples. The calcium values were in the range of 25.39 – 30.91mg/100g while those of magnesium were in the range of 5.60 – 13.60mg/100g. Minerals are known to play important metabolic and physiologic roles in the living system. Magnesium helps maintain muscle and nerve functions, keeps heart rhythm steady and supports a healthy immune system and regulates blood sugar levels (Saris et al., 2000). It has been reported that edible palm weevil contains various minerals like: Calcium 39.58mg/100g; Phosphorus 126.4mg/100g; Magnesium 7.54mg/100g; Iron 12.24mg/100g (Banjo et al., 2006; Ekpo et al., 2006). Processing could also have affected the mineral values of the snack product.

Table 3. Mineral composition of maize-wheat chin-chin enriched with palm weevil

SAMPLES PARAMETERS (mg/100g)	20M5R75W	15M5R80W	10M5R85W	5M5R90W	100M0R0W	LSD
Phosphorus	$332.96^b \pm 0.16$	$331.17^c \pm 0.49$	$323.85^d \pm 1.03$	$317.55^e \pm 0.80$	$376.75^a \pm 0.33$	1.19
Magnesium	$8.80^b \pm 1.39$	$8.8^b \pm 1.39$	$11.20^{ab} \pm 1.38$	$13.60^a \pm 1.38$	$5.60^a \pm 1.39$	2.52
Calcium	$29.39^a \pm 2.31$	$29.39^a \pm 2.31$	26.72 ± 2.32	25.39 ± 2.32	30.91 ± 2.48	-
Potassium	$97.73^a \pm 0.61$	$76.8^a \pm 0.40$	75.2 ± 0.80	74.80 ± 40	116.27 ± 0.12	-
Sodium	$63.1^b \pm 0.95$	$63.63^b \pm 0.42$	$65.07^b \pm 1.10$	$70.3^b \pm 1.67$	$59.0^c \pm 1.04$	2.02

Means in the same row with the same superscript are not significantly different at P<0.05
20M5R75W (20% maize flour : 5% Rhynchophorus phoenicis paste: 75% wheat flour)
15M5R80W (15% maize flour : 5% Rhynchophorus phoenicis paste: 80% wheat flour)
10M5R85W(10% maize flour : 5% Rhynchophorus phoenicis paste: 85% wheat flour)
5M5R90W (5% maize flour : 5% Rhynchophorus phoenicis paste: 90% wheat flour)
100M0R0W (100% maize flour : 0% Rhynchophorus phoenicis paste: 0% wheat flour)

3.2 Functional Properties of Maize – Wheat Composite Flour

Table 4 presents the result of the functional properties of maize-wheat composite flour. There was significant difference in the wettability values of the maize –wheat composite flour blends. The values were in the range of 46.67 – 200. The values were decreasing as the rate of addition of maize flour reduced. The sample 100M0R0W (100% maize flour) had the highest value of 200 secs. Some of the samples high wettability values could be due to its low protein content. It has been reported that the lower the level of denatured protein present, the slower it takes to get wetted or imbibe water (Oti & Akobundu, 2008).There was no significant difference in the swelling index values of the samples. The values were in the range of 1.29 – 1.68. The swelling power of flour granule is an indication of the extent of associative forces within the granules (Moorthy & Ramanujan, 1986). Swelling capacity can also be related to the water absorption index of the starch-based flour during heating. There was no significant difference in the oil absorption capacity of the samples. The values were in the range of 1.04 – 1.45. The reason why the oil absorption capacity of Sample 100M0R0W (100% maize flour) is 1.45 and therefore greater than the ones of the other samples might be that the increase in oil absorption is associated with heat dissociation of the protein and denaturation which is expected to unmask the nonpolar residue from the interior of the protein molecule (Kinsella, 1976).

Table 4. Functional properties of maize- wheat composite flour blends

SAMPLES PARAMETERS	20M5R75W	15M5R80W	10M5R85W	5M5R90W	100M0R0W	LSD
Swelling index(v/v)	1.41±0.02	1.37±0.04	1.32±0.07	1.29±0.02	1.68±0.04	-
Oil absorption capacity(g/ml)	1.39±0.06	1.26±0.05	1.10±0.05	1.04±0.10	1.45±0.05	-
Bulk density(g/ml)	0.95±0.01	0.96±0.01	0.96±0.01	0.97±0.01	0.98±0.01	-
Wettability(secs)	181.67[b]±6.51	90.67[c]±7.23	72.67[d]±3.51	46.67[e]±1.15	200[a]±7.21	10.31

*Means in the same row with the same superscript are not significantly different at P<0.05
20M5R75W (20% maize flour : 5% Rhynchophorus phoenicis paste: 75% wheat flour)
15M5R80W (15% maize flour : 5% Rhynchophorus phoenicis paste: 80% wheat flour)
10M5R85W(10% maize flour : 5% Rhynchophorus phoenicis paste: 85% wheat flour)
5M5R90W (5% maize flour : 5% Rhynchophorus phoenicis paste: 90% wheat flour)
100M0R0W (100% maize flour : 0% Rhynchophorus phoenicis paste: 0% wheat flour)

The bulk density of the flour samples ranged from 0.95 in Sample 20M5R75W (20% maize flour : 5% Rhynchophorus phoneicis paste : 75% wheat flour) to 0.98 in Sample 100M0R0W (100% maize flour). Adengunwa et al. (2014) reported bulk density of 0.76 – 0.83 for millet-wheat composite flour blends while Etudaiye et al. (2015) reported bulk density values of 0.68 – 0.82 for sweet potato –wheat composite flour blends and 0.74 – 0.78 for sweet potato starch – wheat composite blends. The bulk density is generally affected by particle size and the density of flour or flour blend and it is very important in determining the packaging requirement, raw materials handling and application in wet processing in food industry (Ajanaku et al., 2012). It has been reported that bulk density of foods increases with increase in starch content (Bhattacharya & Prakash, 1994). Ojinnaka et al. (2009) in their work on the use of modified cocoyam (Xanthosoma sagittifolium) starch in cookie production reported packed bulk density of 0.67 and 0.62 for starch from native cultivars Ede Uhie and Ede ocha as well as loose bulk density of 0.49 and 0.47 for Ede uhie and Ede ocha respectively. High bulk density of protein material is also important in relation to its packaging (Onimawo et al., 1998). High bulk density of the flours and starches indicate that they would serve as good food thickners in food products (Adebowale et al., 2005) while the low bulk density of the flours and starch samples will be suitable for the formulation of high nutrient density weaning food (Mepba et al., 2007b).

3.3 Sensory Evaluation

Table 5 shows the results of the sensory evaluation of maize- wheat chin-chin enriched with palm weevil. There were no significant differences (P≤0.05) in all the organoleptic properties (appearance, flavor, taste, texture, general acceptability) measured. The values for appearance and texture attributes ranged from 7.39 – 7.76 and 6.83 – 7.50 respectively. The appearance was best for sample 20M5R75W (20% maize flour :5% Rhynchophorus phoneicis paste : 75% wheat flour) with value of 7.76. It was observed that the value in appearance decreased as the level of maize substitution reduced. The samples were well accepted by the members of the panel though sample 20M5R75W (20% maize flour :5% Rhynchophorus phoneicis paste : 75% wheat flour) and 15M5R80W (15% maize flour :5% Rhynchophorus phoneicis paste : 80% wheat flour) were most preferred. Idolo (2010) reported that wheat buns produced at 5%- 15% level of substitution with R.phoenicis were found to be acceptable but observed that with 5% substitution was most acceptable.

Table 5. Mean sensory scores of maize-wheat chinchin enriched with *Rhynchophorus phoenicis*

SAMPLES PARAMETERS	20M5R75W	15M5R80W	10M5R85W	5M5R90W	100M0R0W	LSD
Appearance	7.76±1.24	7.72±1.18	7.50±1.34	7.5±1.42	7.39±1.54	-
Flavour / Aroma	7.28±1.18	6.83±1.47	7.39±1.33	6.56±1.46	7.28±1.71	-
Taste	6.89±1.53	7.56±1.50	6.89±1.71	7.44±1.34	7.56±1.34	-
Texture	7.39±2.00	7.33±1.28	6.83±1.50	7.22±1.52	7.50±1.54	-
General Acceptability	7.78±1.17	7.72±1.07	7.39±1.02	7.39±1.38	7.56±1.46	-

Means in the same row with the same superscript are not significantly different at P<0.05
20M5R75W (20% maize flour : 5% Rhynchophorus phoenicis paste: 75% wheat flour)
15M5R80W (15% maize flour : 5% Rhynchophorus phoenicis paste: 80% wheat flour)
10M5R85W(10% maize flour : 5% Rhynchophorus phoenicis paste: 85% wheat flour)
5M5R90W (5% maize flour : 5% Rhynchophorus phoenicis paste: 90% wheat flour)
100M0R0W (100% maize flour : 0% Rhynchophorus phoenicis paste: 0% wheat flour)

4. Conclusion

The process developed for production of wheat-maize chin-chin enriched with edible palm weevil *(Rhynchophorous phoenicis)* paste at 5% level of substitution could easily be adopted due to its nutritional quality and sensory attributes. Results from the study showed that the chin-chin samples were nutritionally acceptable by the members of the panel. Edible insects can however be incorporated into different food formulations that will be palatable as well as nutritious to consumers especially those that find it difficult to consume edible insects in their forms. More studies can also be carried out on how best to incorporate these edible insects into different food products to create varieties so that their use as sustainable diets will be encouraged since they are nutritionally, economically and ecologically important.

References

Adedire, C. O., & Aiyesanmi, A. F. (1999). Proximate and mineral composition of the adult and immature forms of the variegated grasshopper, *Zonocerusvariegatus*(L) (Acridoidea: Pygomorphidae). *Bioscience Research Communications, 11*(2):121-126.

Adegunwa, M. O., Ganiyu, A. A., Bakare, H. A., & Adebowale, A. A. (2014). Quality evaluation of composite millet-wheat Chin-chin. *Agriculture and Biology Journal of North America, 5*(1), 33-39.

Aiyesanmi, A. F., &Oguntokun, M. O. (1996). Nutrient composition of *Diocleareflexa* seed—an underutilized edible legume. *La RivistaItalianaDelleSostanze Grasse*, LXXIII, 521-523.

Ajanaku, K. O., Ajanaku, C. O., Edobor-osoh, A., &Nwinyi, O. C. (2012). Nutritive value of sorghum Ogi fortified with groundnut journal seed (*Arachis hypogeal*). *Am. J. Food Technol., 79*, 82-88. http://dx.doi.org/10.3923/ajft.2012.82.88

Akubor, P. I. (2004). Protein contents, physical and sensory properties of Nigerian snack foods (cake, chin-chin, puff-puff) prepared from cowpea-wheat flour blends. *Int. J. Food Sci. and Technol., 39*, 419-424. http://dx.doi.org/10.1111/j.1365-2621.2004.00771.x

Aldryhim, Y., & Al-Bukiri, S. (2003). Effect of irrigation on within-grove distribution of red palm weevil *Rhynchophorous ferrugineus. Agric. and Marine Sci., 8*(1), 47-49.

AOAC. (2000). Association of Official Analytical Chemists. Official methods of Analysis (Vol. II; 17th edition) of AOAC International.Washington, DC, USA.

Banjo, A. D., Lawal, O. A., & Songonuga, E. A. (2006). The nutritional value of fourteen species of edible insects in Southwestern Nigeria. *Afr. J. Biotechnol., 5*(3), 298- 301.

Belluco, S., Losasso, C., Maggioletti, M., Alonzi, C., Paoletti, M. G., & Ricci, A. (2013). Edible Insects in a Food Safety and Nutritional Perspective: A Critical Review. *Comprehensive Reviews in Food Science and Food Safety., 12*(3), 296-313. http://dx.doi.org/10.1111/1541-4337.12014

Bhattacharya, S. M., & Prakash. (1994). Extrusion blends of rice and chicken pea flours: A response surfaceanalysis. *J. Food Eng., 21*, 315-330. http://dx.doi.org/10.1016/0260-8774(94)90076-0

Borkovcova M., Bednarova M., Fiser V., Ocknecht P. &Kuchynehmyzemzpestrena. (2009). 1st ed., Lynx: Brno, Czech Republic. 136 p. ISBN 975–80–86787–37–4.

Braide, W., &Nwaoguikpe, R. N. (2011). Assessment of microbiological quality and nutritional values of a processed edible weevil caterpillar (*Rhychophorus phoenicis*) in Port Harcourt, Southern Nigeria. *International Journal of Biological and Chemical Sciences, 5*(2), 410-418. http://dx.doi.org/10.4314/ijbcs.v5i2.72059

Cerda, H., Martinez, R., Briceno, N., Pizzoferrato, L., Hermoso, D., & Paoletti, M. (1999). Cria, Analysis nutricional y sensorial del picudo del cocotero R.P (Coleoptera: Curculionidea), insecto de la dieta traditional indigenaAmazonica. *Ecotropicos, 12*, 25-32.

Ekop, E. A., Udoh, A. I., & Akpan, P. E. (2010). Proximate and anti-nutrient composition of four edible insects in Akwa Ibom State, Nigeria. *World J. Appl. Sci. Technol., 2*(2), 224-231.

Ekpo, K., &Onigbinde, A. (2005). Nutritional potentials of the larva of *Rhynchophorusphoenicis*. Ambrose Alli University, Ekpoma, Nigeria. *Pakistan journal of nutrition, 4*(5), 287-290.

Etudaiye, H. A., Oti, E., Aniedu, C., &Omodamiro, M. R. (2015). Utilization of sweet potato starches and flours as composites with wheat flours in the preparation of confectioneries. *African Journal of Biotechnology, 14*(1), 17-22. http://dx.doi.org/10.5897/AJB12.2651

Falola, A. O., Olatidoye, O. P., Adesala. S. O., &Amusan, M. (2014). Modification and Quality Characteristics of Cocoyam Starch and its Potential for Chin-Chin Production. *Pakistan Journal of Nutrition, 13*(12), 768-773. http://dx.doi.org/10.3923/pjn.2014.768.773

Fasasi, O. S. (2009). Proximate, Antinutritional factors and functional properties of processed pear millet. *Journal of Food Technology, 7*, 92-97.

Fasoranti, J. O., & Ajiboye, D. O. (1993). Some edible insects ofKwara State, Nigeria. *American Entomologist, 39*(2), 113-116. http://dx.doi.org/10.1093/ae/39.2.113

Fasoranti, J. O. (1997). The value of African insects as human food supplements. *J. Sci. Tech., 10*, 1-5.

Fennema R. O. (1996). (ed) Food chemistry,3rd edn. Marcel Dekker, Inc. 270 *Food Chemistry*, (3d edition): 365-369

Food and Agricultural Organization. (2012). *Assessing the Potential of Insects as Food and Feed in Assuring Food Security*.Presented at Tech. Consult. Meet., 23–25 January, FAO, Rome, Italy

Idolo, I. (2010). Nutritional and Quality Attributes of Wheat Buns Enriched with the Larvae of *Rhynchophorus phoenicis*F. *Pakistan Journal of Nutrition, 9*(11), 1043-1046. http://dx.doi.org/10.3923/pjn.2010.1043.1046

Inyang, U. E., & Iduh, A. O. (1996). Influence of pH and salt concentration on protein solubility, emulsifying and foaming properties of sesame protein concentrate. *J.the American Oil Chemists' Society, 73*, 1663-1667. http://dx.doi.org/10.1007/BF02517969

Iwe, M. O. (2002). Handbook of Sensory methods and analysis. Rojoint Comm. Services Ltd. Enugu, Nigeria

Iwe, M. O. (2007). Current trends in sensory evaluation of foods. Rojoint Comm. Services Ltd, Enugu, Nigeria.

Kiin-Kabari, D. B., &Ogbonda, K. H. (2013). Production, Proximate and Sensory Evaluation of *Rhynchophorus Phoenicis* (F)Larva paste. *Journal of Food Studies., 2*(1), 13-18. http://dx.doi.org/10.5296/jfs.v2i1.2938

Kinsella, J. E. (1976). Functional properties of protein foods. *Crit. Rev. Sci. Nutr., 1*, 219-229. http://dx.doi.org/10.1080/10408397609527208

Kinyuru, J. N., Kenji, G. M., & Njoroge, M. S. (2009). Process development, nutrition and sensory qualities of wheat buns enriched with edible termites (*Macrotermes subhylanus)* from lake Victoria region, Kenya. *African Journal of Food Agriculture Nutrition and Development, 9*(8), 1739-1750. http://dx.doi.org/10.4314/ajfand.v9i8.48411

Kure, O. A., Bahago, E. J., & Daniel, E. A. (1998). Studies on the Proximate Composition and Effect of Flour Particle Size on Acceptability of Biscuit Produced from Blends of Soyabeans and Plantain flours. *Namida Tech-Scope J., 3*, 17-21.

López, C., & Shanley, P. (2004). Riches of the forest: For health, life and spirit in Africa,center for International Forestry Research. pp: 1 – 24.

Mepba, H. D., Achinewhu, S. N., & Wachukwu, C. K. (2007a). Microbiological quality of selected street foods in Port Harcourt, Nigeria. *J. Food Safety, 27*, 208. http://dx.doi.org/10.1111/j.1745-4565.2007.00073.x

Mepba, H. D., Eboh, L., & Banigo, D. E. B. (2007b). Effects of processing treatments on the nutritive composition and consumer acceptance of some Nigerian edible leafy vegetables. *Afr. J. Food, Agric. Nutr. Dev., 7*(1), 1-18.

Mitsuhashi, J. (2010). The future use of insects as human food. Proceedings of the forest insects as food: humans bite back, Chiang Mai,Thailand, 19–21 February 2010, Durst, P.B., Johnson D.V., LeslieR.N., Shono K., Eds., RAP Publication: Bangkok, Thailand. pp. 115–122.

Mlcek, J., Rop, O., Borkovcova, M., &Bednarova, M. (2014). A Comprehensive Look at the Possibilities of Edible Insects as Food in Europe – a Review. *Pol. J. Food Nutr. Sci., 64*(3), 147-157. http://dx.doi.org/10.2478/v10222-012-0099-8

Moorthy, S. N., & Ramanujan, T. (1986). Variation in properties of starch in Cassava varieties in relation to age of the crop. *Starch/Starke, 38*(2), 58-61. http://dx.doi.org/10.1002/star.19860380206

Ojinnaka, M. C., Akobundu, E. N. T., & Iwe, M. O. (2009). Cocoyam starch modification effects on functional, sensory and cookies qualities. *Pak. J. of Nutr., 8*(5), 558-567. http://dx.doi.org/10.3923/pjn.2009.558.567

Okezie, B. O., & Bello, A. B. (1988). Physicochemical and functional properties of winged bean flour and isolate compared with soy isolate. *J. Food Sci., 53*, 450-454. http://dx.doi.org/10.1111/j.1365-2621.1988.tb07728.x

Omueti, O., Otegbayo, B., Jaiyeola, O., & Afolabi, O. (2009). Functional properties of complementary diets developed from soybean (*Glycine max.*), groundnut (*Arachis hypogea*) and crayfish (*MacrobrachiumSpp*). *EJEAFChe, 8*(8), 563-573.

Onimawo, I. A., Momoh, A. H., & Usman, A. (1998). Proximate composition and functional properties of four cultivars of bambara groundnut (*Voandzeiasubterranea*). *Plant Foods Human Nutr., 53*, 153-158. http://dx.doi.org/10.1023/A:1008027030095

Onwuka, G. I. (2005). Food analysis andinstrumentation theory and practice. Napthali Prints.Lagos.

Opara, M. N., Sanyigha, F. T., Ogbuewu, I. P., & Okoli, I. C. (2012). Studies on the production trend and quality characteristics of palm grubs in the tropical rainforest zone of Nigeria. *J. Agric. Technol., 8*(3), 851-860

Oti, E., &Akobundu, E. N. T. (2008). Potentials of cocoyam-soybean-crayfish mixtures in complementary feeding. *Nig. Agric. J., 39*, 137-145.

Ozimek, L., Sauer, W. C., Kozikowski, V., Ryan, J. K., Jorgensen, H., & Jelen, P. (1985). Nutritive value of protein extracted from honey bees. *J. Food Sci., 50*, 1327-1329. http://dx.doi.org/10.1111/j.1365-2621.1985.tb10469.x

Prinyawiwatkul, W., Beuchat, L. R., & McWatters, K. H. (1993). Functional property changes in partially defatted peanutflour caused by fungal fermentation and heattreatment. *J. Food Science, 58*, 1318-1323 http://dx.doi.org/10.1111/j.1365-2621.1993.tb06174.x

Sanni, L. O., Adebowale, A. A., & Tafa, S. O. (2006). Proximate, Functional, Pasting and Sensory Qualities of Instant Yam Flour. A Paper Presented at the 14 ISTRC Symposium, Central Tuber Crops the Research Institute, Trivandrum, Kerala State, India

Saris, N. E., Mervaala, E., Karppanen, H., Khawaja, J. A., & Lewenstam, A. (2000). Magnesium:An update on physiological, clinical and analytical aspects. *Clinical ChimicaActa, 294*, 1-26. http://dx.doi.org/10.1016/S0009-8981(99)00258-2

Ukpabi, U. J., & Ndumele, C. (1996). Evaluation of the quality of garri produced in Imo State. *Nig. Food J., 8*, 103-110.

USDA-NAL. (2005). United States Department of Agricultural-NationalAgricultural Library. USDA Nutrient Database for Standard Reference, Release 18. Retrieved from http://www.nal.usda.gov/fnic/foodcomp/gibin/list_nut_edit.pl.

Womeni, H. M., Tiencheu, B., Linder, M., Nabayo, E. M. C., Tenyang, N., Mbiapo, F. T., ... Parmentier, M. (2012). Nutritional value and effect of cooking, drying and storage process on some functional properties of *Rhynchophorusphoenicis*. *Int. J. Life Sci. Pharm. Res., 2*(3), 203-219.

Yang, L., Siriamornpun, S., & Li, D. (2006). Polyunsaturated fatty acidcontent of edible insects in Thailand. *J. Food Lipids, 13*, 277–285. http://dx.doi.org/10.1111/j.1745-4522.2006.00051.x

11

Effect of Temperature on the Intensity of Basic Tastes:
Sweet, Salty and Sour

Keri Lipscomb[1], James Rieck[1] & Paul Dawson[1]

[1] Food, Nutrition and Packaging Sciences, Clemson University, United States

Correspondence: Paul Dawson, Food, Nutrition and Packaging Sciences, Clemson University, United States. E-mail: pdawson@clemson.edu

Abstract

Sensory panels were trained to identify specific concentrations of sucrose, sodium chloride and citric acid as an intensity level value of 6 on a 15-point scale for flavors of sweet, salty and sour, respectively. Trained panels were exposed to a single concentration of each taste singly, in combinations of 2 and all three at 3 temperatures (3°C, 23°C, 60°C) using concentrations previously identified at an intensity level of 6. Panelists determined the perceived intensity of each taste at each temperature in the single and combined treatments. Sweetness was perceived as more intense at 60°C than 23°C and 3°C when tasted alone but not when in combination with other tastes (salty and sour). Salty perceived intensity was not affected by serving temperature while sourness was perceived as more intense at 23°C compared to 3°C and 60°C. In general, perceived sweetness was less suppressed when combined with other tastes than salty and sour.

Keywords: taste intensity, basic tastes, sweet, sour, salty, temperature effect

1. Introduction

Scientists have been curious about the effect of temperature on basic tastes for many years (Hahn & Gunther, 1932). Food is often prepared at one temperature but served and eaten at another. Sweet, sour, salty, and bitter have been considered primary tastes for many years (Meilgaard, 2007), although recently some have proposed a fifth basic taste called umami or savory (Johnson & Wales University, 2003). This study focuses on just three of these tastes: sweet, sour, and salty. Several studies have evaluated the influence of temperature on varying concentrations of a basic taste solution to determine threshold values or perceived intensity, but none have reported using a single concentration with varying temperatures (Calvino, 1986; Pangborn et al., 1970; Paulus & Reisch, 1980).

Bartoshul et al. (1982) evaluated 7 sucrose concentrations served at 4, 12, 20, 28, 36 and 44°C and concluded that the perceived sweetness of sucrose was more affected by higher temperature at lower sucrose concentrations and temperature effects on perceived intensity became negligible as sucrose concentration approached 0.5M. Calvino (1986) reported similar results testing sucrose concentrations of 0 091-1.462 M at temperatures of 7, 37 and 50°C with sweetness being more intense at the higher serving temperature for the lower concentrations while this effect disappeared at about 0.4 -0.5M sucrose concentrations. Schiffman et al. (2000) further verified the higher perceived intensity of sucrose served at higher temperatures with an all female panel using 6, 22 and 50°C serving temperatures and sucrose concentrations of 2.5, 5, 7.5 and 10% sucrose. Schiffman et al. (2000) also tested glucose, fructose, three terpenoid sweeteners, mannitol, sorbitol, alitame, aspartame, acesulfame-K, saccharin, cyclamate, thaumatin, dihydrochalcone and sucralose concluding that serving temperature had little effect on perceived intensity.

No effect of temperatures on salty perception was reported by Pangborn et al. (1970) using temperatures of 0, 22, 37 and 55°C and sodium chloride concentrations of 0.04, 0.08, 0.16, 0.24, 0.32, 0.40, 0.52 and 0.64% and these researchers reported a linear increase for perceived intensity in this range of sodium chloride concentrations. McBurney et al. (1973) reported that sodium chloride was perceived as more intense at 17, 37 and 42°C compared to intermediate serving temperatures of 22, 27 and 32°C, respectively. However in this study one temperature was served per day which may have compromised the results relative to serving temperature effects. Citric acid taste threshold level was highest at 2°C and lowest at 20.5°C with 41°C in between for threshold level (Powers et al., 1971).

Paulus and Reisch (1980) studied the effect of temperature on sweet, salt, sour and bitter. The solutions for the determination of recognition thresholds were 0.5, 1.0, 2.0, 4.0, 8.0 and 16 grams of sucrose, 0.093, 0.185, 0.375, 0.75, 1.5 and 3.0 grams of NaCl, and 0.031, 0.062, 0.125, 0.25, 0.5 and 1.0 grams of citric acid to 1 liter of distilled water for sweet, salty and sour, respectively. Each taste was sampled at 10, 20, 40 and 60°C. The recognition threshold for sucrose was the lowest at 20 and 40°C, increased at 60°C and 10°C. For NaCl, there was an in recognition threshold between 10 and 20°C, a more pronounced increase at 40°C, and no information reported at 60°C. With citric acid the threshold increased between 10 and 20°C, and decreased at 40°C to nearly the same threshold value found at 10°C. Interestingly, Green et al. (1988) observed that the sweetness of saccharin was not affected by temperature (36 vs 20°C) while glucose, fructose and aspartame were perceived as more intense at the higher temperature.

Previous studies using two tastes found bitterness can be suppressed both by sweet tastes (Lawless, 1979; Kroeze & Bartoshuk, 1985; Calvino et al., 1993; Frijters & Schifferstein, 1994) and by sodium salts (Brelsin & Beauchamp, 1995; Green, 2003), and that sweetness can suppress both sourness (Schifferstein & Fritjer, 1991; Frank & Archambo, 1986) and saltiness (Kroeze, 1978; Panghorn, 1962). Studies have also shown that sweetness can be suppressed by tastes that evoke bitterness (Calvino et al., 1990), saltiness (Kroeze, 1979) or sourness (Bonnans & Noble, 1993). Pangborn (1962) reported that combining sucrose with sodium chloride suppressed salty perception while the perceived intensity of sweetness was enhanced by the addition of sodium chloride. Both Pangborn (1961) and Mcbride & Finlay (1990) reported that combining sucrose and citric acid suppressed the intensity of sucrose while Schifferstein and Frijters (1990) found that combining sucrose and citric acid suppressed both sweet and sour intensities. Pangborn and Trabue (1967) looked at the interaction of salt-acid mixtures: NaCl-0.05, 0.15, 0.45 and 1.35% and citric acid-0.005, 0.0125, 0.0313 and 0.078%, served at 20°C. In solution it was found that higher concentrations of citric acid suppressed saltiness while the lower concentrations enhanced saltiness. However, NaCl had an overall suppression effect on the perceived sourness of citric acid. In another study combining citric acid and sodium chloride, Wise and Breslin (2011) found that NaCl suppressed the perceived intensity of citric acid when compared to the perceived intensity of the same concentrations of citric acid without NaCl.

No published studies have focused on the effect of temperature on the interrelationship of basic tastes when tastes are combined, though information on basic taste interrelationships is available. Green et al. (2010) evaluated the predominance of sweetness in combination with other tastes at 2 different temperatures however the objective was to determine if the intensity of sweetness was affected by the difference in temperature and not on how temperature impacted the intensity of all tastes singly and in combination. Therefore, the objective of this research was to determine the effect of serving temperature on the perceived intensity of basic tastes (sweet, salty, or sour) singly and in combination when all tastes were held at a single intensity level.

2. Material and Methods

2.1 Test Solution Preparation

Test solutions were prepared from food grade solutes of sucrose (Wal-Mart Private Label, Wal-Mart Stores, Inc., Bentonville, AK), sodium chloride (salt without iodine, Wal-Mart Private Label, Wal-Mart Stores, Inc., Bentonville, AK),) and citric acid (NOW® Foods, Bloomington, IL) and water (Diamond Springs Water, Richmond, VA, U.S.A.). Screening solutions for the Basic Taste Acuity and Intensity Ranking Tests were prepared using the concentrations outlined by Wheeler et al. (1981). Solutions used during training were prepared using the concentrations provided in the Basic Taste Intensity Level Spreadsheet (Table 1). This spreadsheet was created based on standards of percentages for certain basic taste intensity levels in *Sensory Evaluation Techniques* (Meilgarrd et al., 2007). Solutions were prepared at least 12 hours before the day of use to allow ample time to equilibrate. The solutions were not kept in excess of 96 hours. Solutions were stored in Ball® glass mason jars with plastic screw top lids (Jarden Home Brands, Daleville, IN, U.S.A.) and refrigerated when not in use. The day of screening, solutions were removed from the refrigerator and allowed to equilibrate to 23°C before distribution. The temperature of one sample was measured using a thermocouple and a sample equilibrium took approximately 30 minutes.

Table 1. Basic taste intensity levels and ingredient weights added to 1 liter of water needed to achieve each intensity level

Ingredient	Intensity Level														
	1	2	3	4	5	6	7	8	9	10	11	12	13	14	15
Sucrose	10g	20g	30g	40g	50g	60g	70g	80g	90g	100g	112g	124g	136g	148g	160g
Citric Acid	0.4g	0.5g	0.6g	0.7g	0.8g	0.94g	1.08g	1.22g	1.36g	1.5g	1.6g	1.7g	1.8g	1.9g	2.0g
Sodium Chloride	0.4g	1.5g	2.3g	2.9g	3.5g	3.92g	4.34g	4.76g	5.15g	5.45g	5.75g	6.05g	6.35g	6.65g	7.0g

2.2 Panelist Recruitment and Training

Subjects recruited for the study were ages 18 to 65 in good health, i.e. no individuals with diabetes, hypoglycemia, hypertension, dentures, chronic colds or sinusitis. Subjects were recruited by email and through personal communication. Twenty individuals, 14 female and 6 male, meeting the age and health requirements, completed the screening process before proceeding to training. A University Institutional Review Board approved all panels. After completing a screening questionnaire and meeting minimum requirements for participation in the panels, subjects completed a Basic Taste Recognition Test, an Intensity Ranking of Basic Taste Evaluation, an Exercise in Taste Scaling, then both a Triangle Test to determine if subjects could distinguish between intensities 5 and 6 and an Intensity Ranking Training before data collection commenced. Retention testing was also performed throughout the duration of the basic taste training to determine if panelists were retaining the ability to distinguish between intensity levels. . For example, during salty training a sweet reference and unknown samples were given in addition to the day's salty samples on days 6 and 8. The reference and samples were not identified as sweet, only labeled as REF for reference and 3-digit code numbers for the samples. Throughout sour basic training a salty reference and unknown samples were provided along with the sour samples on day 6 with no prior notice to the subjects. Using the same retention method, a sweet reference and unknown samples were incorporated into the training on day 7 of sour training. A total of 40 hours of training were completed including basic taste acuity testing and screening.

2.3 Facilities and Ballot

The sensory lab consists of 6 privacy booths equipped with Dell Mini P787J notebook computers (Dell, Santa Clara, CA, U.S.A.) on which computerized ballots recorded subject's evaluation of the samples. Each booth also contained bottled water, cups, napkins and expectorant cups. Results were recorded using SIMS 2000 Sensory Evaluation Testing Software (Sensory Computer Systems, Morristown, NJ, U.S.A.). DELETED SENTENCE. In addition to providing scales to record results, the sensory ballots clearly outlined the instructions for evaluating samples and the objectives of the study before the subjects received samples

The sensory evaluation samples were served at 3 temperatures (3, 23, 60°C) with 30 ml being dispensed in 118 ml (4 oz) plastic cups with lids (Solo® Cup Co., Urbana, IL, U.S.A.) affixed with predetermined 3-digit code numbers corresponding to the basic taste concentration using a dispenser (Cole Palmer, U.S.A.). After dispensing, room temperature samples and references were held until they equilibrated to room temperature. Cold samples (3°C) placed in refrigeration until they equilibrated after which they were presented individually to panelists. 60°C samples were held in a Ball® Mason jar fitted with the dispenser (Cole Palmer, U.S.A.). The jar was then placed in an 18 liter (4.75 gallon) clear plastic water bath (Cambro® Manufacturing Co., Huntington Beach, CA, USA) fitted with a Sous Vide Immersion Circulator (PolyScience, Niles, IL, USA) set to 60°C. Thirty mL of the hot solution were dispensed into a 4oz Styrofoam portion cup with a vented lid (Dart® Container Corp., Mason, MI, U.S.A.) for each individual subject immediately before distribution. All temperatures were verified with a -20~110°C mercury thermometer (Barker Diagnostics Inc., Deerfield, IL, U.S.A.).

When building each sensory ballot in the SIMS 2000 software, a rotation plan was automatically created. Each ballot was constructed so that a subject was required to log into the computerized ballot with their panelist number to activate the test. One sample set from the rotation plan was written on a note card and given to a subject when they entered a privacy booth. Subjects were provided with a tray of all reference samples (room temperature) and unsalted crackers upon entry to a booth. The note cards only provided code numbers and gave no indication of the order of sample presentation.

2.4 Selection of Single Intensity Level

Before training began, a preliminary test was performed to determine which single intensity level would be used during sensory evaluations. Subjects participating in the preliminary tests tasted sucrose solutions of intensities 5 through 10. The subjects were asked to taste each intensity level, between which they were instructed to eat a portion of an unsalted cracker and drink bottled water before moving on to another sample. During a round table discussion, subjects indicated intensity levels 6, 7, and 8 to be the most palatable of the sweet solutions. Upon further questioning, some of the subjects expressed difficultly differentiating between intensities 7 and 8. All subjects verbalized they were able to detect a difference between the samples of intensity levels 6 and 7. Considering this feedback, intensity level 6 was chosen as the single intensity level for use during all sensory evaluations in this study.

2.5 Sensory Evaluation

Of the 20 subjects participating in the sensory evaluations, there were14 females and 6 males, with a combined mean age of 27. All were non-smokers who completed the necessary screening procedures and at least 80% of training for each basic taste. One subject was unavailable for all 10 days of salty basic taste training, but completed 100% of the required training for sweet and sour. This subject was excluded from any sensory evaluation containing salt.

Each basic taste and combination test was replicated three times for a total of 21 sensory panels. The order of the sensory panels was randomized using SAS® 9.2 Business Analytics Software (SAS® Institute Inc., Cary, NC, U.S.A.). Panels were scheduled over 4 weeks with panels occurring twice daily with a target time of 11:00 A.M. and 2:00 P.M. Subjects were allowed to schedule specific panel times if unable to participate at the target times. Eight subjects of the total 20 cleared to participate, were used for each panel. A total of 16 subjects were used each day, with 4 panelists available as backup if a scheduled subject could not participate.

2.6 Statistical Analysis

An analysis of variance (ANOVA) was used to determine any significant effects of serving temperature on intensity perception. When serving temperature was found to have a significant effect ($p \leq 0.05$) on perceived intensity, least square mean (LSM) p-values ($p \leq 0.05$) were used to determine which serving temperatures differed in perceived intensity. Standard errors of mean intensity were also determined. The ANOVA, LSM and standard error results for basic tastes and basic taste combinations were performed using SAS® 9.2 Business Analytics Software.

3. Results

3.1 Single Basic Tastes

Serving temperature had a significant effect ($p=0.0005$) on perceived intensity of sweet taste with the 60°C serving temperature perceived as more intense than when sweet taste at the same sucrose concentration was served at room temperature and cold (3°C) (Table 2).

However, there was insufficient evidence to conclude temperature influenced the perceived saltiness of NaCl solutions ($p=0.7746$) (Table 2). While temperature did not affect the perceived saltiness of the samples, it is interesting to note the standard error of the 23°C sample means was the lowest of the 3 samples, and therefore the individual estimations by the subjects were closer to the mean intensity. Furthermore, all serving temperatures were very close to the actual intensity of 6, thus temperature had little influence on perceived saltiness.

Similar to sweet taste, serving temperature had a significant effect on the perceived intensity of sour taste ($p=0.0007$) with the hot and cold serving temperatures being perceived as less sour than samples served at room temperature. With sour taste, the room temperature sample was perceived as more intense (7.0) than the reference (6.0) possibly due to the relative perception of the other temperature samples during testing.

Table 2. Perceived intensity (1 to 15 scale) of sweet, salty and sour tastes, (and standard errors) served singly at three different temperatures

Temperature	Sweet taste	Salty taste	Sour taste
60°C	8.6[a] (0.49)	6.3 (0.57)	5.6[b] (0.43)
23°C	7.3[b] (0.41)	5.7 (0.35)	7.0[a] (0.33)
3°C	6.8[b] (0.47)	6.2 (0.69)	4.5 [b] (0.53)

[ab]means within the same column with different superscripts are significantly different ($p \leq 0.05$).
(n=20).

3.2 Sweet/Salty Combination

When sweet and salty tastes were served together, there was as significant effect due to temperature on sweetness (p=0.039) but not on saltiness (p=0.816) (Table 3). When served with salty taste, sweetness was more intense when served hot (p=0.0089) and when served at room temperature (p=0.0017) compared to when served cold. This differed from when sweet taste was served alone when the hot serving temperature was more intense than both room and cold serving temperatures. Also, the overall perceived intensity of both sweet and salty were lower when the tastes were served together than when served singly.

Table 3. Perceived intensity (1 to 15 scale) and standard errors for sweet, salty and sour tastes served in paired combinations at three different temperatures

Temperature	Sweet and salty tastes together	
	Sweet taste	Salty taste
60°C	5.6^a (0.41)	3.3 (0.51)
23°C	5.9^b (0.37)	3.6 (0.43)
3°C	4.4^b (0.37)	3.7 (0.37)
	Sweet and sour tastes together	
	Sweet taste	Sour taste
60°C	5.1^a (0.51)	4.8 (0.55)
23°C	4.7^{ab} (0.41)	4.2 (0.53)
3°C	3.5^b (0.41)	3.5 (0.53)
	Salty and sour tastes together	
	Salty taste	Sour taste
60°C	4.5^a (0.49)	5.0 (0.53)
23°C	3.3^b (0.45)	5.3 (0.47)
3°C	3.3^b (0.51)	4.5 (0.41)

^{ab}means with different superscripts are significantly different (p≤ 0.05).

(n=24).

3.3 Sweet/Sour Combination

The results of the sweet/sour combination were very similar to those of the sweet/salty combination in that serving temperature did affect (p=0.033) perceived sweetness but did not affect perceived sourness (0.079) (Table 3). Also, like the sweet/salty combination, both hot (p=0.012) and room (p=0.0118) serving temperatures were perceived to be more intense than the cold serving temperature. A significant difference was not found between the hot and room temperature means (p=0.4126).

3.4 Sour/Salty Combination

For the sour/salty taste combination, salty taste intensity was affected (p=0.045) by serving temperature while sour was not (Table 3). Salt served at room g temperature y was perceived as more intense compared to when it was served cold (p=0.034) and hot (p=0.266). This perception differed from when salty was served alone where no difference in perceived intensity was found. As with the previous combinations, overall intensity for both taste were lower than when the tastes were served alone. While the sour intensity was not significantly affected by temperature, it was ranked as more intense than the salty taste.. In single evaluations, citric acid in solution produced a statistical difference due to temperature, whereas the single NaCl evaluations did not.

3.5 Sour/Sweet/Salty Three-way Combination

The sour portion of the three-way combination provided the only significant a p-value (p=0.0455) of the trio (Table 4). The mean sour intensity was different between cold and hot, and cold and room temperature were not detectable. The temperature effect between the hot and room temperature means provided a verifiable difference due to temperature (p=0.0188).

No research was found on the interaction of three of these basic tastes in combination using a temperature treatment or otherwise. As with all combinations involving sucrose in this study, sweetness was the dominant taste in that it lessened the perceived intensity of sour and salty tastes. The perceived intensity of sucrose also increased with temperature, as seen when sweet was evaluated alone, the sweet/sour combination and to a slightly lesser degree, the sweet/salty panels. The difference in intensity estimates between citric acid and sucrose appear more pronounced when combined with NaCl, then when evaluated without. In the trio combination, sour was estimated as more intense than salty at 3 and 60°C but not at 23°C. However, when the two were paired, sour had a higher mean across all temperatures. The mean intensities of sour and salty were estimated to be very close at the two lower temperatures. The largest difference in perceived intensity between the two was observed at 60°C.

Table 4. Perceived intensity and standard errors for sour/sweet/salty basic tastes for sour/sweet/salty basic taste evaluations

	Sour Basic Taste	
Temperature	Mean Estimated Intensity	Standard Error
60°C[a]	2.8[b]	0.47
23°C	4.0[a]	0.49
3°C	3.0[b]	0.51
	Sweet Basic Taste	
60°C	5.1	0.39
23°C	5.3	0.63
3°C	4.1	0.47
	Salty Basic Taste	
60°C	3.2	0.35
23°C	2.6	0.49
3°C	2.8	0.49

[ab]means with different superscripts are significantly different ($p \leq 0.05$).

(n=24).

4. Discussion

4.1 Sweet Basic Taste

The basic taste results for sweet found as the sample temperature increased so did the perceived sweetness of the samples by the subjects. These results are concurrent with the findings of Baroshuk et al. (1981), Calvino (1983), Schiffman et al. (2000) and Talavera et al. (2005) that found a linear relationship between temperature and the perceived sweetness of sucrose solutions. However, a study by Paulus and Reisch (1980) indicated, at least with stimulus and recognition thresholds, that perceived intensity did not increase with an increase in sample temperature. They found intensity was lowest between 20 and 40°C and perceived at its highest at 60°C. The sucrose intensities were higher at 10°C than the mid-range temperatures (20 and 40°C) but not significantly so. These results provide a nonlinear, almost U shaped curve of the effect of temperature on stimulus and recognition thresholds. It is noteworthy to mention at this point that much of the research that will be discussed reporting conflicting results studied taste thresholds and not mid-range or optimal sensory levels as was done in the current study.

4.2 Salty Basic Taste

While the salty evaluation results are similar to the observations of Pangborn et al. (1970) that perceived intensity of sodium chloride solutions changed due to temperature, these results are not supported by some other works. A study by McBurney et al. (1973) established that the threshold for NaCl was higher at temperatures 4 and 42°C and lower at temperatures 22 to 32°C, where they were also very close in range. In contrast, a 1932 study performed by Hahn and Gunther provided evidence that like the sucrose samples in the sweet evaluations, NaCl thresholds rose as the temperature increased. Hahn and Gunther's theory that perception of saltiness decreases as the temperature is increased is supported by a study almost 50 years later by Paulus and Reisch (1980).

The inability of the panel used in this study to distinguish a temperature effect between the three samples could be due to heightened NaCl sensitivity. Of the 20 individuals who participated in the salty sensory evaluations, 50% indicated on the Screening Questionnaire they felt their consumption of salty foods was not significant. Evidence generated from the results of a study observing how long-term reduction of sodium can alter one's taste of salt, indicates low intake over time, can increase the perceived intensity of salt in foods (Bertino, Beauchamp & Engleman, 1982). It is reasonable to speculate that an individual who consumes a reduced amount of dietary sodium could perceive the salt in food as more intense than someone who consumes more sodium.

4.3 Sour Basic Taste

The results of the sour sensory evaluations are not supported by two other studies. In a study of the effect of temperature on the threshold values of several substances including citric acid, Powers et al. (1971), found the threshold of citric acid to be highest at 2°C. Yet panelists found the citric acid threshold to be lower at 20.5 and 41°C. In contrast, a study by Paulus and Reisch (1980) evaluating the effect of temperature on the stimulus and recognition thresholds of citric acid provided no statistically verifiable evidence that temperature had an effect on either threshold. However, the authors do mention that though not statistically significant, there was a tendency of the threshold to increase as the temperature increased.

4.4 Sweet/Salty Combination

Literature on the pairing of sweet/salty combinations provides little insight into the relationship between sucrose and NaCl combination solutions based on temperature, as the research previously conducted does not use temperature as a treatment. Pangborn (1962) found that the addition of sucrose suppressed saltiness at the three higher concentrations (0.36, 1.08 and 3.24%) while no affect was found on saltiness at the lowest concentration (0.12%). The perceived sweetness of the two lower sucrose concentrations (0.75 and 2.25%) was enhanced by the addition of NaCl at all three concentrations. However, the higher concentrations of sucrose (6.75 and 20.25%) were suppressed by all concentrations of NaCl. The 20.25% was suppressed the most, while 6.75% was only slightly suppressed. More research by Pangborn and Chrisp (1964) and Pangborn and Trabue (1964) on the interrelationship of sucrose and NaCl was performed using not only varying concentrations of tastes, but also evaluates the basic taste combinations in canned tomato juice and lima bean puree respectively. However, in an article on "Flavor effects of sodium chloride" by Gillette (1985), the author states that the perception of sweetness is amplified by the addition of salt. While it is evident the sweetness of the solution was rated as more intense by subjects than salty, the mean intensities of perceived sweetness were evaluated as less intense in combination than when evaluated singularly. The saltiness of the combination mimics the observations from the single salty panel in that change due to temperature is insignificant.

4.5 Sweet/Sour Combination

In a report by Horn (1981) on evaluating sweetness and the many factors that influence its perception, it was stated that acidic ingredients such as citric acid can somewhat suppress the sweetness of sucrose. The results of this evaluation appear to reinforce Horn's statement since the same intensity of sucrose solution was perceived as higher when evaluated alone, than when in combination with citric acid. The mean intensities for sweet and sour were estimated relatively close at all temperatures in combination, while the means for the sweet/salty combination were not as similar.

McBride and Finlay (1989, 1990), Pangborn (1961) and Schiffman et al. (2000) used varying concentrations of sucrose and citric acid in combination solutions, however none of the authors researched how temperature would influence the interrelationships. Pangborn (1961) reported for the single-sample presentation, all concentrations (0.00, 0.007, 0.023 and 0.073%) of citric acid suppressed the perceived intensity of sucrose at all concentrations (0.5, 1.8, 5.8 and 20.0%). The interpretation of the pair-sample presentation using the same concentrations above plus additional sucrose (1.0, 2.0, 5.0, 10.0 and 20.0%) and citric acid (0.00, 0.005, 0.010, 0.020 and 0.040%) concentrations provided a bit more insight. Citric acid suppressed the sweetness of sucrose at the lower concentrations than at the higher sucrose concentrations. McBride and Findlay (1990) found that when rating sweetness, only the highest concentration of citric acid (of 0.00, 0.006 or 0.05 M) suppressed the sweetness of sucrose. Similar results were found when subjects rated the acidity of the mixtures. The highest level of sucrose (of 0, 0.172 or 0.8 M) in the solution suppressed the acidity of citric acid indicating that sucrose and citric acid mixtures mutually suppressed one another depending on rating criteria. Schifferstein and Frijters (1990) conducted 3 investigations (total intensity, sweetness and sourness) consisting of various concentrations of both sucrose (0.00, 0.125, 0.250, 0.500 and 1.000 M) and citric acid (0.00, 0.00125, 0.0025, 0.005, 0.010 M). They found that the suppression of citric acid in solution was dependant on both the concentration of the citric acid and sucrose, while the sweetness suppression of sucrose was only dependant on the concentration of citric acid.

4.6 Sour/Salty Combination

Literature by Pangborn and Trabue (1967) and Wise and Breslin (2011), was based on the combination of varying concentrations of citric acid and salt and incorporated no temperature treatment. Pangborn and Trabue (1967) found that higher concentrations of citric acid (of 0.005, 0.0125, 0.0313 and 0.078%) suppressed saltiness while the lower concentrations enhanced saltiness. However, NaCl had an overall suppression effect on the perceived sourness of citric acid. The higher concentrations of citric acid were suppressed the most with the two lowest concentrations showing a slight initial drop in perceived sourness. Wise and Breslin (2011) used citric acid concentrations of 1.67, 5 and 15 mM and for NaCl 130, 280 and 500 mM and concluded that the addition of NaCl suppressed the perceived intensity of citric acid when compared against the perceived intensity of the same concentrations of citric acid alone.

4.7 Sweet/Salty/Sour Combination

No research was found on the interaction of three of these basic tastes in combination using more than 2 temperatures. As with all combinations involving sucrose in this study, it was the dominant taste when in combination with other basic tastes. The perceived intensity of sucrose also increased with temperature, as seen in the single sweet panel, the sweet/sour combination and to a slightly lesser degree, the sweet/salty panels. The difference in intensity estimates between citric acid and sucrose appear more pronounced when combined with NaCl, than when evaluated without NaCl. In the trio combination, sour was estimated as more intense than salty at 3 and 60°C but not at 23°C. However, when the two were paired, sour had a higher mean across all temperatures. The mean intensities of sour and salty were estimated to be very close at the two lower temperatures. The largest difference in perceived intensity between the two was observed at 60°C.

Green (2010) reported that the intensity of binary, tertiary and quaternary mixtures of sucrose, citric acid, sodium chloride and quinine sulfate were additive and that sweetness (sucrose) was the dominant quality, being most resistant to suppression by other tastes and more likely to suppress other tastes. The role of sweetness and bitterness for food choices have been theorized to be related to survival and evolution of man to find high-energy foods (sweet) and avoid potentially poisonous foods (bitter) (Lawless, 1979; Kroese & Bartoshuk, 1985; Gilan, 1982, 1984).

Temperature has been shown to affect perceived intensity of most single tastes. For instance, Bartoshuk et al. (1982) found that weak sucrose solutions (< 0.5M) had an increase in perceived intensity with an increase in temperature but above 0.5M the temperature effect diminished. This finding was substantiated by Calvino (1986) who reported that cool sucrose solutions were judged less sweet than warmer solution up to the 0.4 to 0.5M concentration zone. Moskowitz (1973) estimated the sensing intensity between 25-50°C for glucose, NaCl, citric acid and quinine sulfate and found that all except citric acid were unaffected by temperature. McBurney et al. (1973) reported that the taste thresholds for NaCl (salty), HCl (sour) and quinine sulfate (bitter) were lowest between 22 and 32°C and that thresholds rose above these temperatures in the test range of 17 to 42°C. One study (Schiffman et al., 2000) contradicting temperature effects on sweetness in nearly all sweeteners commercially available was Schiffman et al. (2000).

The current study provided support of published information on the temperature/perceived intensity relationship when taste solutions were set at a single concentration and only temperature was altered. The primary findings of this study can be summarized as follows. Sweet and sour were perceived as more intense at 60°C and 23°C, respectively, compared to the other two temperatures tested. When sweet/salty were tasted together, sweet intensity was less at 3°C compared to the other temperatures tested while temperature had no effect on salty perceived intensity. Ironically, the same temperature effect was seen for salty (lower at 3°C) when combined with sour. For sweet/sour combinations sweet was perceived as more intense at 60°C compared to 3°C. When all three basic tastes were combined, only sour produced a significant difference due to temperature.

Since significant effects for perceived intensity were found due to serving temperature and since combining basic tastes also affects perceived intensities, food formulations and serving temperatures are important factors to consider in the food and beverage industry.

References

Bartoshuk, L. M., Rennert, K., & Stevens, J. C. (1982) Effects of temperature on the perceived sweetness of sucrose. *Physiology and Behavior, 28*, 905-910. http://dx.doi.org/10.1016/0031-9384(82)90212-8

Bertino, M., Beauchamp, G. K., & Engleman, K. (1982) Long-term reduction in dietary sodium alters the taste of salt. *The American Journal of Clinical Nutrition, 36*, 1134-1141.

Bonnans, S., & Noble, A. C. (1993). Effect of sweetener type and of sweetener and acid levels on temporal perception of sweetness, sourness and fruitiness. *Chemical Senses, 18*, 273-283. http://dx.doi.org/10.1093/chemse/18.3.273

Breslin, P. A., & Beauchamp, G. L. (1995). Suppression of bitterness by sodium: variation among bitter taste stimuli in humans. *Physiology & Behavior, 20*, 609-623. http://dx.doi.org/10.1093/chemse/20.6.609

Calvino, A. M. (1986). Perception of sweetness: The effects of concentration and temperature. *Physiology and Behavior, 36*, 1021-1028. http://dx.doi.org/10.1016/0031-9384(86)90474-9

Calvino, A. M., GarciA-Medina M. R., & Cometto-Muniz, J. E. (1990). Interactions in caffeine-sucrose and coffee-sucrose mixtures: evidence of taste and flavor suppression. *Chemical Senses, 15*, 505-519. http://dx.doi.org/10.1093/chemse/15.5.505

Calvino, A. M., Garcia-Medina, M. R., Cometto-Muniz, J. E., & Rodriguez, M. B. (1993). Perception of sweetness and bitterness in different vehicles. *Perception & Pyschophysics, 54*, 751-758. http://dx.doi.org/10.3758/BF03211799

Frank, R. A., & Archambo, G. (1986). Intensity and hedonic judgments of tastemixtures. *Perception & Pyschophysics, 48*, 326-330.

Frijter, J. E., & Schifferstein, H. N. (1994). Perceptual interactions in mixtures containing bitter tasting substances. *Physiology & Behavior, 56*, 1243-1249. http://dx.doi.org/10.1016/0031-9384(94)90372-7

Gilan, D. J. (1982). Mixture suppression: the effect of cooling on the perception of carbohydrate and NaCl. *Perception & Pyschophysics, 32*, 504-510.

Gilan, D. J. (1984). Evidence for peripheral and central processed in taste adaptation. *Perception & Pyschophysics, 35*, 1-4. http://dx.doi.org/10.3758/BF03205918

Gilette, M. (1985). Flavor effects of sodium chloride. *Food Technology, 39*(6), 47-56.

Green, B. G. (2003). Studying taste as a cutaneous sense. *Food Quality & Preference, 14*, 99-109. http://dx.doi.org/10.1016/S0950-3293(02)00071-X

Green, B. G., & Frankman, S. P. (1988). The effect of cooling on the carbohydrate and intensive sweeteners. *Physiology & Behavior, 43*(4), 515-519. http://dx.doi.org/10.1016/0031-9384(88)90127-8

Green, B. G., Lim, J., Osterhoof, F., Blacher, K., & Nachtigal, D. (2010). *Physiology & Behavior, 101*(5), 731-737. http://dx.doi.org/10.1016/j.physbeh.2010.08.013

Hahn, H., & Gunther, H. (1932). Uber die reize und die reizbedingungen des geschmacksinnes. *Pflugers Arch Ges. Physiol., 231*, 48. http://dx.doi.org/10.1007/BF01754527

Horn, H. E. (1981). Evaluating sweetness: Many factors influence perception. *Food Product Development, 15*(4), 28.

Johnson & Wales University. (2003). *Culinary fundamentals*. Thailand: Johnson & Wales University.

Kroeze, J. H. (1978). The taste of sodium chloride: masking and adaption. *Chemical Senses, 3*, 43-449. http://dx.doi.org/10.1093/chemse/3.4.443

Kroeze, J. H. (1979). Masking and adaptation of sugar sweetness intensity. *Physiology & Behavior, 22*, 347-351. http://dx.doi.org/10.1016/0031-9384(79)90097-0

Kroeze, J. H., & Bartoshuk, L. M. (1985). Bitterness suppression as revealed by split-tongue taste stimulation in humans. *Physiology & Behavior, 35*, 779-783. http://dx.doi.org/10.1016/0031-9384(85)90412-3

Lawless, H. T. (1979). Evidence for neural inhibition in bittersweet taste mixture. *J. Comp Perception & Pyschophysics, 93*, 538-547. http://dx.doi.org/10.1037/h0077582

McBride, R. L., & Finlay, D. C. (1989). Perception of taste mixtures by experienced and novice assessors. *Journal of Sensory Studies, 3*, 237. http://dx.doi.org/10.1111/j.1745-459X.1989.tb00447.x

McBride, R. L., & Finlay, D. C. (1990). Perceptual integration of tertiary taste mixtures. Perception & Pyschophysics, *48*(4), 326. http://dx.doi.org/10.3758/BF03206683

McBurney, D. H., Collings, V. B., & Glanz, L. M. (1973). Temperature dependence of human taste responses. *Physiology & Behavior, 11*, 89. http://dx.doi.org/10.1016/0031-9384(73)90127-3

Meilgarrd, M. C., Civille, G. V., & Carr, B. T. (2007). *Sensory evaluation techniques* (4th ed.). Boca Raton, FL: CRC Press Inc.

Moskowitz, H. R. (1972). Effects of solution temperature on taste intensity in humans. *Physiology & Behavior, 10*(2), 289-292. http://dx.doi.org/10.1016/0031-9384(73)90312-0

Pamghorn, R. M. (1961). Taste interrelationships. II. suprathreshold solutions of sucrose and citric acid. *Journal of Food Science, 26*, 648. http://dx.doi.org/10.1111/j.1365-2621.1961.tb00811.x

Panghorn, R. M. (1962). Taste interrelationships III. suprathreshold solutions of sucrose and sodium chloride. *Journal of Food Science, 27*, 495.

Panghorn, R. M., & Chrisp, R. B. (1964). Taste interrelationships VI. sucrose, sodium chloride and citric acid in canned tomato juice. *Journal of Food Science, 29*, 726.

Panghorn, R. M., Chrisp, R. B., & Bertolero, L. L. (1970). Gustatory, salivary, and oral thermal responses to solutions of sodium chloride at four temperatures. *Perception & Pyschophysics, 8*(2), 69. http://dx.doi.org/10.3758/BF03210177

Panghorn, R. M., & Trabue, I. M. (1964). Taste interrelationships V. sucrose, sodium chloride, and citric acid in lima bean puree. *Journal of Food Science, 29*, 233. http://dx.doi.org/10.1111/j.1365-2621.1964.tb01724.x

Panghorn, R. M., & Trabue, I. M. (1967). Detection and apparent taste intensity of salt-acid mixtures in two media. *Perception & Pyschophysics, 2*(11), 503. http://dx.doi.org/10.3758/BF03210255

Paulus, K., & Reisch, A. M. (1980). The influence of temperature on the threshold values of primary tastes. *Chemical Senses, 5*(1), 11. http://dx.doi.org/10.1093/chemse/5.1.11

Powers, J. J., Howell, D. A., & Vacinek, S. J. (1971). Effect of temperature on threshold values for citric acid, malic acid and quinine sulphate---energy of activation and extreme-value determination. *Journal of the Science of Food and Agriculture, 22*, 543. http://dx.doi.org/10.1002/jsfa.2740221012

Schifferstein, H. N., & Frijters, J. E. (1990). Sensory integration in citric acid/sucrose mixtures. *Chemical Senses, 15*(1), 87. http://dx.doi.org/10.1093/chemse/15.1.87

Schifferstein, H. N., & Frijters, J. E. (1991). The effectiveness of different sweeteners in suppressing citric acid sourness. *Perception & Pyschophysics, 49*, 1-9. http://dx.doi.org/10.3758/BF03211610

Schiffman, S. S., Sattely-Miller, E. A., Graham, B. G., Bennett, J. L., Booth, B. J., & Desai, N. (2000). Effect of temperature, pH, and ions on sweet taste. *Physiology & Behavior, 68*, 469. http://dx.doi.org/10.1016/S0031-9384(99)00205-X

Talavera, K., Yasumatsu, K., Voets, T., Droogmans, G., Shigemura, N., & Ninomiya, Y. (2005). Heat activation of TRPM5 underlies thermal sensitivity of sweet taste. *Nature, 438*, 1022. http://dx.doi.org/10.1038/nature04248.

Wheeler, J. B., Hoersch, H. M., Mahy, H. P., & Kleinberg, A. S. (Eds.). (1981). *Guide-lines for the selection and training of sensory panel members, ASTM STP 758*. Philadelphia, PA: American Society for Testing and Materials.

Wise, P. M., & Breslin, P. A. S. (2011). Relationship among taste qualities assessed with response-context effects. *Chemical Senses, 36*, 581. http://dx.doi.org/10.1093/chemse/bjr024

12

Antioxidant and Prebiotic Activity of Selected Edible Wild Plant Extracts

Poloko Stephen Kheoane[1], Clemence Tarirai[1], Tendekayi Henry Gadaga[2], Carmen Leonard[1] & Richard Nyanzi[3]

[1]Department of Pharmaceutical Sciences, Tshwane University of Technology, Pretoria, South Africa

[2]Department of Environmental Health Science, University of Swaziland, Mbabane, Swaziland

[3]Department of Biotechnology, Tshwane University of Technology, Pretoria, South Africa

Correspondence: Clemence Tarirai, Department of Pharmaceutical Sciences, Tshwane University of Technology, 175 N. Mandela Drive, Pretoria, 0001, South Africa. E-mail: tariraic@tut.ac.za

Abstract

Edible wild plants were investigated as potential sources of antioxidants and prebiotics to benefit human health. Antioxidant activity, ascorbic acid and total dietary fibre contents were determined in edible wild plants from Lesotho, Swaziland and South Africa. Pure probiotic strains of *Bifidobacterium animalis* subsp. *animalis* (ATCC 25527), *Lactobacillus rhamnosus* (TUTBFD) and *Lactobacillus acidophilus* (ATCC 314) were cultured in broth containing edible wild plant extracts to assess their prebiotic activity. *Cyperus esculantus* had the highest arscobic acid content of 603±64.1 mg/100 g edible dry plant material followed by *Rosa rubiginosa* (500.8±48.8 mg/100 g). The two plants had IC_{50} of 10.7±0.2 μg/mL and 47.8±0.2 μg/mL for DPPH inhibition, respectively. Forty percent (40%) (n=30) of the edible wild plants had significant (p<0.01) total antioxidant activity (IC_{50}<60 μg/mL) and high ascorbic acid content (>200 mg/100 g). *Nasturtium officinale* reported the highest yield for soluble fibre (25%) while *Hypoxis hirsute* had the highest total dietary fibre content (7.3%). *Rorippa nudiuscula* enhanced the growth of *B. animalis* significantly (p=0.001), 8-fold more than inulin. *Chenopodium album* and *Urtica dioica* stimulated the growth of *L. rhamnosus* significantly (p=0.0001) than inulin, respectfully, while *Tragopogon porrifolius* significantly (p=0.0001) stimulated the growth of *L. acidophilus* than inulin. It was concluded that the investigated edible wild plants from southern Africa have antioxidant and prebiotic properties that may be beneficial to human health.

Keywords: antioxidant, Bifidobacterium, dietary fibre, edible wild plant, Lactobacillus, prebiotic, probiotic

1. Introduction

Edible wild plants provide energy, vitamins and minerals in the diet (Boedecker, Termite, Assogbadjo, Van Damme & Lachat, 2014). Reactive oxygen species, including superoxide anions, hydroxyl radicals, and hydrogen peroxide are generated in specific organelles of the cell under normal physiological conditions (Haraguchi, 2001). Excessive production of these reactive oxygen species and free radical mediated reactions are associated with degenerative diseases such as aging, cancer, coronary heart disease, and Alzheimer's disease (Sun, Wang, Fang, Gao & Tan, 2004). Consumption of sufficient amounts of fruits, vegetables and tubers leads to reduced incidences of heart disease, cancers and other degenerative diseases (Kaur & Kapoor, 2001). These foods provide an optimal mix of beneficial phytochemicals such as natural antioxidants, fibres and other biotic compounds (Kaur & Kapoor, 2001). According to Halliwell (1996), dietary compounds such as vitamins E and C are strong antioxidants while the antioxidant capacity of plant pigments (e.g. carotenoids), plant phenolics and flavonoids may involve other biological mechanisms such as increasing the expression of endogenous antioxidants. Some edible wild plant foods are also sources of vitamins and minerals. For example, a study by Bwembya, Thwala, Silaula and Otieno (2014) reported that Swazi wild edible vegetables like *Solanum nigrum* (Umsobo), *Corchorius olitorus* (Ligusha), *Momordica involucrate* (Inkhakha), *Amaranthus spinosus* (Imbuya) and *Bidens pilosa* (Chuchuza) are sources of nutrients such as pro-vitamin A, iron, calcium and zinc. Additionally, *Acacia senegal* (Gum Arabic) functionally increases the absorption of coenzyme Q10 (Ozaki et al., 2010).

Prebiotics are non-digestible food ingredients naturally present in some plant materials that stimulate the growth and/or activity of 'friendly' probiotic bacteria in the digestive system in ways claimed to be beneficial to health (Gibson & Roberfroid, 1995). Prebiotic molecules can be divided into three categories, namely short-chain, long-chain and full spectrum prebiotics (Roberfroid, 2007). Inulin, oligofructose and lactulose have prebiotic effect and have been part of the human diet for centuries (Franck, 2002). The human intestine is densely populated with microorganisms that influence human biological processes as well as drug absorption (Kullberg, 2008). Some of these microorganisms originate from fermented foods or the environment and are known to have benefical effects on human health; hence the term probiotics (Holzapfel & Schillinger, 2002; Lei & Jakobsen, 2004). Microorganisms which are commonly used as probiotics belong to the heterogeneous group of lactic acid bacteria (Lactobacillus, Enterococcus) and to the genus Bifidobacterium (Food and Agriculture Organization of the United Nations and the World Health Organization [FAO/WHO], 2001; Ghosh, 2012; Holzapfel, Haberer, Geisen, Björkroth & Schillinger, 2001). The beneficial bacteria types include the well-known probiotic strains *B. animalis* subsp. *animalis*, *L. rhamnosus* and *L. acidophilus* (Jungersen et al., 2014; Weese & Anderso, 2002). Alvarez-Olmos and Oberhelman (2001) and Tien et al. (2006) described the mechanism of action for probiotics against gastrointestinal pathogens as based principally on: i) competition for nutrients and sites of adhession, ii) production of antimicrobial metabolites, iii) changing the gastrointestinal environmental conditions and, iv) modulation of the immune response of the host.

Probiotic bacteria have specific nutrient requirements and some of them are selectively stimulated by non-digestible carbohydrate molecules such as oligofructose (Gibson & Roberfroid, 1995; Quigley, 2010; Roberfroid, 2007). Previous studies reported that extracts from herbal products and dietary supplements act as prebiotics or bifidogenic factors by modulating the balance of human gut microbes (Roberfroid et al., 2010; Wang, Qi, Wang & Li, 2011). Probiotics have shown several benefits in human health (Parvez, Malik, Ah Kang & Kim, 2006). Some of the health benefits of probiotics include: i) cancer prevention (Baldwin et al., 2010), ii) reduction of *Helicobacter pylori* infection (Gotteland, Brunser & Cruchet, 2006; Pacifico et al., 2014), iii) reduction of cholesterol and triacylglycerol plasma concentrations (Jones, Martoni, Parent & Prakash, 2011), iv) beneficial effects on mineral metabolism (Lamberti et al., 2011), v) relief from constipation (Malaguarnera et al., 2012), vi) reduction of allergic symptoms (Matsuda et al., 2012), and vii) relief from irritable bowel syndrome (Thomson, Chopra, Clandinin & Freeman, 2012).

Antioxidants and prebiotics can be extracted from plants (e.g. ascorbic acid and inulin, respectively), by enzymatic hydrolysis (e.g. oligofructose from inulin), by synthesis (e.g. by trans-glycosylation reactions) from monosaccharides and disaccharides such as sucrose (fructooligosaccharides) and lactose (trans-galactosylated oligosaccharides or galactooligosaccharides) (Crittenden & Playne, 1996). Inulin and lactose oligosaccharides are the most studied prebiotics and have been recognized as dietary fibres in most countries (Moshfegh, Friday, Goldman & Chugahuja, 1999). Inulin is known to occur naturally in some plant species such as *Cichorium intybus* (Chicory), Jerusalem artichokes and to a lesser extent in onion, garlic, banana, asparagus, tomatoes and leek (Crittenden & Playne, 1996; Kassim, Baijnath & Odhav, 2014; Roberfroid, 2007). Among cereal grains, wheat is the best source of prebiotics, providing about 2.5 g fructooligosaccharides (FOS) and 2.5 g inulin (fructopolysaccharide) per 100 g raw bran (Van Loo, Coussement, De Leenheer, Hoebregs & Smits, 1995). Wheat is the major food source of naturally occurring inulin and FOS, providing about 70% of these compounds in American diets (Moshfegh et al., 1999).

In Lesotho, Swaziland, South Africa and southen Africa in general, fewer people in urban areas now use edible wild plants for nutrition due to the availability of domesticated food plants and fast foods (Hart & Vorster, 2006) while rural communities may depend more on these plants. Arguably, some domesticated food plants have low ascorbic acid content and antioxidant activity as compared to the edible wild plants (Legwaila, Mojeremane, Modisa, Mmolotsi & Rampart, 2011). Studies on edible wild plants from other geographical locations around the world largely focus on the proximate and micronutrient content analysis as well as agronomic improvement of wild and semi-wild crop varieties (Uprety et al., 2012). Meanwhile, southern African indigenous knowledge on the value of edible wild plants is rapidly eroding due to limited local scientific evidence to substantiate their benefit to human health. This paper reports on the ascorbic acid content, total antioxidant capacity and prebiotic activity of edible wild plants from Lesotho, Swaziland and the Limpopo province of South Africa to motivate for their benefits to human health, rational use and conservation.

2. Materials and Methods

2.1 Materials

The 2, 2-diphenyl-1-picrylhydrazyl (DPPH) dye and the ascorbic acid reference standard were purchased from Sigma-Aldrich (Johannesburg, South Africa). The 96 well UV/VIS microplates and HPLC grade methanol were purchased from Anatech (Johannesburg, South Africa). Edible wild plants were sourced from the Lesotho, Swaziland and the Limpopo province of South Africa. De Man, Rogosa, Sharpe (MRS) broth, MRS agar, and petri dishes were purchased from Separations Scientific (Johannesburg, South Africa). Inulin, lactulose, α-amylase, protease, amyloglucosidase, *L. acidophilus* (ATCC 25527, Lot: 1092-05-1, France) and *B. animalis* subsp. *animalis* (ATCC 25527, Lot: 1092-05-1, France) were purchased from Sigma-Aldrich (Johannesburg, South Africa) while *L. rhamnosus* (TUTBFD) species was received from the Department of Biotechnology and Food Technology, Tshwane University of Technology. Diatomaceous earth flour (celite) and fritted-crucibles (porosity number 4 (14 μm)) all other materials were of analytical grade and were used as received.

2.2 Collection of Edible Wild Plants

Fruits, vegetables, roots, and tubers were among the samples collected. The fruits collected were *Opuntia megacantha*, *Rubus cuneifolius*, and *Rosa rubiginosa*. The vegetables were *Sonchus dregeanus*, *Rorippa nudiuscula*, *Amaranthus retroflexus*, *Chenopodium album*, *Urtica dioica*, *Wahlengergia androsacea*, *Lepidium capense*, *Tragopogon porrifolius*, *Sonchus oleraceus*, *Sonchus asper*, *Sisymbrium capense*, *Nemesia fruticans*, *Nasturtium officinale*, *Berkheya purpurea*, *Hypochaeris radicata*, *Sisymbrium thelungii*, *Sonchus integrifolius*, *Solanum retroflexum*, *Momordica foetida*, *Corchorus tridens*, *Amaranthus hybridus*, *Amarathus spinosus*, *Bidens pilosa*, and *Momordica involucrate*. *Cyperus esculentus* bulbs, *Hypoxis hirsute* and *Tragopogon porrifolius* taproot and *Discorea minutiflora* tubers were also collected.

2.3 Preparation of Plant Extracts

Local people in the areas from which the edible wild plants were collected usually chop the edible wild plants before consumption. Of interest was the simulation of these conditions and simple methods used for the determination of the amount of ascorbic acid left after the appropriate preparation method involving homogenizing, grinding with mortar and pestle, or shredding with a knife and natural drying. The fruits were homogenized in 200 mL of distilled water using a Kenwood kitchen food blender (Game, Pretoria, South Africa). The homogenate was frozen and lyophilized using a VirTis Benchtop K free-drier (Thermo Electron, Germiston, South Africa) at -50 °C for 48 hours. Plant leaves were air-dried at room temperature for 7 days and ground with a motor and pestle. Roots were shredded using a sharp knife then ground and ultra-sonicated for 24 hours at 25 °C in 500 mL distilled water, followed by filtration using Whatman filter paper pore size 0.45 μm (Sigma-Aldrich, Johannesburg, South Africa) into a volumetric flask. The remaining plant root material was ultra-sonicated with 300 mL distilled water at 30 °C with stirring for a further 12 hours. The two extract portions were combined to make 800 mL, which was frozen and lyophilized. The resultant powders were micronized (250 μm sieve mesh) and stored in labelled moisture-free airtight amber glass vials until further use (Tarirai, Viljoen & Hamman, 2012).

2.4 Determination of Ascorbic Acid Content

A 5 g amount of corn starch was added to 250 mL of distilled water in a beaker, which was heated with continuous stirring until the corn starch dissolved. The corn starch solution (5 mL) and distilled water (250 mL) were mixed in a 500 mL beaker to which four drops of a 0.05 M iodine solution were added and mixed thoroughly. The final iodine-corn starch dye solution was blue in color. The blue iodine-corn starch dye was standardized by volume using the mass of the ascorbic acid standard. The standard was prepared by dissolving ascorbic acid (5 mg) in 10 mL of distilled water. The ascorbic acid solution was poured into a burette and titrated into a 5 mL volume of the blue iodine-corn starch dye to a colorless endpoint that persisted for at least 15 seconds. The amount of ascorbic acid required to turn the 5.0 mL volume of the blue iodine-corn starch dye to colorless was calculated by using equation (1):

$$M_1 = VM/V_1 \qquad\qquad\qquad (1)$$

Where M_1 (mg) is the amount of ascorbic acid contained in a titrated volume, V_1 (mL), used to reduce 5 mL of the dye from blue to colorless. M (mg) is the known amount of ascorbic acid that was weighed off and dissolved into the total volume (V, mL) of ascorbic acid stock solution i.e. M plus mass of distilled water.

Edible wild plant extracts were prepared by dissolving a 1 g of crude plant extract powder in 20 mL of distilled water (i.e. 50 mg/mL before filtration). The mixture was left for 30 minutes at room temperature and then filtered using Whatman filter paper pore size 0.2 μm (Sigma-Aldrich, Johannesburg, South Africa). For the qualitative determination of the presence of ascorbic acid in a plant sample, the aqueous plant extracts were slowly titrated into the standardised 5 mL volume of iodine-corn starch solution until the blue solution became clear or the plant extract solution (10 mL) got depleted. For the quantitative assay of ascorbic acid, the volume of the plant extract required to turn the blue iodine-corn starch dye to colorless was recorded. The concentration of ascorbic acid in the plant extract aliquot titrated against 5 mL standardized dye was proportionally determined using equation (2):

$$M_2 = V_1 M_1 / V_2 \tag{2}$$

Where M_2 is the amount of ascorbic acid (mg) in the aliquot of crude plant extract contained in a V_2 (mL) of the plant extract aliquot titrated into 5 mL of the dye. M_1 is the mass of ascorbic acid in the volume V_1 of ascorbic acid solution used to reduce 5 mL of dye from blue to colorless from equation (1).

The amount of ascorbic acid in the total volume of crude plant extract prepared was obtained using equation (3):

$$M_t = V_2 M_2 / V_t \tag{3}$$

Where M_t (mg) is the total amount of ascorbic acid in the total stock volume (V_t, mL) of crude plant extract prepared. M_2 (mg) is the amount of ascorbic acid in an aliquot (V_2, mL) of crude plant extract titrated, to endpoint, into 5 mL of dye from equation (2).

The ascorbic acid content of each plant extract was determined using equation (4) and was expressed as milligrams of ascorbic acid per 100 grams of plant material:

$$Ascorbic\ acid\ content = (M_t \times 100) / M_s \tag{4}$$

Where M_t (g) is the total amount of ascorbic acid in a known amount (M_s, g) of crude plant extract used to prepare the plant extract stock solution.

2.5 Evaluation of Total Antioxidant Capacity

About 100 g of each freeze-dried, ground and micronized plant material was soaked in 1 L of high performance liquid chromatography (HPLC) grade methanol 98% v/v for 5 days and was stirred every 18 hours using a sterilized glass rod at room temperature. The final extracts were passed through Whatman filter paper pore size 0.2 μm (Sigma-Aldrich, Johannesburg, South Africa). The filtrates obtained were concentrated under vacuum using a Rotavapor R-200 (Buchi Labortechnik AG, Flawil, Switzerland) at 40 °C until the methanol evaporated. The samples were wetted with 10 mL of distilled water, lyophilized and stored at 4 °C for further use (Khalaf, Shakya, Al-Othman, El-Agbar & Farah, 2008).

Stock solutions were prepared by dissolving 5 mg of dry filtered plant extract in 1 mL of 98% v/v HPLC grade methanol. Working solutions of the extracts were appropriately diluted from the stock solutions to obtain concentrations of 10, 50, 100, 250, and 500 μg/mL. Ascorbic acid was used as a reference across a 10-500 μg/mL concentration range. The 2,2-diphenyl-1-picrylhydrazyl (DPPH) dye was prepared in HPLC grade methanol to obtain a 0.002% w/v DPPH dye solution.

A 150 μL aliquot of each concentration of filtered plant extract solution and 150 μL of ascorbic acid solution were separately plated (in triplicates) in a 96-well microplate followed by 150 μL of DPPH solution added to each well. The microplate was placed in a SpectrostarNano UV/Vis microplate reader (BMG LabTech, Ortenberg, Germany) and shaken for 10 seconds to mix its contents. The assays were kept in the dark for 30 minutes and their optical densities or UV/Visible absorbance was measured at 517 nm. Methanol (150 μL) with DPPH solution (0.002%, 150 μL) was used as a blank. The optical density was recorded and the % inhibition of DPPH activity was calculated using equation (5):

$$\%\ inhibition\ of\ DPPH\ activity = (A_{test\ sample} / A_{blank}) \times 100 \tag{5}$$

Where A = optical density of the test sample or the blank (Khalaf, Shakya, Al-Othman, El-Agbar & Farah, 2008). The radical scavenging effect of test samples was measured as a decrease in the absorbance of DPPH over time. The purple color of the reaction mixture changed to yellow and the UV/Visible absorbance at 517 nm decreased in the presence of increasing antioxidant activity (Tung, Wu, Kuo & Chang, 2007). Plots (not shown) of the % DPPH inhibition (y-axis) and the concentration of plant extract (x-axis) were used to extrapolate the concentration (IC_{50}, ug/mL) at which 50% inhibition of DPPH occurred for each plant extract.

2.6 Extraction of Prebiotics

The prebiotic carbohydrates (soluble fibre) were extracted according to an adaptation of the method described by Iwata, Hotta and Goto (2009). The powders were extracted in duplicate for each sample. The dried powder samples were soaked in 50% v/v ethanol and left at room temperature for 3 days. The samples were then filtered through a Whatman No. 1 filter paper. The filtrates were freeze-dried (VirTis Benchtop K, Model 2KBTES, SP industries) at -50 °C for 18 hours) after concentrating using a rotary evaporator (Model R-200 Buchi Rotavapor, Netherlands) at 200 rpm at 50 °C for 20 minutes). The freeze-dried extracts of edible wild plants were preserved in airtight glass containers until further used. The % yield of the plant extracts was determined using equation (6) as described by Wichienchot et al. (2011):

$$Yield\ (\%)\ =\ (weight\ dried\ extract\ (g)\ x\ 100)\ /\ initial\ amount\ (g) \qquad (6)$$

2.7 Determination of Total Dietary Fibre Content

Digestion in acid and enzyme media was used to separate the digestible fibre from the indigestible material. The acid digestion test was done according to the method(s) described by Iwata et al. (2009) and Wichienchot et al. (2011). Solutions (10% w/v) of the dried extracted carbohydrates were made in distilled water by dissolving a 250 mg ethanol-based sample of the crude plant extract from Section 2.3 in 2.25 mL distilled water. The mixture was incubated with a 225 µL HCl buffer of pH 1 at 37 °C for 4 hours. The reaction was terminated by neutralization using a 225 µL of 1 N sodium hydroxide solution. The acid-digested samples were diluted with a 12.5 mL phosphate buffer (20 mM) at pH 6.0 and the final volume was 15.2 mL. The pH of this final sample was adjusted to 6.0.

For enzymatic digestion, the acid-digested solutions were treated according to a modified AOAC method 985.29 (AOAC International, 1995). The acid-digested samples were incubated with 30 µL of 3000 units/mL α-amylase at 50 °C for 60 minutes in a water bath. The samples were cooled to room temperature and the pH was adjusted to 7.5. This was followed by digestion with 30 µL of 50 mg/mL protease at 60 °C for 30 minutes in the water bath. The samples were cooled to room temperature and their pH was adjusted to 4.0. The samples were further incubated with 75 µL of amyloglucosidase at 60 °C for 30 minutes in the water bath. The pH of the samples was measured using a SensION^{TM} $^{+}$PH31 meter (HACH, Serial no.: 415006, Spain) throughout the experiment. When the enzymatic digestion sequence was completed, 60 mL of 95% v/v ethanol previously heated to 60 °C, was added to each sample. The solutions were left to precipitate over 12 hours at room temperature and then filtered using fritted-crucibles, porosity number 4 (14 µm) covered with diatomaceous flour (celite) (particle size 16 µm) bed under vacuum. The fritted-crucibles were first cleaned and 0.5 g diatomaceous flour was weighed off in each crucible. The flour was wetted with 78% v/v ethanol followed by vacuum application to get rid of the ethanol thus forming a thin bed of diatomaceous flour over the base of each crucible. The fritted-crucibles containing the diatomaceous flour bed were then heated at 130 °C in an oven. After heating, the crucibles were cooled in a desiccator, weighed off and used to filter the precipitated enzymatic digested samples.

The residues inside the fritted-crucibles were dried over 12 hours. After the drying process, each fritted-crucible with its contents was weighed and its mass was recorded. One duplicate set of the residues was analysed for protein content using a TruMac$^{®}$ N Nitrogen determinator Version 1.20 (Part number: 200-739, LeCo, USA). The second duplicate set of residue was used to determine ash content by incinerating in the oven at 320 °C for 8 hours. The residue weight, an average of duplicate samples, was calculated by subtracting the mass of the crucible plus diatomaceous flour from the mass of crucible plus diatomaceous flour plus sample precipitate. This residue weight, minus the protein and ash residue weights, represented the weight of insoluble dietary fibre and high molecular weight soluble dietary fibre. Protein content of the residue was calculated using equation (7):

$$Protein\ content\ =\ N\ x\ 6.25 \qquad (7)$$

Where N is the nitrogen content of the residue and 6.25 is the protein conversion factor.

Total dietary fibre content was calculated using equation (8):

$$Total\ dietary\ fibre\ =\ weight\ (residues)\ -\ (protein\ +\ ash) \qquad (8)$$

The percentage total dietary fibre was calculated using equation (9):

$$\%\ Total\ dietary\ fibre\ =\ (Total\ diatary\ fibre\ x\ 100)/weight\ of\ test\ sample \qquad (9)$$

Where weight residue = average of duplicates, and weight test portion = average of duplicates.

2.8 Assessment of Prebiotic Activity

2.8.1 Cultivation of Bacteria Cultures

Pure strains of *B. animalis* subsp. *animalis* (ATCC 25527, Lot: 1092-05-1, France) and *L. acidophilus* (ATCC 314, Lot: 885-34-11, France) were activated by aseptically inoculating them into sterile MRS broth (5 mL). Aliquots (1.5 mL) of each of the activated cultures were separately transferred to fresh MRS broth (15 mL) in centrifuge tubes. Each treatment was done in triplicate. *B. animalis* and *L. rhamnosus* containing tubes were incubated anaerobically (Anaerocult gas mixture, Merk, SA) for 48 hours at 37 °C, while the *L. acidophilus* was incubated in a 5% carbon dioxide atmosphere at 37 °C for 48 hours. After incubation, the tubes were centrifuged at 277 xg in a 5417C/R centrifuge (Epperndorf, Germany) for 15 minutes to obtain a pellet of bacteria. The broth was aseptically decanted and resuspended in MRS broth (5 mL) containing 20% glycerol. The pellets were then stored at below -20 °C until required for further tests.

2.8.2 Reactivating Frozen Bacterial Cultures

A loopful of thawed bacterial culture was streaked onto MRS agar (Biolab, South Africa). *B. animalis* and *L. rhamnosus* were incubated anaerobically at 37 °C for 48 hours, while *L. acidophilus* was incubated in a 5% carbon dioxide atmosphere at 37 °C as previously described in section 2.8.1.

2.8.3 Preparation of Test Solutions and Inoculation of Bacterial Cultures

A solution (2.5 mL) containing 5 mg/mL of plant extract for each of the selected edible wild plants was transferred to MRS broth (10 mL) in triplicates. The tubes were thoroughly mixed using a vortex mixer and sterilized at 121 °C for 15 minutes. After cooling, the tubes were inoculated with a loopful of *B. animalis*, *L. rhamnosus*, or *L. acidophilus*. *B. animalis* and *L. rhamnosus* were incubated anaerobically 37 °C for 24 hours as previously described in section 2.8.2, while *L. acidophilus* was incubated under 5% carbon dioxide.

2.8.4 Enumeration of Bacteria

The bacterial macroculture counts for inoculated test broth were enumerated on MRS agar using the drop plate technique (Herigstad, Hamilton & Heersink, 2001; Munsch-Alatossava, Rita & Alatossava, 2007; Naghili et al., 2013). *B. animalis* and *L. rhamnosus* were incubated anaerobically at 37 °C for 72 hours, while *L. acidophilus* was incubated in 5% carbon dioxide/oxygen.

2.9 Statistical Analysis of Data

Ascorbic acid content (mg/100 g of dry edible plant material), total antioxidant activity (ug/mL) and the log (bacterial count/mL) were determined in the presence of plant extracts. One-way analysis of variance (ANOVA) was peformed using MS Excel-*XLSAT* (Microsoft, Redmond, USA) to determine the differences in ascorbic acid content, antioxidant and prebiotic activity between plant extracts and also in comparison with the controls. Significant differences were taken for values of $p < 0.01$.

3. Results and Discussion

3.1 Percentage Yield of Soluble Dietary Fibre

The results for the percentage yields of ethanolic extracts of edible wild plants are presented in Table 1. *Rosa rubiginosa* had the highest percentage yield of 49.4% while *Discorea minutiflora* species had the lowest percentage yield of 7.8%.

3.2 Antioxidant Activity of Edible Wild Plants

Cyperus esculentus showed the highest potency ($IC_{50} = 10.7 \pm 0.2$ µg/mL) and corresponding high percentage inhibition of DPPH (data not shown) followed by *Berkheya purpurea* ($IC_{50} = 11.9 \pm 0.1$ µg/mL) while *Discorea minutiflora* ($IC_{50} = 338.6 \pm 0.3$ µg/mL) had the lowest capacity (Table 1). The antioxidant activity of the edible wild plants showed a trend similar to that observed for the values of ascorbic acid content (Table 1) for 87% (n=30) of the edible wild plants. This suggests that the antioxidant activity observed in the plant extracts was partly due to ascorbic acid.

Other antioxidants previously reported in edible wild plants and herbs include tocopherols and carotenoids (Choi, Jeong & Lee, 2007); tannins (Lee, Koo & Min, 2004); phenols and flavonoids (Nair, Nagar & Gupta, 1998). These phytochemicals partly contributed to the total antioxidant activity reported as indicated by the significantly ($p < 0.01$) low IC_{50} values obtained for those plant extracts which had low ascorbic acid content notably *Rorippa nudiuscula* ($IC_{50} = 22.6 \pm 0.2$ ug/mL, but had 77.0 ± 3.0 mg/100 g ascorbic acid) and *Sonchus integrifolus* ($IC_{50} = 19.1 \pm 0.3$ ug/mL with 80.2 ± 8.4 mg/100 g ascorbic acid). Conversely, *Amaranthus hybridus* ($IC_{50} = 233.0 \pm 1.0$ ug/mL) and *Momordica involucrate* ($IC_{50} = 146.6 \pm 0.3$ ug/mL), which scored fairly high in ascorbic acid content showed weak antioxidant activities possibly due to low content of tocopherols, carotenoids, tannins, phenolics and flavonoids. The content of these phytochemicals in the edible wild plant extracts need further investigation. Additionally, all the three *Sonchus* subspecies included in the currect study consistently showed high ascorbic acid content and potent antioxidant activity.

Ozen (2010) reported that DPPH inhibition by aqueous wild edible plant extracts (100 µg/mL) from Turkey ranged from 48% to 78%. *Amaranthus retroflexus* aqueous extract caused a 69% DPPH inhibition while the methanolic extract (100 µg/mL) of the same plant elicited 43% in the present study. *Momordica foetida* leaves showed 45% inhibition of the superoxide anion at 175 µg/mL in an Italian study conducted by Acquaviva et al. (2013) while inhibition of DPPH activity by the same plant was 55% at 175 µg/mL in this study. Gacche, Kabaliye, Dhole and Jadhav (2010) reported that the DPPH scavenging activity of 1 mg/mL common vegetable extracts ranged from 20% to 67% while that of edible wild plant extracts at 500 µg/mL ranged from 53% to 100% in the present study.

Table 1. Percentage Yield of Soluble Fiber, Total Antioxidant Capacity and Ascorbic Acid Content of the Edible Wild Plant Extracts

Description of plant sample				% yield of soluble fibre	Total antioxidant activity*	Ascorbic acid content*
Plant botanical name	Plant common English name	Plant vernacular name#	Part of plant used	(mean±sd, n=2)	IC$_{50}$, µg/mL (mean±sd, n=3)	mg/100 g (mean±sd, n=3)
Ascorbic acid	Vitamin C				3.5±0.2[a]	
Rubus Cuneifolius	Wild berries/ Sand bramble	Monokots'oai (Ss)	Ripe fruits	39.5±3.2	30.2±0.3[c]	298.5±24.3[e]
Opuntia megacantha	Prickly pears	Torofeiea	Ripe fruits	47.0±4.8	348.9 ± 0.4	NT*
Sonchus dregeanus	Thistle	Leharasoana (Ss)	Young leaves	16.4±1.6	18.1±0.2[b]	128.5±6.5
Rorippa nudiuscula	-	Papasane (Ss)	Young leaves	20.0±2.1	22.6±0.2[b]	80.2±8.4
Rosa rubiginosa	Rosehip	'Morobei (Ss)	Ripe fruits	49.4±4.9	47.8±0.2[e]	500.8±48.8[f]
Amaranthus retroflexus	Red pigweed	Thepe (Ss)	Young leaves	19.3±0.7	139.3±0.6	132.8±12.1
Chenopodium album	Wild spinach	Seruoe (Ss)	Young leaves	19.6±0.8	77.8±0.2	240.5±16.5[b]
Urtica dioica	Sting nettle plant	Bobatsi (Ss)	Young leaves	13.3±0.2	89.0±0.1	205.5±13.8[a]
Wahlengergia androsacea	Hare-bell	Tenane (Ss)	Young leaves	12.2±0.1	121.8±0.4	273.0±19.9[d]
Lepidium capense	Cape pepper cress/weed	Qhela (Ss)	Young leaves	12.7±0.3	122.2±1.0	160.8±16.1
Tragopogon porrifolius	Salsify	Moetse-oa-pere (Ss)	Young leaves, taproot	13.1±0.4	33.4±0.6[c]	240.5±16.5[b]
Sonchus oleraceus	Sowthistle	Leshoabe (Ss)	Young leaves	20.1±4.6	35.2±0.4[d]	121.4±10.6
Sonchus asper	Sowthistle	Leshoabe (Ss)	Young leaves	15.0±2.1	54.2±0.4[f]	287.0±38.3[d]
Sisymbrium capense	-	Tlhako ea khomo (Ss)	Young leaves	13.9±0.8	181.5±0.5	NT*
Nemesia fruticans	Wild nemesia	Malana a konyana (Ss)	Young leaves	20.2±3.5	48.5±0.5[c]	261.5±19.9[c]
Nasturtium officinale	Watercress	Semetsing/Selae (Ss)	Young leaves	25.0±4.7	82.6±0.4	123.0±8.4
Berkheya purpurea	Purples	Sehlohlo (Ss) (Ss)	Young leaves	20.3±2.1	11.9±0.1[a]	298.5±24.3[e]
Hypochaeris radicata	Cat's ear	Lepheo-la-khoho (Ss)	Young leaves	15.3±1.8	42.2±0.3[d]	155.0±8.6

Note. #Ss=Sesotho, V=Venda, Sw=Swati. *NT = Not tested (no ascorbic acid based on preliminary qualitative test, data not shown). +Different letters indicate significant statistical difference (p<0.01) for total antioxidant activity and ascorbic acid content values of edible wild plant extracts.

Table 2. Percentage Yield of Soluble Fiber, Total Antioxidant Capacity and Ascorbic Acid Content of the Edible Wild Plant Extracts (continued)

Description of plant sample				% yield of soluble fibre	Total antioxidant activity*	Ascorbic acid content*
Plant botanical name	Plant common English name	Plant vernacular name#	Part of plant used	(mean±sd, n=2)	IC$_{50}$, µg/mL (mean±sd, n=3)	mg/100 g (mean±sd, n=3)
Sisymbrium thelungii	Wild mustard	Sepaile (Ss)	Young leaves	22.2±2.6	101.0±0.5	121.4±10.6
Cyperus esculentus	Sedge/ Earth Almond	Monakalali (Ss)	Bulb	22.8±1.9	10.7±0.2[a]	603.0±64.1[f]
Hypoxis hirsute	-	Lentsikitlane (Ss)	Taproot	9.6±0.1	334.5±0.5	NT*
Sonchus integrifolius	Thistle	Sentlhokojane (Ss)	Young leaves	20.8±0.9	19.1±0.3[b]	77.0±3.0
Solanum retroflexum	Wonderberry/nightshade	Muxe	Young leaves	14.8±0.4	198.9±0.2	172.1±5.6
Momordica foetida	-	Nngu (V)	Young leaves	15.5±0.4	155.1±0.3	116.0±4.9
Corchorus tridens	Wild jute plant	Delele (V)	Young leaves	10.4±0.3	157.5±0.5	68.5±7.0
Amaranthus hybridus	Pigweed	Thebe (Ss)	Young leaves	18.7±0.8	233.0±1.0	212.4±8.4[a]
Amarathus spinosus	Pigweed	Imbuya (Sw)	Young leaves	9.8±0.3	121.1±0.6	109.8±4.1
Bidens pilosa	Black jack	Chuchuza (Sw)	Young leaves	11.5±0.3	193.2±0.4	66.1±0.9
Momordica involucrate	Bitter gourd	Inkhakha (Sw)	Young leaves	23.8±1.6	146.6±0.3	224.1±6.5[a]
Discorea minutiflora	Yam	Emadhumbe (Sw)	Tubers	7.8±0.2	338.6±0.3	20.0±1.0

Note. #Ss=Sesotho, V=Venda, Sw=Swati. *NT = Not tested (no ascorbic acid based on preliminary qualitative test, data not shown). +Different letters indicate significant statistical difference (p<0.01) for total antioxidant activity and ascorbic acid content values of edible wild plant extracts.

3.3 Ascorbic Acid Content of the Edible Wild Plants

A total of 27 (90%) edible wild plants contained ascorbic acid ranging from 20.0±1.0 to 603±64.1 mg per 100 g edible portion of dry plant material (Table 2). *Cyperus esculantus* had the highest ascorbic acid content of 603±64.1 mg/100 g followed by *Rosa rubiginosa* at 500.8±48.8 mg per 100 g.

Local people in the areas from which the plant samples were collected usually chop the edible wild plants before consumption. These conditions were simulated to determine the amount of ascorbic acid left after such simple, but harsh preparation procedures like homogenizing, grinding with mortar and pestle, or shredding with a knife. Despite these harsh treatments, the ascorbic acid content of *Rosa rubiginosa* in this study correlates well with previous reports. For example, Ropciuc, Cenusa, Caprita and Cretescu (2011) reported ascorbic acid content ranging between 347.1 and 621.3 mg/100 g for *Rosa rubiginosa*, which varied with altitude, type of soil and humidity. Ascorbic acid content of *Rosa rubiginosa* was at 500.8±48.8 mg per 100 g in the current study.

Conversely, Pradhan, Manivannan, and Tamang (2015) reported the ascorbic acid content of 44.00±0.17 mg/100 g for *Amaranthus viridis*, 7.0±0.1 mg/100 g for *Urtica dioica*, and 3.0±0.1 mg/100 g for *Chenopodium album* while the ascorbic acid of *Amaranthus* species in the present study ranges from 109.8 to 212.4 mg/mL, 205.5 mg/mL for *Urtica dioica*, and 240.5 mg/mL for *Chenopodium album*. Difference in results from the two studies is explained in terms of the different geographical locations (i.e. India and Southern Africa), which inherently give rise to different effects from altitude, climate, and soil type/fertility on plant composition as well as the methods of plant preparation and handling.

A frequency distribution of the edible wild plants investigated in this study, categorized by ascorbic acid content, antioxidant activity and the perceived potential societal value, is presented in Table 3. Approximately 87% (n=30) of the edible wild plants had adequate (>60 mg/100 g) ascorbic acid content according to the Food and Drug Administration [FDA] (2016). *Cyperus esculentus* and *Rosa rubiginosa* and ten other edible wild plants had significantly (p<0.01) high ascorbic acid content (>200 mg/100 g edible plant material) and potent total antioxidant activity (IC_{50}<60 ug/mL). *Cyperus esculentus*, *Rosa rubiginosa* and the *Sonchus* species are potential commercial sources of ascorbic acid and antioxidants.

Many rural communities in southern Africa depend on edible wild plants as famine foods. People even preserve these edible wild plants by chopping, then air and/or sun drying them for later use during the dry seasons (i.e. April to August). Plants tend to lose their ascorbic acid when their tissues are ground without being immersed in 5% metaphosphoric acid. However, the present study reports adequate amounts of ascorbic acid of >60 mg/100 g edible portion of plant material as stipulated by the FDA (FDA, 2016) were retained in 87% (n=30, Table 3) of the edible wild plants prepared by indigenous southern African methods.

Table 3. Classification of Edible Wild Plants Based on their Potential Antioxidant Value

Potential value of edible wild plants	Ascorbic acid content[*] mg/100 g of plant extract	Total antioxidant activity[**] IC_{50} (µg/mL) of plant extract	% (n=30)
No/low value category Edible wild plants with no or low value with respect to ascorbic acid content and total antioxidant activity.	0	>300	10
[**]*Limited value category* Generally meets the FDA daily value (DV) for ascorbic acid (60 mg), but has limited antioxidant value.	≤60	201-300	3
Average value category Average ascorbic acid and moderate total antioxidant activity. Edible wild plants with potential for use both as daily sources of ascorbic acid and antioxidants as well as for therapeutic purposes.	61-200		47
[+]*High value category* Edible wild plants with high ascorbic acid content and high total antioxidant capacity. Edible wild plants that potentially have high therapeutic and high commercial value e.g. *Rosa rubiginosa* fruit and *Cyperus esculentus* bulb.	≥201	≤60	40

Note. [*]A daily value (DV) of 60 mg for dietary ascorbic acid consumption (FDA, 2016) was used as reference for *high value*. [**]IC_{50} = Concentration (ug/mL) of plant extract (ascorbic acid and phenolic) that induced 50% inhibition of 0.002% w/v DPPH activity.

Notably, vegetables are usually cooked before consumption, unless used in a salad, and this can affect the ascorbic acid content in the final dish. Paul and Ghosh (2012) reported that degradation of pure ascorbic acid and pomegranate fruit (*Punica granatum*) increased with an increase in temperature (during cooking), but the degradation of pure ascorbic acid was higher than that of the pomegranate fruit. The plant cell materials, including insoluble starch and celluloses fibers, in pomegranate and edible wild plants prepared would shield the ascorbic acid from overheating thus slowing the degradation process. Future studies can explore the ascorbic acid content of final dishes for those edible wild plants from this study, which are prepared through cooking.

In comparison, Venkatachalam, Rangasamy and Krishnan (2014) determined the ascorbic acid content of cultivated common vegetables and fruits from southern India. The ascorbic acid content of cultivated common vegetables such as mulberries, papaya, red grapes, guava, tomato, red onion, red cauliflower, carrot, and beetroot ranged from 10.83 to 68.71 mg/100 g. The values are low when compared to those of edible wild plants investigated in the present study. These findings support the viewpoint that edible wild plants are arguably better sources of ascorbic acid than cultivated plants (Legwaila et al., 2011). Most cultivated plants are genetically modified cultivars, and theoretically, they tend to grow faster before taking up enough nutrients from the soil. On the other hand, edible wild plants slowly grow in natural habitats thus gaining sufficient nutrients by their late maturity state.

3.4 Total Dietary Fibre Content of Edible Wild Plants

Table 4 presents the results for total dietary fibre content in the edible wild plant extracts. The total dietary fibre in the edible wild plants ranged from 7.97 mg/g to 73.48 mg/g with *Hypoxis hirsute* having the highest amount of total dietary fibre (7.3%) followed by *Amarathus spinosus* with 5.5%, and *Momordica involucrate* with 4.8%. The lowest amount of total dietary fibre (0.8% or less) was observed for *Sisymbrium capense*. *Rorippa nudiuscula*, *Sonchus dregeanus*, *Rosa rubiginosa*, *Urtica dioica*, *Berkheya purpurea*, *Sisymbrium thelungii*, *Cyperus esculentus*, *Corchorus tridens*, *Amaranthus hybridus* and *Bidens pilosa* also had higher amounts of total dietary fibre.

An ideal prebiotic must escape acidic and mammalian enzyme digestion in the upper gastrointestinal tract so that it can be released in the lower gastrointestinal tract and be used by the beneficial microorganisms in the colon, mainly bifidobacteria and lactobacilli (Gibson, Probert, Loo, Rastall & Roberfoid, 2004). Prebiotics are non-digestible carbohydrates like dietary fibre which are selectively fermented by probiotics (Iwata et al., 2009). The total dietary fibre content of edible wild plants indicates the undigested portion of plant material that can reach the lower gastrointestinal tract, where bifidobacteria and lactobacilli are densely populated. Edible wild plants with high total dietary fibre content are expected to stimulate the growth of probiotics better than edible wild plants with low amounts of total dietary fibre. However, there are several other factors than total dietary fibre alone, such as reducing sugars, which can be contributing to the stimulatory effects of edible wild plants.

3.5 Prebiotic Activity of Edible Wild Plants

3.5.1 Effects of Edible Wild Plant Extracts on the Proliferation of *B. Animalis* Subsp. *Animalis* (ATCC 25527)

The colonies of *B. animalis* were observed as white, glistering and medium in size. The bacteria were short, rod-shaped and gram positive. The stimulation of *B. animalis* growth by edible wild plants ranged from log (cfu/mL)=7.00±0.01 for *Discorea minutiflora* to log (cfu/mL)=8.15±0.05 in the presence of *Rorippa nudiuscula* after 48 hours as presented in Table 4. Twenty-two edible wild plants stimulated the growth of *B. animalis* much better as compared to the broth (negative control, log (cfu/mL)=7.52±0.07 and inulin (positive control, log (cfu/mL)=7.75±0.05. Other plants that markedly stimulated the growth of *B. animalis* better than the positive control were *Rubus cuneifolius*, *Sonchus dregeanus*, *Amaranthus retroflexus*, *Chenopodium album*, *Urtica dioica*, *Wahlengergia androsacea*, *Lepidium capense*, *Nemesia fruticans*, *Berkheya purpurea*, *Taraxacum officinale*, *Sisymbrium thelungii*, *Cyperus esculentus*, *Sonchas integrifolius*, *Solanum retroflexum*, *Corchorus tridens*, *Amaranthus hybridus* and *Amarathus spinosus*.

Most of these plants stimulated the net growth of *B. animalis* several times better than inulin (positive control). For example, *Rorippa nudiuscula* enhanced the growth of *B. animalis* 11 times more than inulin, *Sonchus dregeanus* by 9-fold more, with *Sisymbrium thelungii* at 7-fold, and *Urtica dioica* was 6-fold more effective than inulin. The observed high growth stimulatory effects of these edible wild plants could be due to the presence of high total dietary fibre, reducing sugars or inulin. *Rorippa nudiuscula* had high total dietary fibre content, which correlates with its higher stimulatory effects on the growth of *B. animalis* and so were *Sisymbrium thelungii* and *Urtica dioica*. Kassim et al. (2014) reported that *Solanum nigrum* (leaves), *Momordica balsamina* (leaves), *Amaranthus spinosus* (leaves), *Amaranthus hybridus* (leaves), *Amaranthus dubius* (leaves), *Sonchus oleraceus* (leaves and roots), *Taraxacum officinale* (leaves and roots) had a higher growth stimuli on probiotics. The same edible wild plants from different geographical locations also showed high stimulation on growth of *B. animalis* in the present study. Sreenivas and Lele (2013) reported that gourd family vegetables had significant prebiotic ability on *B. breve*. Similarly, in the present study, *Momordica foetida* stimulated the growth of *B. animalis*. However, *Momordica involucrate*, which had appreciable total dietary fibre content, inhibited the growth of all three probiotics indicating that the edible wild plants probably possesses some antimicrobial activity.

3.5.2 Effects of Edible Wild Plant Extracts on the Proliferation of *L. Rhamnosus* (TUTBFD)

The colonies of *Lactobacillus rhamnosus* were white, glistering and of medium size. The bacteria were gram positive rods. Log colony counts of *L. rhamnosus* ranged from log (cfu/mL)=8.29±0.03 for *Amaranthus hybridus* to log (cfu/mL)=8.85±0.01 in the presence of *Urtica dioica* as shown in Table 4. Twenty-three edible wild plants stimulated the growth of *L. rhamnosus* better than both the broth (negative control, log (cfu/mL)=8.51±0.01 and the positive control i.e. inulin, log (cfu/mL)=8.58±0.01. Other plants that demonstrated stimulatory effects on the growth of *L. rhamnosus* include *Rorippa nudiuscula*, *Rosa rubiginosa*, *Chenopodium album*, *Lepidium capense*,

Table 4. Effects of Soluble Fibres from Edible Wild Plants on the Growth of *Bifidobacterium animalis* subsp. *animalis* (ATCC 25527), *Lactobacillus rhamnosus* (TUTBFD) and *Lactobacillus acidophilus* (ATCC 314)

Test sample	Total dietary fibre[*]	Log colony forming units per mL ($\log_{10}$ CFU/mL) after 48 hours[**+]		
	mg/g plant extract	*Bifidobacterium animalis*[#]	*Lactobacillus rhamnosus*[##]	*Lactobacillus acidophilus*[#]
Inulin (control)		7.75±0.05[a]	8.58±0.01[a]	6.46±0.02[a]
Broth (baseline)		7.52±0.07	8.51±0.01	5.72±0.12
Rubus cuneifolius	21.24±0.38	7.84±0.06	8.45±0.03	6.63±0.04
Opuntia megacantha	20.95±0.12	7.72±0.12	8.56±0.02	6.10±0.05
Sonchus dregeanus	44.30±2.36	7.98±0.03[c]	8.40±0.02	6.20±0.03
Rorippa nudiuscula	37.00±5.17	8.15±0.05[d]	8.75±0.01[b]	6.48±0.04
Rosa rubiginosa	39.97±1.66	7.46±0.15	8.65±0.02	6.47±0.02
Amaranthus retroflexus	21.69±1.80	7.79±0.10	8.59±0.01	6.15±0.05
Chenopodium album	23.09±0.61	7.77±0.07	8.82±0.01[c]	5.59±0.11
Urtica dioica	34.68±4.88	8.01±0.06[c]	8.85±0.01[c]	0 (bactericidal)
Wahlengergia androsacea	32.21±0.53	7.84±0.06	8.57±0.02	6.41±0.06
Lepidium capense	14.74±1.04	7.73±0.05	8.72±0.01	6.86±0.01
Tragopogon porrifolius	21.78±1.44	7.59±0.11	8.73±0.01	7.05±0.01[b]
Sonchus oleraceus	19.25±1.69	7.42±0.10	8.58±0.04	6.87±0.04
Sonchus asper	19.96±0.92	7.66±0.10	8.79±0.01[b]	6.91±0.02[b]
Sisymbrium capense	7.97±1.35	7.56±0.07	8.48±0.04	5.76±0.15
Nemesia fruticans	22.59±2.72	7.95±0.05[b]	8.49±0.03	6.63±0.08
Nasturtium officinale	16.86±0.81	7.66±0.10	8.77±0.01	6.94±0.02[b]
Berkheya purpurea	31.21±2.13	7.77±0.07	8.64±0.02	6.62±0.03
Hypochaeris radicata	18.88±0.40	7.95±0.05[b]	8.53±0.01	6.65±0.04
Sisymbrium thelungii	41.77±2.34	8.02±0.06[c]	8.76±0.01[b]	6.78±0.04
Cyperus esculentus	36.13±0.95	7.84±0.06	8.66±0.01	6.83±0.03
Hypoxis hirsute	73.48±3.48	7.46±0.15	8.71±0.01	NT[$]
Sonchus integrifolius	27.52±1.01	7.98±0.03[c]	8.71±0.01	5.46±0.15
Solanum retroflexum	19.58±0.46	7.86±0.09	8.74±0.01[b]	6.81±0.02
Momordica foetida	25.56±0.19	7.59±0.11	8.63±0.01	6.82±0.02
Corchorus tridens	39.44±0.67	7.75±0.05	8.68±0.01	5.46±0.15
Amaranthus hybridus	34.79±0.33	7.77±0.07	8.29±0.03	6.62±0.04
Amaranthus spinosus	55.29±1.82	7.82±0.04	8.41±0.02	6.31±0.03
Bidens pilosa	39.83±1.25	7.46±0.15	8.65±0.01	6.75±0.03
Momordica involucrate	47.93±0.91	7.42±0.10	8.37±0.03	5.90±0.05
Dioscorea minutiflora	21.50±0.74	7.00±0.01	8.63±0.02	6.99±0.08[b]

Note. Values are mean± sd, n=2[*] and n=3[**]. [+]No further growth was observed at 72 hours. [#]Conversion factor for colony forming units $=1\times10^5$, or [##]1×10^3. [$]NT = Not tested (plant material was insufficient), Values with different superscript letters indicate significant statistical difference (p<0.01) in growth stimulation by the edible wild plants per probiotic strain.

Tragopogon porrifolius, Sonchus asper, Nasturtium officinale, Berkheya purpurea, Sisymbrium thelungii, Cyperus esculentus, Hypoxis hirsute, Sonchas integrifolius, Solanum retroflexum, Momordica foetida, Corchorus tridens, Bidens pilosa and *Discorea minutiflora.*

Urtica dioica had the highest prebiotic response on *L. rhamnosus* of 7-fold more than inulin (positive control). The total dietary fibre content of *Urtica dioica* was high (3.5%) making it a better prebiotic candidate. However, *Urtica dioica* had inhibitory effects on *L. acidophilus*, which indicates selective and varied responses by the bacterial strains to the edible wild plant extracts. Sreenivas and Lele (2013) studied the prebiotic activity of gourd family vegetable fibres using *Lactobacillus fermentum*. They reported that gourd family vegetables had significant prebiotic ability. In the present study, *Momordica foetida* also stimulated the growth of probiotics. Petkova and Denev (2013) reported that *Sonchus oleraceus* had high content of both prebiotic fructooligosaccharrides and inulin. This explains the high growth stimulatory effects of all *Sonchus* species that we observed in this study.

3.5.3 Effects of Edible Wild Plant Extracts on the Proliferation of *L. Acidophilus* (ATCC 314)

The *L. acidophilus* bacteria cells were very tiny, non-spore forming gram-positive rods. The colonies were tiny greyish-white in colour and non-glistering. Twenty-five edible wild plants showed prebiotic activity on *L. acidophilus* compared to the broth (negative control, Table 4). Twenty plants (*Rubus Cuneifolius, Sonchus dregeanus, Wahlengergia androsacea, Lepidium capense, Tragopogon porrifolius, Sonchus asper, Nemesia fruticans, Nasturtium officinale, Berkheya purpurea, Taraxacum officinale, Sisymbrium thelungii, Cyperus esculentus, Hypoxis hirsute, Solanum retroflexum, Momordica foetida, Amaranthus hybridus, Bidens pilosa,*

Discorea minutiflora and *Sonchas integrifolius*) possessed much higher prebiotic activity than both positive control (inulin) and the negative control.

Tragopogon porrifolius species had the highest prebiotic activity with log (cfu/mL)=7.05±0.01 after 48 hours while *Sonchus integrifolius* and *Corchorus tridens* with log (cfu/mL)=5.46±0.15 had the lowest prebiotic activity among the group. *Tragopogon porrifolius* and *Discorea minutiflora* stimulated the growth of *L. acidophilus* by 5-fold more than inulin (positive control). *Tragopogon porrifolius* had high total dietary fibre content while *Discorea minutiflora* had a mixture of total dietary fibre and high indigestible protein content. Five edible wild plants (i.e. *Rorippa nudiuscula*, *Rosa rubiginosa*, *Amarathus spinosus*, *Opuntia megacantha* and *Sisymbrium capense*) had low prebiotic activity than inulin (positive control). Notably, *Urtica dioica* inhibited the growth of *L. acidophilus* possibly by selective bactericidal mechanisms confirming a previous report by Modarresi-Chahardehi, Ibrahim, Fariza-Sulaiman and Mousavi (2012).

4. Limitations of the Study

The simulated methods used to prepare the plant extracts possibly result in loss of ascorbic acid and therefore the actual ascorbic acid content was not reported. The effect of final dish preparation methods especially boiling and cooking of edible wild plant material needs to be investigated. The DPPH method determined the total antioxidant activity related to many free radicals or reactive oxygen species. The antioxidant capacity reported herein is therefore a compound value, which cannot be attributed to one type of molecule although it was mostly correlated to ascorbic acid content. The individual composition of other antioxidants in these edible wild plants may be determined prior to *in vivo* studies.

5. Conclusion

Eighty seven percent (87%) (n=30) of the edible wild plants met the FDA recommended daily value for ascorbic acid. *Cyperus esculentus* and *Rosa rubiginosa* had the highest content of ascorbic acid and correspondingly highly potent antioxidant activities while the Sonchus family consistently showed high ascorbic acid content and potent antioxidant activity. The antioxidant capacity of edible wild plants partly depends on their ascorbic acid content. Antioxidant activity observed for plant extracts with low ascorbic acid content suggest the presence of other types of antioxidants in the edible wild plants. *Rubus Cuneifolius*, *Lepidium capense*, *Berkheya purpurea*, *Sisymbrium thelungii*, *Cyperus esculentus*, and *Solanum retroflexum* stimulated the growth of all the three probiotics strains. Other edible wild plants demonstrated prebiotic activity on at least two of the probiotics while those that stimulated the growth of *B. animalis* and *L. rhamnosus* did not necessarily show the same trends with *L. acidophilus*. Probiotics may have selective nutritional preferences and different sensitivities to non-prebiotic phytochemicals in certain edible wild plants possibly due to different metabolic disposition of both the probiotics and the prebiotics.

Acknowledgements

This study was funded by the National Research Foundation (NRF) through the NRF Thuthuka Grant (Rating track) UID: 87848, the Tshwane University of Technology (TUT) Emerging Researcher Grant and the South African Medical Research Council Self-Initiated Research Grant of Prof Tarirai. Opinions expressed and conclusions arrived at, are those of the authors and are not necessarily to be attributed to the sponsors.

Conflict of Interest Statement

The authors hereby declare that they have no conflicts of interest.

References

Acquaviva, R., Di Giacomo, C., Vanella, L., Santangelo, R., Sorrenti, V., … Iauk, L. (2013). Antioxidant activity of extracts of *Momordica Foetida* Schumach. et Thonn. *Molecules, 18*, 3241-3249. http://dx.doi.org/10.3390/molecules18033241

Alvarez-olmos, M. I., & Oberhelman, R. A. (2001). Probiotic agents and infectious diseases: A modern perspective on a traditional therapy. *Clinical Infectious Diseases, 32*, 1567-1576. http://dx.doi.org/10.1086/320518

AOAC Official Method 985.29. (1995). Total dietary fibre in foods. Gaithersburg, MD, USA: AOAC International. Retrieved from https://secure.megazyme.com/files/Booklet/K-TDFR_DATA.pdf

Baldwin, C., Millette, M., Oth, D., Ruiz, M. T., Luquet., FM., & Lacroix, M. (2010). Probiotic *Lactobacillus acidophilus* and *Lactobacillus casei* mix sensitize colorectal tumoral cells to 5-fluorouracil-induced apoptosis. *Nutrition and Cancer, 62*, 371-378. http://dx.doi.org/10.1080/01635580903407197

Boedecker, J., Termote, C., Assogbadjo, A. E., Van Damme, P., & Lachat, C. (2014). Dietary contribution of wild edible plants to women's diets in the buffer zone around the Lama forest, Benin – an underutilized potential. *Food Security, 6*, 833-849. http://dx.doi.org/10.1007/s12571-014-0396-7

Bwembya, G. C., Thwala, J. M., Silaula, S. M., & Otieno, D. A. (2014). *Nutritional analysis of some major micronutrients in common Swazi foods*. Abstract. Retrieved from http://www.uniswa.sz/sites/default/files/research/urc/docs/Nutritional%20analysis%20of%20some%20majo r%20micronutrients%20in%20common%20Swazi%20foods.pdf

Choi, Y., Jeong, H. S., & Lee, J. (2007). Antioxidant activity of methanolic extracts from some grains consumed in Korea. *Food Chemistry, 103*, 130-138. http://dx.doi.org/10.1016/j.foodchem.2006.08.004

Crittenden, R. G., & Playne, M. J. (1996). Production, properties and applications of food-grade oligosaccharides. *Trends in Food Science and Technology, 7*, 353-361. http://dx.doi.org/10.1016/S0924-2244(96)10038-8

Food and Agriculture Organization of the United Nations and World Health Organization [FAO/WHO] (2001) FAO food and nutrition paper 85. Probiotics in food health and nutritional properties and guidelines for evaluation: Report of a joint FAO/WHO expert consultation on evaluation of health and nutritional properties of probiotics in food including powder milk with live lactic acid bacteria. Cordoba, Argentina. Retrieved from http://www.fao.org/3/a-a0512e.pdf.

Food and Drug Administration (FDA). 2016. *Guidance for industry: A food labelling guide (14. Appendix F: Calculate the percent daily value for the appropriate nutrients)*. The Code of Federal Regulations (CFR) annual. Retrieved from http://www.ecfr.gov/cgi-bin/text-idx?SID=10896471be7fb6ff7aae0acf00081a82&mc=true&node=pt21.2.10 1&rgn=div5#se21.2.101_19 .

Franck, A. (2002). Technological functionality of inulin and oligofructose. *British Journal of Nutrition, 87*, S287-S291. http://dx.doi.org/10.1079/BJN/2002550

Gacche, R. N., Kabaliye, V. N., Dhole, N. A., & Jadhav, A. D. (2010). Antioxidant potential of selected vegetables commonly used in diet in Asian subcontinent. *Indian Journal of Natural Products and Resources, 1*(3), 306-313.

Ghosh, A. R. (2012). Appraisal of probiotics and prebiotics in gastrointestinal infections. *Webmed Central Gastroenterology, 3*, 1-27. http://dx.doi.org/10.9754/journal.wmc.2012.003796

Gibson, G. R., & Roberfroid, M. B. (1995). Dietary modulation of the human colonic microbiota: Introducing the concept of prebiotics. *Journal of Nutrition, 125*, 1401-1412.

Gibson, G. R., Probert, H. M., Loo, J. V., Rastall, R. A., & Roberfroid, M. B. (2004). Dietary modulation of the human colonic microbiota: Updating the concept of prebiotics. *Nutritional Research Reviews, 17*, 259-275. http://dx.doi.org/10.1079/NRR200479

Gotteland, M., Brunser, O., & Cruchet, S. (2006). Systematic review: Are probiotics useful in controlling gastric colonization by *Helicobacter pylori*? *Alimentary Pharmacology and Therapeutics, 23*, 1077-1086. http://dx.doi.org/doi:10.1111/j.1365-2036.2006.02868.x

Halliwell, B. (1996). Antioxidants in human health and disease. *Annual Reviews in Nutrition, 16*, 33-50.

Haraguchi, H. (2001). Antioxidative plant constituents. *In bioactive compounds from natural sources, first published*. New York: Taylor and Francis Inc.

Hart, T. G. B., & Vorster, H. J. (2006). *Indigenous knowledge on the South African landscape – Potentials for agricultural development. Urban, rural and economic development programme*. Occasional paper No 1. Cape Town: HSRC Press.

Herigstad, B., Hamilton, M., & Heersink, J. (2001). How to optimize the drop plate method for enumerating bacteria. *Journal of Microbiol Methods, 44*(2), 121-129. http://dx.doi.org/10.1016/S0167-7012(00)00241-4

Holzapfel, W. H., & Schillinger, U. (2002). Introduction to pre- and probiotics. *Food Research International, 35*, 109-116. http://dx.doi.org/10.1016/S0963-9969(01)00171-5

Holzapfel, W. H., Haberer, P., Geisen, R., Björkroth, J., & Schillinger, U. (2001). Taxonomy and important features of probiotic microorganisms in food and nutrition. *American Journal of Clinical Nutrition, 73*, 365S-373S.

Iwata, E., Hotta, H., & Goto, M. (2009). The screening method of a bifidogenic dietary fibre extracted from inedible parts of vegetables. *Journal of Nutrition Science and Vitaminology, 55*, 385-388. http://dx.doi.org/10.3177/jnsv.55.385

Jones, M. L., Martoni, C. J., Parent, M., & Prakash, S. (2011). Cholesterol-lowering efficacy of a microencapsulated bile salt hydrolase-active Lactobacillus reuteri NCIMB 30242 yoghurt formulation in hypercholesterolaemic adults. *British Journal of Nutrition, 9*, 1-9. http://dx.doi.org/10.1017/S0007114511004703

Jungersen, M., Wind, A., Johansen, E., Christensen, J. E., Stuer-Lauridsen, B., & Eskesen, D. (2014). The science behind the probiotic strain *Bifidobacterium animalis subsp. lactis* BB-12[®]. *Microorganisms, 2*, 92-110. http://dx.doi.org/10.3390/microorganisms2020092

Kassim, M. A., Baijnath, H., & Odhav, B. (2014). Effect of traditional leafy vegetables on the growth of lactobacilli and bifidobacteria. *International Journal of Food Science and Nutrition, 68*, 977-980. http://dx.doi.org/10.3109/09637486.2014.945155

Kaur, C., & Kapoor, H. C. (2001). Antioxidants in fruits and vegetables – the millennium's health. *International Journal of Food Science and Technology, 36*, 703-725.

Khalaf, N. A., Shakya, A. K., Al-Othman, A., El-Agbar, Z., & Farah, H. (2008). Antioxidant activity of some common plants. *Turkish Journal of Biology, 32*, 51-55.

Kullberg, M. C. (2008). Immunology: Soothing intestinal sugars. *Nature, 453*, 602-604. http://dx.doi.org/10.1038/453602a

Lamberti, C., Mangiapane, E., Pessione, A., Mazzoli, R., Giunta, C., Lipi, D. & Raychaudhuri, U. (2012). Role of nutraceuticals in human health. *Journal of Food Science and Technology*, 49:173-183. http://dx.doi.org/10.1007/s13197-011-0269-4

Lee, J., Koo, N., & Min, D. B. (2004). Reactive oxygen species, aging and antioxidative neutraceuticals. *Comprehensive Reviews in Food and Food Security, 3*, 21-33.

Legwaila, G. M., Mojeremane, W., Modisa, M. E., Mmolotsi, R. M., & Rampart, M. (2011). Potential of traditional food plants in rural household food security in Botswana. *Journal of Horticulture and Forestry, 3*(6), 171-177.

Lei, V., & Jakobsen, M. (2004). Microbiological characterisation and probiotic potential of koko and koko sour water; African spontaneously fermented millet porridge and drink. *Journal of Applied Microbiology, 96*, 384-397.

Malaguarnera, M., Vacante, M., Antic, T., Giordano, M., Chisari, G., Acquaviva, R., ... Galvano, F. (2012). *Bifidobacterium longum* with fructo-oligosaccharides in patients with non-alcoholic steatohepatitis. *Digestive Diseases and Sciences, 57*, 545-553. http://dx.doi.org/10.1007/s10620-011-1887-4

Matsuda, A., Tanaka, A., Pan, W., Okamoto, N., Oida, K., Kingyo, N., ... Matsuda, H. (2012). Supplementation of the fermented soy product ImmuBalance[TM] effectively reduces itching behaviour of atopic NC/Tnd mice. *Journal of Dermatological Science, 67*, 130-139. http://dx.doi.org/10.1016/j.jdermsci.2012.05.011

Maundu, P. M., Ngugi, G. W., & Kabuye, C. H. S. (1999). *Traditional Food Plants of Kenya*. Kenya Resource Centre for Indigenous Knowledge, National Museum of Kenya, Nairobi, 270-288.

Modarresi-Chahardehi, A., Ibrahim, D., Fariza-Sulaiman, S., & Mousavi, L. (2012). Screening antimicrobial activity of various extracts of *Urtica dioica*. *Revista de Biologia Tropica, 60*(4), 1567-1576. http://dx.doi.org/10.15517/rbt.v60i4.2074

Moshfegh, A. J., Friday, J. E., Goldman, J. P., & Chugahuja, J. K. (1999). Presence of inulin and oligofructose in the diets of Americans. *Journal of Nutrition, 129*, 1470S-1411S.

Munsch-Alatossava, P., Rita, H., & Alatossava, T. (2007). A faster and more economical alternative to the standard plate count (SPC) method for microbiological analyses of raw milk. In Mendez-Vilas, A. (ed.) *Communicating current research and educational topics and trends in applied microbiology (Microbiology Series: Volume 1)*. Badajoz: Formatex.

Naghili, H., Tajik, H., Mardani, K., Rouhani, S. M. R., Ehsani, A., & Zare, P. (2013) Validation of drop plate technique for bacterial enumeration by parametric and nonparametric tests. *Veterinary Research Forum, 4*, 179-183.

Nair, S., Nagar, R., & Gupta, R. (1998). Antioxidant phenolics and flavonoids in common Indian foods. *Journal of the Association of Physicians of India, 46*, 708-710.

Ozaki, A., Muromachi, A., Sumi, M., Sakai, Y., Morishita, K., & Okamoto, T. (2010). Emulsification of coenzyme Q10 using gum arabic increases bioavailability in rats and human and improves food-processing suitability. *Journal of Nutritional Science and Vitaminology Tokyo, 56*(1), 41-47.

Ozen, T. (2010). Antioxidant activity of wild edible plants in the Black Sea region of Turkey. *Grasas Yaceites, 61*(1), 86-94.

Pacifico, L., Osborn, J. F., Bonci, E., Romaggioli, S., Baldini, R., & Chiesa, C. (2014). Probiotics for the treatment of *Helicobacter pylori* infection in children. *World Journal of Gastroenterology, 20*(3), 673-683. http://dx.doi.org/10.3748/wjg.v20.i3.673

Parvez, S., Malik, K. A., Ah Kang, S., & Kim, H. Y. (2006) Probiotics and their fermented food products are beneficial for health. *Journal of Applied Microbiology, 100*, 1171-1185. http://dx.doi.org/10.1111/j.1365-2672.2006.02963.x

Paul, R., & Ghosh, U. (2012). Effect of thermal treatment on ascorbic acid content of pomegranate juice. *Indian Journal of Biotechnology, 11*, 309-313.

Petkova, N., & Denev, P. (2013). Evaluation of fructan content of the taproots of *Lactucaserriola L.* and *Sonchus oleraceus L. Science Bulletin, 17*, 117-122.

Pradhan, S., Manivannan, S., & Tamang, J. P. (2015). Proximate, mineral composition and antioxidant properties of some wild leafy vegetables. *Journal of Scientific and Industrial Research, 74*, 155-159.

Quigley, E. M. M. (2010). Prebiotics and probiotics: Modifying and mining the microbiota. *Pharmacology Research, 61*, 213-218. http://dx.doi.org/10.1016/j.phrs.2010.01.004

Roberfroid, M. B. (2007). Prebiotics: The concept revisited. *Journal of Nutrition*, 137:830S-837S.

Roberfroid, M. B., Gibson, G. R., Hoyles, L., Mccartney, A. L., Rastall, .R, Rowland, I., ... Meheust, A. (2010). Prebiotic effects: Metabolic and health benefits. *British Journal of Nutrition, 104*, S1-S63. http://dx.doi.org/10.1017/S0007114510003363

Ropciuc, S., Cenusa, R., Caprita, R., & Cretescu, I. (2011). Study on the ascorbic acid content of rose hip fruit depending on stationary conditions. *Animal Science and Biotechnologies, 44*(2), 129-132.

Sreenivas, K. M., & Lele, S. S. (2013). Prebiotic activity of gourd family vegetable fibres using *in vitro* fermentation. *Food Bioscience, 1*, 26-30. http://dx.doi.org/10.1016/j.fbio.2013.01.002

Sun, C., Wang, J., Fang, L., Gao, X., & Tan, R. (2004). Free radical scavenging and antioxidant activities of EPS2, an exopolysaccharide produced by a marine filamentous fungus Keissleriella sp. YS 4108. *Life Sciences, 75*, 1063-1073.

Tarirai, C., Viljoen, A. M., & Hamman, J. H. (2012). Effects of dietary fruits, vegetables and a herbal tea on the *in vitro* transport of cimetidine: Comparing the Caco-2 cell model with porcine jejunum tissue. *Pharmaceutical Biology, 50*(2), 254-263.

Thomson, A. B. R., Chopra, A., Clandinin, M. T., & Freeman, H. (2012). Recent advances in small bowel diseases: Part I. *World Journal of Gastroenterology, 18*, 3336-3352. http://dx.doi.org/10.3748/wjg.v18.i26.3336

Tien, M. T., Girardin, S. E., Regnault, B., Le Bourhis, L., Dillies, M. A,, Coppée, J. Y., ... Pédron, T. (2006). Anti-inflammatory effect of *Lactobacillus casei* on Shigella-infected human intestinal epithelial cells. *Journal of Immunology, 176*, 1228-1237. http://dx.doi.org/10.4049/jimmunol.176.2.1228

Tung, Y., Wu, J., Kuo, Y., & Chang, S. (2007). Antioxidant activities of natural phenolic compounds from *Acacia confusa* bark. *Bioresource Technology, 98*, 1120-1123.

Uprety, Y., Poudel, R. C., Shrestha, K. K., Rajbhandary, S., Tiwari, N. N., Shrestha, U. B., & Asselin, H. (2012). Diversity of use and local knowledge of wild edible plant resources in Nepal. *Journal of Ethnobiology and Ethnomedicine, 8*, 1-16.

Van Loo, J., Coussement, P., De Leenheer, L., Hoebregs, H., & Smits, G. (1995). On the presence of inulin and oligofructose as natural ingredients in the Western diet. *Crititical Reviews in Food Science and Nutrition, 35*, 525-552. http://dx.doi.org/10.1080/10408399509527714

Venkatachalam, K., Rangasamy, R., & Krishman, V. (2014). Total antioxidant activity and radical scavenging capacity of selected fruits and vegetables from South India. *International Food Research Journal, 21*(3),

1039-1043.

Wang, H., Qi, L., Wang, C., & Li, P. (2011). Bioactivity enhancement of herbal supplements by intestinal microbiota focusing on the ginsenosides. *American Journal of Chinese Medicine, 39*, 1103-1115. http://dx.doi.org/10.1142/S0192415X11009433

Weese, J. S., & Anderson, E. C. (2002). Preliminary evaluation of *Lactobacillus rhamnosus* strain GG, a potential probiotic in dogs. *Canadian Veterinary Journal, 43*, 771-774.

Wichienchot, S., Thammarutwaski, P., Jongjareonrak, A., Chanuwan, W., Hmadhlu, P., Hongpattarakere, T., … Ooraiku, B. (2011). Extraction and analysis of prebiotics from selected plants from Southern Thailand. *Songklanakarin Journal of Science and Technology, 33*, 517-523.

13

Postharvest Practices of Maize Farmers in Kaiti District, Kenya and the Impact of Hermetic Storage on Populations of *Aspergillus* Spp. and Aflatoxin Contamination

Angeline W Maina[1], John M Wagacha[1], Francis B Mwaura[1], James W Muthomi[2] & Charles P Woloshuk[3]

[1]School of Biological Sciences, University of Nairobi, Kenya

[2]Department of Plant Science and Crop Protection, University of Nairobi, Kenya

[3]Department of Botany and Plant Pathology, Purdue University, United States

Correspondence: Angeline W Maina, School of Biological Sciences, University of Nairobi, Kenya. E-mail: angewanjiku@gmail.com

Abstract

Aflatoxin contamination in maize by *Aspergillus* spp. is a major problem causing food, income and health concerns. A study was carried out in Kaiti District in Lower Eastern Kenya to evaluate the effect of three months storage of maize in triple-layer hermetic (PICS™) bags on the population of *Aspergillus* spp. and levels of aflatoxin. Postharvest practices by maize farmers including time of harvesting, drying and storage methods were obtained with a questionnaire. *Aspergillus* spp. in soil and maize were isolated by serial dilution-plating and aflatoxin content was measured using Vicam method. Maize was mostly stored in woven polypropylene (PP) and sisal bags within granaries and living houses. *Aspergillus flavus* L-strain was the most predominant isolate from soil (Mean = 8.4 x10^2 CFU/g), on the harvested grain (4.1 x 10^2 CFU/g) and grain sampled after three months of storage (1.1 x 10^3 CFU/g). The type of storage bag significantly ($P \leq 0.05$) influenced the population of members of *Aspergillus* section *Flavi*, with *A. flavus* (S and L strains) and *A. parasiticus* being 71% higher in PP bags than in PICS bags. Total aflatoxin in maize sampled at harvest and after three months storage ranged from <5 to 42.7 ppb with 55% lower aflatoxin content in PICS bags than in PP bags. After storage, the population of *Aspergillus* section *Flavi* was positively correlated with aflatoxin levels. The results of this study demonstrate that PICS bags are an effective management option for reducing population of toxigenic *Aspergillus* spp. and aflatoxin in stored maize.

Key words: aflatoxin, *Aspergillus* spp., hermetic bag, maize, polypropylene bag

1. Introduction

Maize (*Zea mays* L.) is a staple food crop in Kenya accounting for about 40% of daily calories with per capita consumption of 98 kilograms per annum (Muiru, Charles, Kimenju, Njoroge, & Miano, 2015). With about 90% of the rural households in Kenya depending on maize, this grain dominates all national food security considerations (Ouma & De-Groote, 2011). An important challenge for maize production in Kenya is contamination with aflatoxin, which poses negative health effect to humans and animals and causes huge economic losses (Okoth et al., 2012). Aflatoxins are toxic secondary metabolites produced by several *Aspergillus* spp. Aflatoxins have been associated with stunted growth in children, immune-system suppression, cancer and even death in humans (Strosnider et al., 2006). In animals, aflatoxin contaminated feeds have been associated with aflatoxicosis, impaired growth, immunosuppression, liver and kidney tumors in rodents and reduced quality of milk and milk products because of the presence of aflatoxin M1, a derivative of aflatoxin B1 (Lizárraga-Paulín, Moreno-Martínez, & Miranda-Castro, 2011).

Numerous recurrent cases of aflatoxicosis as a result of consumption of maize contaminated with aflatoxin have been reported in Kenya since 1981 (Wagacha & Muthomi, 2008). The worst outbreak occurred in 2004 where 317 patient cases were recorded with 125 deaths mainly in lower Eastern regions of Kenya (Center for Disease Control and Prevention [CDC], 2004). The 2004 outbreak was attributed to inappropriate harvest time, early rains and poor post-harvest storage of maize under moist conditions (Muthomi, Njenga, Gathumbi, & Chemining'wa, 2009; Lewis et al., 2005). In 2009, 31,000 and 1,213 bags of maize contaminated with aflatoxins were condemned in Mbeere, Embu County and Bura Irrigation Scheme in Tana River County, respectively (Nyaga, 2010). Due to deleterious health effects associated with aflatoxin consumption, the Kenya Bureau of Standards (KEBS) set the limit for total aflatoxin level in maize at 10 μg/kg (Kenya Bureau of Standards [KEBS], 2007). Despite the existence of such regulations, contamination of food by aflatoxins is still a major challenge in Kenya.

Despite its importance in the country's economy, maize is liable to infections by mycotoxin producing fungi at pre-harvest, harvest and post-harvest stages (Atanda et al., 2011). Colonization of maize by *Aspergillus* spp. is a major challenge in maize production due to aflatoxin contamination (Wagacha & Muthomi, 2008). *Aspergillus* species are ubiquitous in the environment such as soil, air and debris and are widely distributed in tropical and sub-tropical environments (Klich, 2002). They grow as saprophytes in the soil which serves as the main reservoir of their propagules and as a source of primary inocula (Scheidegger & Payne, 2003). Factors such as high temperatures and moisture, drought stress and delayed harvesting predispose maize to infection by *Aspergillus* spp. and consequently aflatoxin contamination (Atanda et al., 2011). Aflatoxins, the most prevalent mycotoxins in Kenya are mainly produced by *A. flavus* and *A. parasiticus* (Okoth et al., 2012) and other less common *Aspergillus* species such as *A. nomius*, *A. bombycis* and *A. parvisclerotigenus* (Reiter, Zentek, & Razzazi, 2009). There are two major morphotypes of *Aspergillus flavus*; S and L strains (Cotty, 1994). *Aspergillus flavus* S-strain produces small numerous dark sclerotia and is considered the most toxigenic since it produces high levels of B type aflatoxins (Cotty & Cardwell, 1999). *Aspergillus flavus* L-strain produce yellow to bright green colonies with fewer sclerotia and less B-aflatoxins (Probst, Bandyopadhyay, Price, & Cotty, 2011). *Aspergillus parasiticus* which is distinguished by dark green colonies and rough conidia (Atehnkeng et al., 2008) produces aflatoxin B1 (AFB1), B2 (AFB2), G1 (AFG1) and G2 (AFG2) (Reiter et al., 2009).

In maize production, post-harvest practices, such as proper drying and storage, are key areas along the maize value chain to maintain grain quality, quantity and safety (Akowuah, Mensah, Chan, & Roskilly, 2015). However, poor storage practices often adopted by smallholder farmers play a major role in fungal growth and aflatoxin contamination (Wagacha et al., 2013). Most smallholder farmers use different packaging/storage materials that include polypropylene bags and giant woven baskets for packing and storing their maize grains (Hell, Cardwell, Setamou, & Poehling, 2000; Shabani, Kimanya, Gichuhi, Bonsi, & Bovell-Benjamin, 2015). These packaging/storage materials provide optimal conditions for fungal growth and aflatoxin contamination (Hell et al., 2000). Moreover, insect pest disseminate spores of *Aspergillus* spp. in storage increasing the chances of aflatoxin contamination (Hell at al., 2008). However, application of chemical insecticides to protect grains against storage pests and other pathogens has yielded minimal results. Moreover, lack of suitable and affordable grain storage technologies often compel majority of the smallholder farmers to sell their produce immediately after harvest (Gitonga, De Groote, & Tefera, 2015).

A modern method that entails the use of Purdue Improved Crop Storage (PICS™) triple-layer hermetic bags for grain storage has been developed and is gaining favor among many farmers given its advantages when compared with the traditional storage methods (Hell, Mutegi, & Fandohan, 2010). This technology relies on creation of bio-generated atmospheres that hinder survival of microorganisms including fungi (Anankware, Fatunbi, Afreh-Nuamah, Obeng-Ofori, & Ansah, 2012). As a result of fungal, insect and grain respiration within the hermetic bags, oxygen levels drop significantly while carbon dioxide levels increase (Baoua, Amadou, Baributsa, & Murdock, 2014). This creates an unfavorable atmosphere for survival of these organisms within the enclosed system. Triple-bagging technology is therefore sustainable, cost effective and effectively maintains high quality maize grains for longer period of time (Anankware et al., 2012). A frequently asked question by maize producers is what impact the hermetic bags have on mycotoxigenic fungi. The objective of this study was therefore to evaluate the effect of PICS bags on the population of *Aspergillus* spp. and levels of aflatoxin in maize grains stored for three months. The population and incidence of *Aspergillus* spp. and total aflatoxin levels of maize stored in PICS and woven polypropylene (PP) bags were compared after three months storage under farmers' storage conditions.

2. Materials and Methods

2.1 Description of the Study Area

The study was conducted in Mukuyuni and Kilala Locations of Kaiti District, Makueni County of Lower Eastern Kenya. Kaiti District lies between latitude 1° 45' 00" S and longitude 37° 42' 00" E. The area is semi-arid to arid with a temperature range between 18°C to 24°C in the cold seasons and 24°C to 33°C in the hot days (Table 1). The rainfall pattern is bi-monsoonal with the long but unreliable rains in March to May and the more reliable short rains in October to December (Makueni County Integrated Development Plan [MCIDP], 2013). The area receives an annual rainfall of between 800-1200 mm and has an elevation of 600m to 1900m above the sea level (MCIDP, 2013). Residents of Kaiti District rely on subsistence and mixed farming as their major source of livelihood. Maize is the primary dietary staple and the main crop produced. The selection of the study area was based on previous reports of re-current aflatoxicosis outbreaks (Lewis et al., 2005).

Table 1. Monthly temperature (°C), precipitation (mm) and relative humidity (%) data of Kaiti District for the year 2015

Month	Minimum temperature (°C)	Maximum temperature (°C)	Precipitation (mm)	Minimum RH (%)	Maximum RH (%)
January	17.6	30.4	0.0	36.7	85.3
February	18.3	31.6	0.2	35.5	87.8
March	18.5	31.2	0.4	35.6	87.8
April	19.5	29.3	9.3	51.0	95.1
May	18.7	27.5	0.7	56.6	94.9
June	17.5	26.5	0.0	55.3	94.4
July	17.5	26.4	0.0	53.0	95.0
August	17.6	25.9	0.1	50.6	93.4
September	17.9	28.4	0.0	42.2	91.1
October	20.0	29.4	0.5	43.5	92.0
November	20.1	28.5	6.2	58.4	96.3
December	19.8	29.1	5.3	54.6	94.8

Source: (awhere, Inc, 2015); RH – Relative humidity

2.2 Field Survey and Sampling

Field survey and sampling were conducted between October 2015 and January 2016. A field survey involving 30 maize farms selected randomly was carried out in October 2015; 15 farms in Mukuyuni and 15 farms in Kilala administrative Locations of Kaiti District, Makueni County. A semi-structured questionnaire was administered to the farmers to collect data on maize production, handling and storage. Questions were designed to determine the time of harvesting, length of maize drying period, storage practices, farmers' knowledge on aflatoxin, type of storage structures, problems associated with storage and duration of grain storage before consumption, selling or planting. Soil samples were collected at planting time from 15 maize fields in each of the study locations. In each farm, a minimum of five sampling points at least 5 m apart were identified randomly. Approximately 100 g of soil was collected from the top 5 cm horizon at each sampling point, thoroughly mixed to make a composite sample from which a 500 g sample was drawn. Soil was put in a zip lock, plastic bag, transported to the laboratory within 72 h of sampling. The soil was air dried on laboratory benches for five days and stored in Kraft bags at room temperature (23 ± 2°C) until mycological analysis.

Dry maize grain samples were collected after harvest from 30 farmers; 15 of whom were randomly selected from each of the two Locations. The sampled maize grains were harvested from the same fields where soil had been sampled at the time of planting. From each household, shelled grains were randomly taken from different parts of the storage bag or container. The incremental sample was thoroughly mixed to form a composite sample from which 1 kg was drawn. A further 30 maize grain samples of 6 kg each were obtained from the same farmers after harvest and divided into two equal portions for storage. The samples were separately stored in woven polypropylene bags (PP) and PICS bags for three months in farmers' storage structures. Sampling from PICS bags and PP bags entailed thoroughly mixing the 3 kg grain sample and drawing a 1 kg sub-sample. The collected maize grain samples were placed in Kraft bags and transported to the laboratory within 72 h of sampling and stored at 4°C until mycological analysis.

2.3 Isolation and Enumeration of Aspergillus Spp. from Soil and Maize Grains

Isolation and enumeration of *Aspergillus* spp. from soil and ground maize grains was carried out aseptically using serial dilution and spread plate technique on potato dextrose agar (PDA) medium amended with antibiotics: 50 mg penicillin/L, 50 mg tetracycline/L and 50 mg streptomycin/L (Muthomi, 2001). One kilogram of maize grain sample was mixed thoroughly and ground using a dry mill kitchen blender (BL335, Kenwood, UK). The sample was divided into two equal sub-samples for microbial and aflatoxin analysis. *Aspergillus* spp. were isolated and enumerated from soil and ground maize grain samples by suspending 1g of sample in 9 mL of sterile distilled water, which was thoroughly shaken and serially diluted up to 10^{-2} of the original concentration.

A 100 μL aliquot of each suspension was plated onto PDA medium amended with antibiotics and incubated for 5 days at 25°C. The isolation and enumeration of *Aspergillus* spp. was carried out in triplicates for each soil and maize grain sample. Fungal colonies from soil and maize were identified and classified and colony counts of *Aspergillus* species expressed in colony forming units per gram of soil (CFU/g) as follows:

$$CFU/g \ sample = \frac{Number \ of \ colonies \ of \ a \ fungal \ species}{Amount \ plated \times Dilution \ factor} \qquad (1)$$

The incidence of each fungal species was calculated as follows:

$$Incidence \ (\%) = \frac{Number \ of \ isolates \ of \ a \ fungal \ species}{Total \ number \ of \ fungal \ species} X \ 100 \qquad (2)$$

2.4 Identification of Aspergillus Species

Single colonies of *Aspergillus* spp. identified in PDA medium amended with antibiotics were sub-cultured onto 5/2 media (5% V8 juice and 2% agar, pH 5.2) (Atehnkeng et al., 2008). The cultures were incubated at 31°C for 5 days. *Aspergillus* spp. were identified based on colony colour, shape, pigmentation, texture and pattern of growth (Klich, 2002). Isolates that produced numerous small dark sclerotia on 5/2 media were identified as *A. flavus* S-strain, while those with yellow to bright green colony without sclerotia were identified as *A. flavus* L-strain. Isolates that formed dark green colonies on 5/2 media and produced rough conidia were considered *A. parasiticus* (Atehnkeng et al., 2008). Colonies that were black on the top surface, while the underside remained pale, were identified as *A. niger*. Microscopic examination of *Aspergillus* spp. was done with modified Riddell slides (Riddell, 1950). All prepared slides were examined using a light microscope (x1000 magnification) (LEICA DM500, Leica Microsystems, Switzerland) and images recorded using a camera (LEICA ICC 50, Leica Microsystems, Wetzler, Germany) mounted on the microscope. Microscopic characteristics used in identification of *Aspergillus* spp. were conidial heads, seriation, conidia size, shape and roughness as described by Klich (2002).

2.5 Determination of Aflatoxin Levels in Maize Grains

Detection of aflatoxin levels in maize grains was performed using VICAM (Milford, MA, USA) protocol (Vicam, 2013; Herrman, Lee, Jones, & McCormick, 2014). Five grams of each ground maize sample was placed in an extraction tube and 30 mL of Agua premix added. The mixture was vortexed for 5 min and filtered through a 24 cm fluted filter paper (VICAM, Watertown, USA). A hundred microlitre of the Afla-V diluent was transferred to a strip test vial and 100 μL of the sample extract added and vortexed for two minutes. A hundred microlitre of the mixture was transferred to the Afla-V strip test at a flow rate of one drop per second vertically into the circular opening (Vicam, 2013). The strip tests were allowed to develop for five minutes on a flat surface. Afla-V strip tests were inserted into the Vertue reader (VICAM, Watertown, USA) for quantification of total aflatoxin in parts per billion (Vicam, 2013). The limits of detection were between 5ppb and 100ppb.

2.6 Data Analysis

Data on the population and incidence of *Aspergillus* spp. in soil and maize grains were analyzed with the Analysis of Variance (ANOVA) PROC ANOVA procedure of GENSTAT version 15. Frequency data that was not normally distributed were transformed to arcsine before analysis whereas the colony forming units data that was not normally distributed were transformed as $log_{10}(x+1)$. Least significant difference (LSD) was used to assess the significance of differences between treatment means at 95% confidence level.

3. Results

3.1 Maize Production Practices in Kaiti District

A majority (96.7%) of farmers determined the maize-harvesting stage by visual observation techniques (Figure 1A). The crop was considered ready for harvesting when leaves started drying, changed color from green to yellow, drooping of the cobs and by pricking kernels. Only 6.7% of the farmers in Mukuyuni used number of days that the crop was in the field to determine the appropriate maize harvesting stage. Farmers harvested their maize by hand after which they removed husks from maize cobs before drying. All farmers dried their maize on cobs in the sun immediately after harvest for a duration ranging from one to four weeks. Most (36.7%) of farmers dried their maize for two weeks while 23.3% dried it for four weeks. The rest (20%) of the farmers dried their maize for one and three weeks, respectively (Figure 1B). A majority (56.7%) of farmers stored their maize in a granary while 43.3% of farmers stored their produce inside the family living house (Figure 1C). The granaries were mainly raised wooden structures with iron sheet roofing (improved granaries). Most (83.3%) of the farmers packed their shelled grain in woven polypropylene bags while only 16.7% used sisal bags to store their maize (Figure 1D).

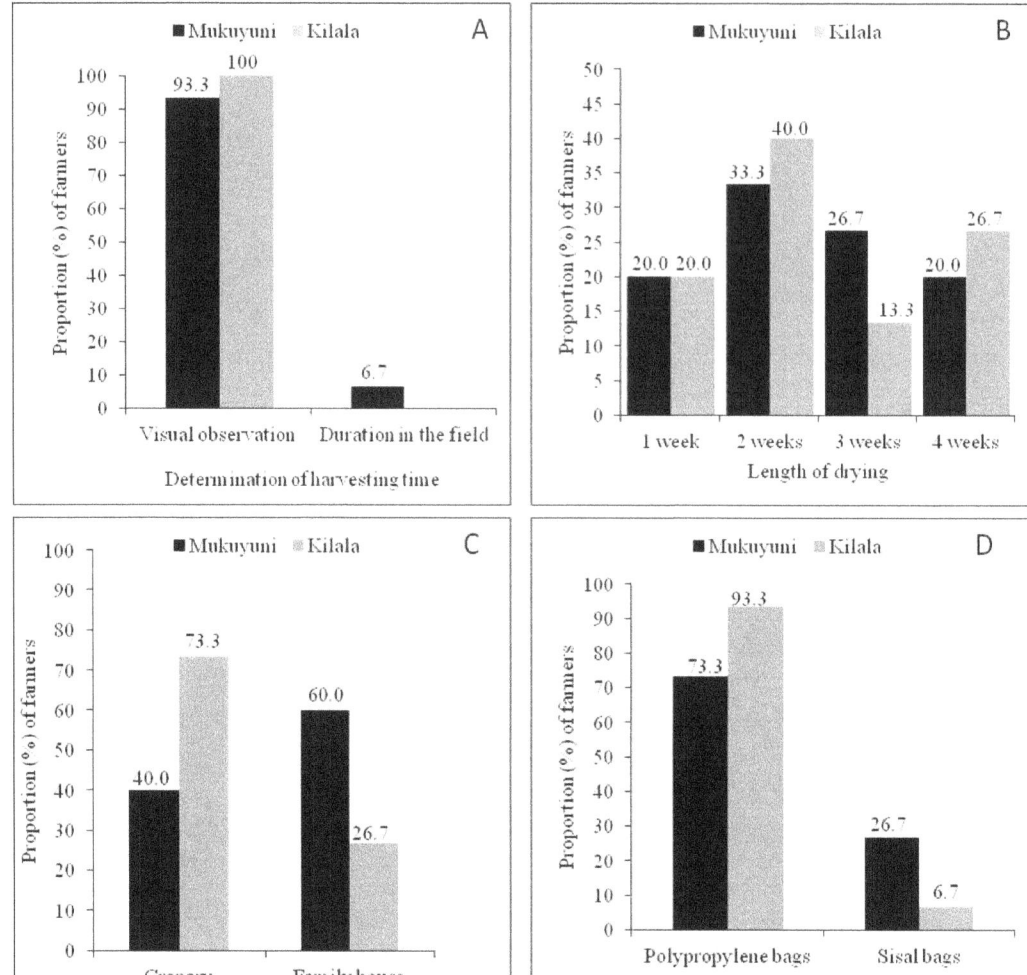

Figure 1. Methods used by farmers to determine when maize is ready for harvesting (A), length of drying period (B), type of storage structures (C) and storage methods (D) in Kaiti District.

The common storage challenges encountered by maize farmers in decreasing importance were insect damage, mold damage, rodent damage and lack of storage bags (Figure 2A). All farmers shelled their maize by hand and treated the grains with chemical insecticides before storage. Most farmers treated their maize to control insects, mainly weevils, with commercial insecticides, however, a few used traditional storage protectants such as ash (Figure 2B). Some farmers used traps to control rodents while all farmers cleaned their storage structures before storage of a new crop. About 87% of farmers in Mukuyuni Location considered mycotoxins a major concern compared to 80% of farmers in Kilala Location (Figure 2C). The majority (70%) of farmers obtained information on good farming practices and aflatoxin contamination from agricultural extension workers while a few relied on their own knowledge (16.7%) (Figure 2D).

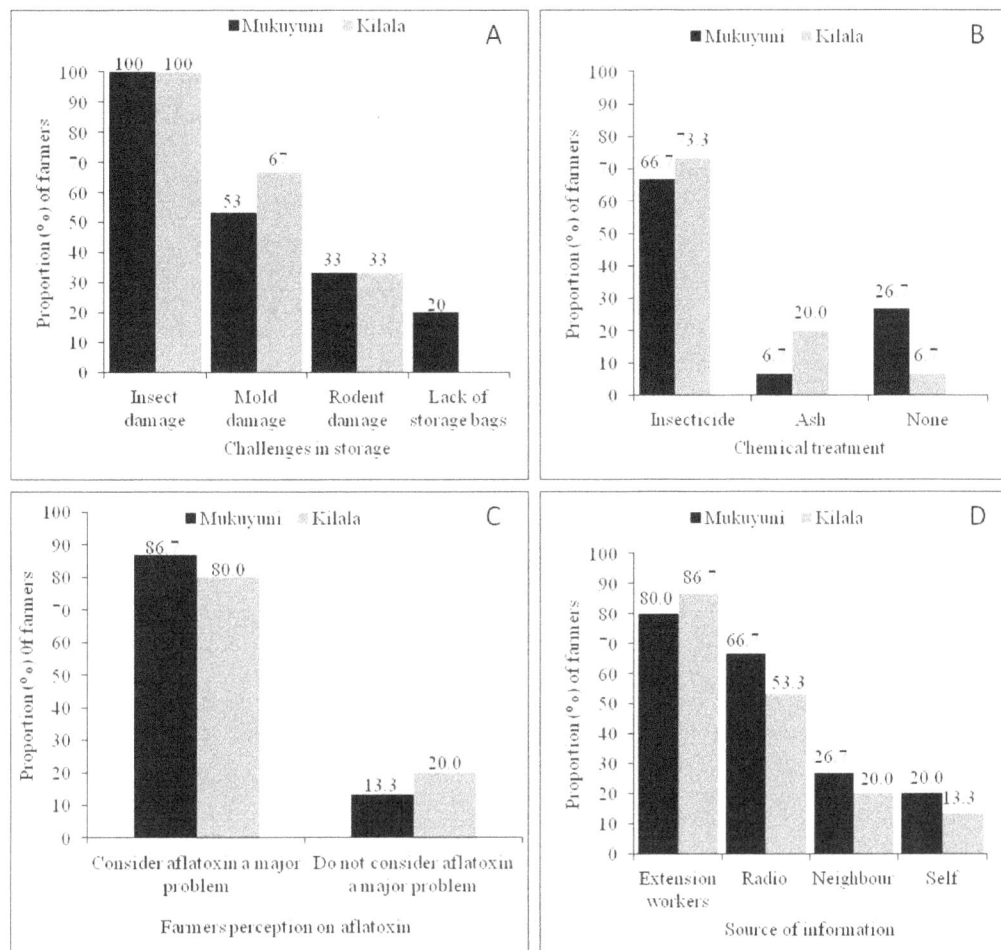

Figure 2. Challenges encountered by farmers during post-harvest storage of maize (A), grain treatment during storage (B), perceptions on aflatoxins in maize (C) and sources of information (D) in the Kaiti District.

3.2 Population and Incidence of Aspergillus Spp. In Soil

Aspergillus species isolated from the soil were: *A. flavus* L-strain, *A. flavus* S-strain, *A. parasiticus* and *A. niger* (Table 2). Among the members of *Aspergillus* section *Flavi*, *A. flavus* L-strain was the most predominant (Range = 0 - 6.0 x 10^3 CFU/g soil), followed by *A. parasiticus* (Range = 0 - 4.0 x 10^3 CFU/g soil) and *A. flavus* S-strain (Range = 0 - 3.0 x10^3 CFU/g soil). Population of *Aspergillus* spp. in soil varied significantly (p ≤ 0.05) between Kilala and Mukuyuni Locations (Table 2). Incidence of *A. flavus* L-strain was higher in soil samples from Mukuyuni while the incidence of *A. parasiticus* was significantly higher (p ≤ 0.05) in soil samples from Kilala Location. There was no significant variation (p ≥ 0.05) in the incidence of *A. flavus* S-strain in soil sampled from the two Locations.

Table 2. Population and incidence of *Aspergillus* spp. in soil sampled from maize fields in Kaiti District

Aspergillus spp.	Mukuyuni		Kilala	
	Population (CFU/g)	Incidence (%)	Population (CFU/g)	Incidence (%)
A. flavus S-strain	5.3 x 10^2 ± 137.1[bc]	10.1 ± 3.1[b]	4.4 x 10^2 ± 121.3[b]	6.6 ± 1.9[b]
A. flavus L-strain	1.0 x 10^3 ± 165.1[ab]	25.6 ± 4.8[a]	6.9 x 10^2 ± 141.5[b]	16.0 ± 3.9[b]
A. parasiticus	4.9 x 10^2 ± 147.8[c]	7.1 ± 2.1[b]	4.2 x 10^2 ± 121.0[b]	6.2 ± 2.0[b]
A. niger	1.4 x 10^3 ± 273.3[a]	35.0 ± 5.7[a]	3.2 x 10^3 ± 430.6[a]	60.1 ± 5.7[a]
Mean	855.6	19.4	1188.9	22.2
LSD (p ≤ 0.05)	509.3	11.6	679.1	10.4
CV (%)	18.5	7.7	5.5	8.7

Means followed by the same letter(s) within columns are not significantly different (Fisher's protected LSD at p ≤ 0.05). LSD - Least significant difference; CV - Coefficient of variation.

3.3 Efficacy of PICS Storage Bags on the Population of Aspergillus Spp. in Maize

Four *Aspergillus* spp. were commonly isolated from maize grains sampled at harvest and after three months of storage in PICS and PP bags (Table 3). The population of *Aspergillus* spp. in maize sampled at harvest and three months after storage was significantly different ($p \leq 0.05$) and in decreasing order: *A. flavus* L-strain (Mean = 8.7 x 10^2 CFU/g), *A. flavus* S-strain (Mean = 6.4 x10^2 CFU/g), *A. parasiticus* (Mean = 3.4 x10^2 CFU/g) and *A. niger* (Mean = 3.3 x10^2 CFU/g). The population of the aforementioned *Aspergillus* spp. was significantly ($p \leq 0.05$) higher in maize sampled after storage compared to samples collected at harvest. The population of *Aspergillus* spp. in maize increased up to three fold from 24.9% at harvest to 74.7% after storage. The type of storage bag significantly ($p \leq 0.05$) influenced the population of members of *Aspergillus* section *Flavi* - *A. flavus* (S and L strains) and *A. parasiticus* – which was 71% higher in maize stored in PP bags than in PICS bags. Overall, the population of *A. flavus* L-strain was 35% higher than that of *A. flavus* S-strain.

Table 3. Population (CFU/g) of *Aspergillus* spp. in maize grains sampled at harvest and three months after storage in PP and PICS bags in Kaiti District

Location	Bag type	AFL	AFS	AP	AN
Mukuyuni	At harvest [a]	5.1 x $10^2 \pm 167.0^b$	3.3 x $10^2 \pm 100.5^a$	4.4 x $10^1 \pm 31.0^b$	8.9 x $10^1 \pm 42.9^b$
	PICS bag	1.1 x $10^3 \pm 270.0^{ab}$	8.2 x $10^2 \pm 242.6^a$	2.9 x $10^2 \pm 87.8^b$	2.2 x $10^2 \pm 89.3^b$
	PP bag	1.4 x $10^3 \pm 435.2^a$	1.2 x $10^3 \pm 553.8^a$	8.0 x $10^2 \pm 163.9^a$	4.7 x $10^2 \pm 170.0^a$
	Mean	1022.2	785.2	377.7	251.9
	LSD ($P \leq 0.05$)	871.1	996.2	300.7	319.9
	CV%	29.9	19.9	46.7	10.2
Kilala	At harvest [a]	3.1 x $10^2 \pm 109.3^b$	1.8 x $10^2 \pm 65.82^b$	1.6 x $10^2 \pm 63.2^b$	2.4 x $10^2 \pm 78.9^b$
	PICS bag	8.0 x $10^2 \pm 184.2^{ab}$	3.8 x $10^2 \pm 106.8^b$	1.6 x $10^2 \pm 83.8^b$	3.6 x $10^2 \pm 96.2^{ab}$
	PP bag	1.0 x $10^3 \pm 277.0^a$	9.6 x $10^2 \pm 203.4^a$	6.2 x $10^2 \pm 172.0^a$	6.0 x $10^2 \pm 147.0^a$
	Mean	718.5	503.7	311.1	400.0
	LSD ($P \leq 0.05$)	568.8	386.6	326.6	312.3
	CV%	18.9	24.3	28.6	19.2

Means followed by the same letter(s) within columns in each location are not significantly different (Fisher's protected LSD at $p \leq 0.05$). LSD - Least significant difference; CV - Coefficient of variation, [a] – Maize grains sampled at harvest. AFL - *A. flavus* L-strain, AFS - *A. flavus* S-strain, AP - *A. parasiticus*, AN - *A. niger*, PICS - Purdue improved crop storage bags, PP - woven polypropylene bags, CFU - colony forming units.

3.4 Efficacy of PICS Storage Bags on the Incidence of Aspergillus Spp. in Maize

Table 4. Incidence (%) of *Aspergillus* spp. in maize grains sampled at harvest and three months after storage in PP and PICS bags in Kaiti District

Location	Bag type	AFL	AFS	AP	AN
Mukuyuni	At harvest [a]	20.0 ± 5.5^a	15.1 ± 4.6^a	2.2 ± 1.6^b	7.1 ± 3.8^a
	PICS bag	34.2 ± 6.4^a	20.5 ± 5.3^a	10.3 ± 3.6^b	10.5 ± 4.4^a
	PP bag	27.2 ± 5.7^a	13.5 ± 4.8^a	32.6 ± 6.2^a	7.8 ± 3.2^a
	Mean	27.1	16.4	15.1	8.5
	LSD ($P \leq 0.05$)	16.6	13.8	12.0	10.7
	CV (%)	3.3	28.8	10.0	12.8
Kilala	At harvest [a]	13.3 ± 4.3^a	6.7 ± 2.4^b	8.3 ± 3.6^{ab}	13.9 ± 4.6^a
	PICS bag	26.1 ± 5.5^a	13.5 ± 4.2^{ab}	4.9 ± 2.7^b	24.2 ± 6.3^a
	PP bag	21.4 ± 4.9^a	24.6 ± 5.2^a	15.6 ± 4.4^a	16.2 ± 4.5^a
	Mean	20.3	15.1	9.6	18.2
	LSD ($P \leq 0.05$)	13.7	11.5	10.2	14.5
	CV (%)	15.9	23.6	17.7	32.0

Means followed by the same letter(s) within columns in each location are not significantly different (Fisher's protected LSD at $p \leq 0.05$). LSD - Least significant difference; CV - Coefficient of variation, [a] – Maize grains sampled at harvest. AFL - *A. flavus* L-strain, AFS - *A. flavus* S-strain, AP - *A. parasiticus*, AN - *A. niger*, PICS- Purdue improved crop storage bags, PP- woven polypropylene bags.

The incidence of *Aspergillus* section *Flavi* isolated from maize grain sampled at harvest and after three months of storage in decreasing order was: *A. flavus* L-strain (Mean incidence = 23.7%), *A. flavus* S-strain (15.6%) and *A. parasiticus* (12.2%) (Table 4). *Aspergillus flavus* L-strain was the most prevalent in harvested (Mean incidence = 16.7%) and stored maize grains (Mean incidence = 27.2%). The incidence of the aforementioned *Aspergillus* spp. was significantly ($p \leq 0.05$) different and 84.8% higher in maize obtained after three months of storage compared to samples collected at harvest. The type of storage bag had a significant influence ($p \leq 0.05$) on the incidence of members of *Aspergillus* section *Flavi* - *A. flavus* (S and L strains) and *A. parasiticus*. The incidence of *A. flavus* S-strain was 12% higher in maize stored in PP bags than in PICS bags while there was no significant variation ($p \geq 0.05$) in the incidence of *A. flavus* L-strain stored in the two bag types.

3.5 Efficacy of PICS Bags on Aflatoxin Levels in Maize

The levels of total aflatoxin in maize grains sampled at harvest and after three months of storage in PICS and PP bags ranged from < 5 ppb to 42.7 ppb (Table 5). The percentage of maize grains sampled at harvest that met different thresholds for total aflatoxin set by various regulatory bodies was as follows: ≤ 4 ppb set by the European Commission (36.7%), ≤ 10 ppb set by the Kenya Bureau of Standards (96.7%) and ≤ 20 ppb set by the US Food and Drug Administration (96.7%). Maize grains stored in PP bags were more contaminated with total aflatoxin (Mean = 4.7 ppb) than grains stored in PICS bags (Mean = 2.1 ppb). Storage of maize in PICS bags reduced aflatoxin contamination by 55% as compared to PP bags. Overall, 50% and 90% of the maize grains stored in PP and PICS bags, respectively met the EC standards for total aflatoxin (≤ 4 ppb). Likewise, 96.7% and 100% of the maize grains stored in PP and PICS bags, respectively met the threshold set by KEBS (≤ 10 ppb) and FDA (≤ 20 ppb) (Table 5).

Table 5. Mean proportion (%) of aflatoxin contamination level categories for maize sampled at harvest and three months after storage in PICS and PP bags in Kaiti District

Location	Bag type	≤ 4	> 4 -10	> 10 - 20	>20	Range (ppb)	Aflatoxin level (ppb)[b]
Mukuyuni	At harvest[a]	66.7	33.3	0.0	0.0	0-4.8	2.2
	PICS bag	100.0	0.0	0.0	0.0	0-3.7	1.5
	PP bag	60.0	40.0	0.0	0.0	0-4.9	2.9
Kilala	At harvest[a]	6.7	86.7	0.0	6.7	0-28.8	7.6
	PICS bag	80.0	20.0	0.0	0.0	0-5.8	7.7
	PP bag	40.0	53.3	0.0	6.7	0-42.7	6.6
	Mean	58.9	38.9	0.0	2.2	0-42.7	4.8

[a] – Maize grains sampled at harvest; [b] – Mean aflatoxin concentration

3.6 Correlation between the Population of Aspergillus Section Flavi and Aflatoxin Levels in Maize

There was a positive significant correlation ($p \leq 0.05$) between the population of *A. parasiticus* and aflatoxin levels in grains sampled after storage (Table 6). However, there was no significant correlation ($p \geq 0.05$) between the population of *A. flavus* (L and S strains) and aflatoxin levels during the two sampling regimes.

Table 6. Correlation between the population of *Aspergillus* section *Flavi* and aflatoxin levels in maize grains sampled at harvest and after three months of storage

	Aspergillus spp.	*A. flavus* S-strain	*A. flavus* L-strain	*A. parasiticus*	Aflatoxin
At harvest	*A. flavus* S-strain	1			
	A. flavus L-strain	0.71**	1		
	A. parasiticus	0.26[ns]	0.25[ns]	1	
	Aflatoxin	-0.18[ns]	-0.10[ns]	0.01[ns]	1
Three months storage	*A. flavus* S-strain	1			
	A. flavus L-strain	0.07[ns]	1		
	A. parasiticus	-0.17[ns]	0.01[ns]	1	
	Aflatoxin	0.04[ns]	0.11[ns]	0.43*	1

**Correlation coefficient significant at $p \leq 0.01$; *correlation coefficient significant at $p \leq 0.05$; ns - not significant.

4. Discussion

Most smallholder farmers do not harvest maize based on physiological maturity, but employ traditional practices such as observing the dried tassels of cobs and drooping of cobs to determine readiness of maize for harvesting (Akowuah et al., 2015). The aforementioned practices were employed by farmers in Kaiti District to determine the harvesting time of maize. A study by Hell et al. (2000) reported that farmers in Benin left maize in the field for 2 to 3 weeks after physiological maturity before harvest. Akowuah et al. (2015) observed that majority (69%) of farmers in Ghana harvested their maize beyond the physiological maturity period, thus late harvesting was identified as a common practice. These techniques are not accurate and therefore, harvested maize may still have high moisture content, thereby making the grains highly susceptible to fungal growth and aflatoxin contamination (Hell et al., 2008). The most suitable time for maize harvesting is at physiological maturity (Kaaya, Harris, & Eigel, 2006; Hell & Mutegi, 2011). Kaaya et al. (2006) observed that delayed harvesting was positively correlated to increase in aflatoxin levels by about four times by the third week after the recommended harvesting time and more than seven times when maize harvest was delayed for four weeks. Thus, timely harvesting and adequate drying of crops can play a role in reducing fungal contamination of crops (Bankole & Adebanjo, 2003; Atanda et al., 2011). However, lack of storage space, unpredictable weather, labour constraint, theft of the produce, rodent damage and other animals compel farmers to harvest at inappropriate time (Bankole & Adebanjo, 2003).

Most of the farmers used woven polypropylene (PP) bags to store maize, while a few used sisal bags. Maize was stored either inside the family house or the granary. Gachara (2015) observed that most farmers in Eastern and Rift valley regions of Kenya first packed their maize in polypropylene bags after which they stored it in the granaries. Studies in Zambia (Kankolongo, Hell, & Nawa, 2009) and Tanzania (Shabani et al., 2015) reported that smallholder farmers stored their maize grains in polypropylene bags inside the family living house. A study by Gitonga et al. (2015) showed that most (60%) smallholder farmers used space in the family living house and improved granaries (17%) to store maize after harvest. Fandohan, Gnonlonfin, Marasas and Wingfield (2006) reported that in most Sub-Saharan African countries, maize is generally stored in cob form either in wooden granaries, under the roofs of farmers' houses, or on floor in houses. In the current study, storage of maize in polypropylene bags and family house might have contributed to high population of *Aspergillus* species. Storage structures differ in their ability to protect grains from fungal contamination. In a study in West Africa, Hell et al. (2000) reported that some types of farmers' storage structures provided conditions that were more conducive to fungal infection and aflatoxin contamination than others. Kaaya et al. (2006) also reported that storage of maize in improved granaries in Uganda was related to reduced aflatoxin contamination. However, use of improved granaries by smallholder farmers to store maize is uncommon due to risk of theft of the produce.

The most common storage problems reported by maize farmers in Kaiti District were infestation by insects and mold damage. Studies carried out by Hell et al. (2008) in Benin and Shabani et al. (2015) in Tanzania, reported that insects and rodents were common maize storage problems. Storage pests, in particular *Cathartus quadricollis* and *Sitophilus zeamais*, play an important role in the contamination of foods with fungi, especially those that produce toxins (Lamboni & Hell, 2009). Pest infestation is largely due to improper post-harvest and storage conditions and the level of insect damage influences the extent of mycotoxin contamination (Atanda et al., 2011).

Application of chemical insecticides to control storage pests was extensively practiced by maize farmers involved in this study an observation consistent with previous reports (Hell et al., 2000; Kaaya et al., 2006; Shabani et al., 2015). Application of insecticides is a significant factor in *A. flavus* and *A. parasiticus* management (García & Heredia, 2006). A previous study by Plasencia (2004) reported that control of insect populations with insecticides in maize storage environments reduced *A. flavus* and *A. parasiticus* contamination. Insecticides are considered too expensive for subsistence farmers and therefore, farmers in Kaiti District did not use the recommended application rates. Use of wrong insecticides might have also contributed to high occurrence of pests during storage. In addition, application of insecticides is labour intensive since farmers have to apply them after every three months.

Aspergillus flavus L-strain, *A. flavus* S-strain, *A. parasiticus* and *A. niger* were isolated from soil. In a similar study, Karanja (2013) isolated different members of *Aspergillus* section *Flavi* from soil sampled from Eastern Kenya. Horn (2003) reported that soil serves as a reservoir for *A. flavus* and *A. parasiticus* that produce different aflatoxin types in agricultural commodities. Under adverse environmental conditions, *Aspergillus* spp. form sclerotia that allow the fungus to survive as saprophytes for extended periods in the soil, maize residue and maize cobs (Wagacha & Muthomi, 2008). Likewise, the sclerotia survive in soil and produce conidiophores and conidia during subsequent seasons to infect the crop via silk (Scheidegger & Payne, 2003). The propagules in the soil and crop debris act as the primary source of contamination with *Aspergillus* spp. infecting maturing maize crops (Atehnkeng et al., 2008). Thus, elimination of *Aspergillus* spp. inoculum sources such as infected debris from the previous harvest may prevent infection of the subsequent crops (Strosnider et al., 2006).

Aspergillus flavus S-strain, *A. flavus* L-strain, *A. parasiticus* and *A. niger* were isolated from the maize grains sampled at harvest. A similar diversity has been reported in maize sampled from Eastern (Murithi, 2014) Kenya at harvest. The high population of *Aspergillus* spp. in maize sampled at harvest could be attributed to the dry weather conditions associated with high temperatures in Makueni County which predispose maize to the fungi at pre-harvest stages in the field (Okoth et al., 2012). Also, improper harvest practices such as delayed harvesting and throwing cobs on the ground during and after harvesting might have contributed to *Aspergillus* spp. contamination of maize. Similar spectrum of the aforementioned *Aspergillus* spp. was also observed in maize grains sampled three months after storage in PP and PICS bags. This could be explained by the occurrence of correspondingly high population of *Aspergillus* spp. resident in maize sampled at harvest which influences the population in storage.

Of all the *Aspergillus* section *Flavi* isolated in maize grains sampled at harvest and after storage in the current study, *A. flavus* L-strain was the most prevalent while *A. parasiticus* was the least predominant species. The current findings on the abundance of *A. flavus* L-strain contradict those of Okoth et al. (2012) who reported that *A. flavus* S-strain was the most predominant in maize obtained from Makueni region of Kenya. A previous study in Eastern Kenya by Probst, Njapau and Cotty (2007) observed a higher incidence of *A. flavus* S-strain (71.8%), followed by *A. flavus* L -strain (28.2%) and lastly *A. parasiticus* (2.1%). Predominance of *A. flavus* L-strain in the current study could be attributed to the fact that the climatic conditions during the study period were not harsh to favour *A. flavus* S-strain over other members of *Aspergillus* section *Flavi*. The high occurrence of *Aspergillus* section *Flavi* in this study indicates a risk of aflatoxin poisoning when conditions are favorable as observed by Probst et al. (2007) since *Aspergillus* section *Flavi*, have been reported as the main contaminants of maize from Eastern regions, Kenya (Okoth et al., 2012).

In this study, the population of *Aspergillus* spp. in maize increased from harvest to sampling after three months of storage in PP and PICS bags. This agrees with reports by Domenico et al. (2016) who observed an increase in the population of *Aspergillus* spp. in maize after three months of storage with a progressive increase until nine months. A previous study by Hell, Cardwell and Poehling (2003) also reported higher frequencies of *A. flavus* in stored maize in Benin compared to maize obtained at harvest. In this study, the population of *Aspergillus* spp. was 71.1% higher in maize stored in woven PP bags than in PICS bags at the third month of storage. Factors conducive for fungal growth such as high moisture content, high relative humidity and aeration of the grains in woven polypropylene bags might have contributed to high population of *Aspergillus* spp. Previous studies have reported that fungal counts in peanuts stored in aerated bags were higher than in hermetically sealed bags after 90 days of storage (Navarro, Navarro, & Finkelman, 2012). Domenico et al. (2016) reported an increase in the population of *Aspergillus* spp. in conventional bags and hermetic bags with respective peaks at three and six months, followed by stability in counts. Furthermore, Viebrantz, Radunz and Dionello (2016) reported that although initial growth of *Aspergillus* spp. was observed, for hermetic and non- hermetic systems, the growth as well as the final incidence was lower in the hermetic system, indicating that low oxygen rates reduced the growth of microorganisms.

The population of *Aspergillus* spp. increased up to two-fold in PICS bags compared to up to three-fold increase in PP bags after three months storage. Maize grains stored in PICS bags were less contaminated with *Aspergillus* spp. than from the woven PP bags possibly due to reduced oxygen (hypoxia) and elevated CO_2 levels (hypocarbia) (International Food Policy Institute [IFPRI], 2010) within hermetic storage that hinders the development of the pathogens (Baoua et al., 2014). Bartosik, Cardoso and Rodríguez (2008) observed that increasing CO_2 concentration from 3% to 30% (even with O_2 concentrations of 21%) resulted in a reduction of fungal counts. Similarly, Hocking (2003), found that growth of *Aspergillus* spp. was significantly reduced but not prevented by storage in an atmosphere of 0.1% O_2 and 21% CO_2. Other studies by Giorni, Battilani, Pietro and Magan (2008) reported that treatment with 25% carbon dioxide reduced *A. flavus* development; however concentration of at least 50% carbon dioxide was required to reduce aflatoxin formation in maize. Changes in internal gas composition due to decrease of oxygen and increase of carbon dioxide within the hermetic bags affect the oxidative metabolism thus fungi and other microorganisms do not generate metabolic water for support of growth and cellular integrity (Mutungi, Affognon, Njoroge, Baributsa, & Murdock, 2014). Low O_2 and high CO_2 concentration reduces the rate of growth of fungi, degree of sporulation, respiratory rate and their ability to attack grain tissues (Moreno-Martinez et al., 2000). Thus hermetic bag protects maize grains better than conventional storage bags (Baoua et al., 2014) and can therefore be used for effective grain storage.

There were varying levels of aflatoxin contamination in maize grains sampled at harvest and three months after storage. Sixty three percent of maize grains collected at harvest had total aflatoxin levels above the acceptable limits set by the European Commission (≤ 4 ppb) while only 3.3% exceeded the limits set by the Kenya Bureau of Standards (≤ 10 ppb) and the US Food and Drug Administration (≤ 20 ppb). Mwihia *et al.* (2008) reported that 35.5% of maize sampled from Makueni in Eastern Kenya at harvest had aflatoxin levels above FDA maximum limit of 20 ppb, while Muthomi et al. (2009) reported aflatoxin levels of up to 160 μg kg^{-1} in maize samples from areas with high prevalence of *A. flavus* in Eastern Kenya. The high levels of aflatoxin contamination in maize sampled at harvest indicated that *A. flavus* infection of maize had already occurred in the field prior to or during harvest. Moreover, drought stress, delayed harvesting, improper harvest practices and spreading of maize on the ground for drying before storage might have increased the risk of maize contamination with *Aspergillus* spp. (Wagacha & Muthomi, 2008).

Maize stored in woven PP bags was 33.4% more contaminated with aflatoxin compared to samples stored in PICS bags. The high aflatoxin levels in PP bags could be attributed to high temperature, relative humidity and moisture conditions (Wagacha et al., 2013) that promote fungal growth and aflatoxin contamination. A study by Domenico et al. (2016) observed that the mean levels of total aflatoxins in kernels stored in conventional bags, hermetic bags and silos were 85.4, 85 and 91.2 μg/kg, respectively. Overall, 90% and 100% of maize samples stored in PICS bags in this study met the Kenyan regulatory threshold of ≤ 10 ppb and FDA standard of ≤ 20 ppb for total aflatoxins. Storage of maize in PICS bags effectively lowered aflatoxin levels by 55.3% after three months of storage which could be attributed to low O_2 content and elevated CO_2 levels in hermetic bags (Moreno-Martinez et al., 2000), which limit fungal growth and toxin production. Hocking (2003) reported that carbon dioxide enrichment hinders aflatoxin formation in the substrate. Other findings demonstrated that the ability of *A. flavus* to produce aflatoxin in groundnuts was substantially reduced with the increase in CO_2 and decrease in O_2 concentrations (Bartosik et al., 2008). Magan & Aldred (2007) observed that aflatoxin production decreased by 25% when CO_2 was elevated to 20% although it had no visible effect on growth and sporulation of fungi.

In this study, there was a significant positive correlation between the population of *A. parasiticus* and aflatoxin levels in maize sampled three months after storage. The positive correlation could be as a result of the presence of sub-optimal storage conditions (high humidity, moisture and ambient temperature) favorable for growth of *A. parasiticus* and aflatoxin contamination. However, Probst et al. (2007) observed no correlation between aflatoxin content and the incidence of *A. parasiticus* in maize. Other studies by Hell et al. (2000) observed a positive correlation between aflatoxin levels and *Aspergillus* spp. in stored maize flour in Benin. In this study therefore, *A. parasiticus* greatly contributed to total aflatoxin in stored maize.

5. Conclusion

The high population of *Aspergillus* spp. in maize stored in polypropylene bags implies a serious health concern to consumers when temperature, relative humidity and storage conditions are favorable for aflatoxin contamination. From the study, the triple-layer hermetic bags effectively suppressed the growth of *Aspergillus* spp. Likewise, hermetic bags provided conditions that were unfavorable for aflatoxin contamination. It is therefore evident that hermetic bags provide good protection against fungal growth and aflatoxin contamination of stored maize grains. Moreover, storage of maize in hermetic bags offers a cost effective and chemical free-method that will enable farmers to preserve high quality grains. Therefore, adoption of hermetic storage technology by smallholder farmers will provide an effective option for managing fungal and aflatoxin contamination of maize grains as well as maintaining grains of high quality.

Acknowledgements

This study was funded by Purdue University through GFS Faculty Seed Grant Number 207780. The first author would like to thank the World Federation of Scientists for support through a one year research internship at the International Centre of Insect Physiology and Ecology (ICIPE), Nairobi Kenya.

Conflict of Interests

The authors have not declared any conflict of interests.

References

Akowuah, J. O., Mensah, L. D., Chan, C., & Roskilly, A. (2015). Effects of practices of maize farmers and traders in Ghana on contamination of maize by aflatoxins: Case study of Ejura-Sekyeredumase Municipality. *African Journal of Microbiology Research*, *9*(25), 1658-1666. http://dx.doi.org/ 10.5897/AJMR2014.7293

Anankware, P. J., Fatunbi, A. O., Afreh-Nuamah, K., Obeng-Ofori, D., & Ansah, A. F. (2012). Efficacy of the multiple-layer hermetic storage bag for biorational management of primary beetle pests of stored maize. *Academic Journal of Entomology*, *5*(1), 47-53.

Atanda, S. A., Pessu, P. O., Agoda, S., Isong, I. U., Adekalu, O. A., Echendu, M. A., & Falade, T. C. (2011). Fungi and mycotoxins in stored foods. *African Journal of Microbiology Research*, *5*(25), 4373-4382.

Atehnkeng, J., Ojiambo, P. S., Donner, M., Ikotun, T., Sikora, R. A., Cotty, P. J., & Bandyopadhyay, R. (2008). Distribution and toxigenicity of *Aspergillus* species isolated from maize kernels from three agro-ecological zones in Nigeria. *International Journal of Food Microbiology*, *122*(1), 74-84. http://dx.doi.org/10.1016/j.ijfoodmicro.2007.11.062

aWhere, Inc. (2015). WeatherTerrain® daily surfaced weather data. Retrieved August 2, 2016, from http://me.awhere.com

Bankole, S. A., & Adebanjo, A. (2003). Mycotoxins in food in West Africa: Current situation and possibilities of controlling it. *African Journal of Biotechnology*, *2*(9), 254-263. http://dx.doi.org/10.5897/AJB2003.000-1053

Baoua, I. B., Amadou, L., Baributsa, D., & Murdock, L. L. (2014). Triple bag hermetic technology for post-harvest preservation of Bambara groundnut (*Vigna subterranea* (L.) Verdc.). *Journal of Stored Products Research*, *58*, 48-52. http://dx.doi.org/10.1016/j.jspr.2014.01.005

Bartosik, R., Cardoso, L., & Rodríguez, J. (2008, September). Early detection of spoiled grain stored in hermetic plastic bags (silo-bags) using CO_2 monitoring. In *Proceeding of 8th International Conference on Controlled Atmosphere and Fumigation in Stored Products* (pp. 550-554).

Centers for Disease Control and Prevention (CDC). (2004). Outbreak of aflatoxin poisoning--Eastern and Central provinces, Kenya, January-July 2004. *Morbidity and Mortality Weekly Report*, *53*(34), 790-793.

Cotty, P. J. (1994). Influence of field application of an atoxigenic strain of *Aspergillus flavus* on the populations of *A. flavus* infecting cotton bolls and on the aflatoxin content of cottonseed. *Phytopathology*, *84*(11), 1270-1277.

Cotty, P. J., & Cardwell, K. F. (1999). Divergence of West African and North American communities of *Aspergillus* section *Flavi*. *Applied and Environmental Microbiology*, *65*(5), 2264-2266.

Domenico, A. S. D., Busso, C., Hashimoto, E. H., Frata, M. T., Christ, D., & Coelho, S. R. M. (2016). Occurrence of *Aspergillus* sp., *Fusarium* sp., and aflatoxins in corn hybrids with different systems of storage. *Acta Scientiarum. Agronomy*, *38*(1), 111-121. http://dx.doi.org/10.4025/actasciagron.v38i1.25621

Fandohan, P., Gnonlonfin, B., Marasas, W. F. O., & Wingfield, J. (2006). Impact of indigenous storage systems and insect infestation on the contamination of maize with fumonisins. *African Journal of Biotechnology*, *5*(7), 546-552.

Gachara, G. W. (2015). *Post-harvest fungi diversity and level of aflatoxin contamination in stored maize: Cases of Kitui, Nakuru and Trans-Nzoia Counties in Kenya.* (Master of Science Thesis, Kenyatta University, Kenya).

García, S., & Heredia, N. (2006). Mycotoxins in Mexico: Epidemiology, management, and control strategies. *Mycopathologia*, *162*(3), 255-264. *http://dx.doi.org/10.1007/s11046-006-0058-1*

Giorni, P., Battilani, P., Pietro, A., & Magan, N. (2008). Effect of a_W and CO_2 level on *Aspergillus flavus* growth and aflatoxin production in high moisture maize postharvest. *International Journal of Food Microbiology*, *122*, 109-113. http://dx.doi.org/10.1016/j.ijfoodmicro.2007.11.051

Gitonga, Z., De Groote, H., & Tefera, T. (2015). Metal silo grain storage technology and household food security in Kenya. *Journal of Development and Agricultural Economics*, *7*(6), 222-230. http://dx.doi.org/10.5897/JDAE2015.0648

Hell, K., & Mutegi, C. (2011). Aflatoxin control and prevention strategies in key crops of Sub-Saharan Africa. *African Journal of Microbiology Research*, *5*, 459-466.

Hell, K., Cardwell, K. F., & Poehling, H. M. (2003). Relationship between management practices, fungal infection and aflatoxin for stored maize in Benin. *Journal of Phytopathology*, *151*(11-12), 690-698. http://dx.doi.org/10.1046/j.1439-0434.2003.00792.x

Hell, K., Cardwell, K. F., Setamou, M., & Poehling, H. M. (2000). The influence of storage practices on aflatoxin contamination in maize in four agroecological zones of Benin, West Africa. *Journal of Stored Products Research*, *36*(4), 365-382. http://dx.doi.org/10.1016/S0022-474X(99)00056-9

Hell, K., Fandohan, P., Bandyopadhyay, R., Kiewnick, S., Sikora, R., & Cotty, P. J. (2008). Pre-and postharvest management of aflatoxin in maize: an African perspective. *Mycotoxins: detection methods, management, public health, and agricultural trade/edited by John F. Leslie, Ranajit Bandyopadhyay, Angelo Visconti*, pp.

219-229.

Hell, K., Mutegi, C., & Fandohan, P. (2010). Aflatoxin control and prevention strategies in maize for Sub-Saharan Africa. *Julius-Kühn-Archives*, (425), 534. http://dx.doi.org/10.5073/jka.2010.425.388

Herrman, T. J., Lee, K. M., Jones, B., & McCormick, C. (2014). Aflatoxin sampling and testing proficiency in the Texas grain industry. *International Journal of Regulatory Science*, 2(1), 7-13.

Hocking, A. D. (2003, June 25-27). Microbiological facts and fictions in grain storage. In *Proceedings of the Australian Postharvest Technical Conference* (pp. 55-58). Canberra: CSIRO.

Horn, B. W. (2003). Ecology and population biology of aflatoxigenic fungi in soil. *Journal of Toxicology: Toxin Reviews*, 22(2-3), 351-379. http://dx.doi.org/10.1081/TXR-120024098

International Food Policy Institute (IFPRI). (2010). Aflatoxin in Kenya. An overview. Aflacontrol project Note1. Retrieved February 3, 2016, from http: // www. *programs.ifpri.org/afla/afla.asp.*

Kaaya, A. N., Harris, C., & Eigel, W. (2006). Peanut aflatoxin levels on farms and in markets of Uganda. *Peanut Science*, 33(1), 68-75. http://dx.doi.org/10.3146/0095-3679(2006)33[68:PALOFA]2.0.CO;2

Kankolongo, M. A., Hell, K., & Nawa, I. N. (2009). Assessment for fungal, mycotoxin and insect spoilage in maize stored for human consumption in Zambia. *Journal of the Science of Food and Agriculture*, 89(8), 1366-1375. http://dx.doi.org/10.1002/jsfa.3596

Karanja, L. W. (2013). *The occurrence of Aspergillus section Flavi in soil and maize from Makueni, County, Kenya* (Master of Science Thesis, University of Nairobi, Kenya).

Kenya Bureau of Standards (KEBS). (2007). Kenya Standard KS 694-1:2007. In: Shelled peanuts (*Arachis Hypogaea* Linn.) Specification. Part 1: Raw peanuts for Table Use. Kenya Bureau of Standards Documentation Centre, Nairobi, Kenya.

Klich, M.A. (2002). *Identification of common Aspergillus species*. Utrecht, the Netherlands: Centraal bureau voor Schimmel cultures, pp. 116.

Lamboni, Y., & Hell, K. (2009). Propagation of mycotoxigenic fungi in maize stores by post-harvest insects. *International Journal of Tropical Insect Science*, 29(01), 31-39. http://dx.doi.org/10.1017/S1742758409391511

Lewis, L., Onsongo, M., Njapau, H., Schurz-Rogers, H., Luber, G., Kieszak, S., Nyamongo, J., Backer, L., Dahiye, A. M., Misore, A., Decoct, K., & Rubin, C. (2005). Aflatoxin contamination of commercial maize products during an outbreak of acute aflatoxicosis in Eastern and Central Kenya. *Environmental Health Perspectives*, 113, 1763-1767.

Lizárraga-Paulín, E. G., Moreno-Martínez, E., & Miranda-Castro, S. P. (2011). *Aflatoxins and their impact on human and animal health: An emerging problem*. INTECH Open Access Publisher.

Magan, N., & Aldred, D. (2007). Post-harvest control strategies: Minimizing mycotoxins in the food chain. *International Journal of Food Microbiology*, 119, 131-139. http://dx.doi.org/10.1016/j.ijfoodmicro.2007.07.034

Makueni County Integrated Development Plan (MCIDP). (2013). First County integrated development plan 2013-2017. REPUBLIC OF KENYA. Retrieved May 28, 2016, from www.kenyampya.com/userfiles/Makueni%20CIDP%20sept2013(1).pdf.

Moreno-Martinez, E., Jiménez, S., & Vázquez, M. E. (2000). Effect of *Sitophilus zeamais* and *Aspergillus chevalieri* on the oxygen level in maize stored hermetically. *Journal of Stored Products Research*, 1(36), 25-36. http://dx.doi.org/10.1016/S0022-474X(99)00023-5

Muiru, W.M., Charles, A.K., Kimenju, J.W., Njoroge, K., & Miano, D.W. (2015). Evaluation of reaction of maize germplasm to common foliar diseases in Kenya. *Journal of Natural Sciences Research*, 5(1), 140-145.

Murithi, M. E. (2014). *Prevalence of Fusarium and Aspergillus species in maize grain from Kitui, Machakos and Meru and use of near infra-red light sorting to remove fumonisins and aflatoxin contaminated grain in Kenya* (Master of Science Thesis, University of Nairobi, Kenya).

Muthomi, J.W. (2001). *Comparative studies on virulence, genetic variability and mycotoxin production among isolates of Fusarium species infecting wheat* (Doctoral dissertation, University of Nairobi, Kenya).

Muthomi, J.W., Njenga, L.N., Gathumbi, J.K., & Chemining'wa, G.N. (2009). The occurrence of aflatoxins in maize and distribution of mycotoxin-producing fungi in Eastern Kenya. *Plant Pathology Journal*, 8(3), 113-119.

Mutungi, C. M., Affognon, H., Njoroge, A. W., Baributsa, D., & Murdock, L. L. (2014). Storage of mung bean (*Vigna radiata* [L.] Wilczek) and pigeonpea grains (*Cajanus cajan* [L.] Millsp) in hermetic triple-layer bags stops losses caused by *Callosobruchus maculatus* (F.)(Coleoptera: Bruchidae). *Journal of Stored Products Research*, *58*, 39-47. http://dx.doi.org/10.1016/j.jspr.2014.03.004

Mwihia, J.T., Straetmans, M., Ibrahim, A., Njau, J., Muhenje, O., Guracha, A., Gikundi, S., Mutonga, D., Tetteh, C., Likimani, S., & Breiman, R.F. (2008). Aflatoxin levels in locally grown maize from Makueni District, Kenya. *East African Medical Journal*, *85*(7), 311-317. http://dx.doi.org/10.4314/eamj.v85i7.9648

Navarro, H., Navarro, S., & Finkelman, S. (2012). Hermetic and modified atmosphere storage of shelled peanuts to prevent free fatty acid and aflatoxin formation. *Integrated Protection of Stored Products*, *81*, 183-92.

Nyaga, P.N. (2010). Report on aflatoxin contamination in maize. Ministry of Agriculture. Retrieved September 7, 2015, from http://www.fao.org/fileadmin/user_upload/drought/docs/KENYAY%20FOOD%20SECURITY%20PRESE NTATION(AFLATOXIN)%20JUNE%202010.pdf.

Okoth, S., Nyongesa, B., Ayugi, V., Kang'ethe, E., Korhonen, H., & Joutsjoki, V. (2012). Toxigenic potential of *Aspergillus* species occurring on maize kernels from two agro-ecological zones in Kenya. *Toxins*, *4*(11), 991-1007. http://dx.doi.org/10.3390/toxins4110991

Ouma, J. O., & De Groote, H. (2011). Maize varieties and production constraints: Capturing farmers perceptions through participatory rural appraisals (PRAs) in Eastern Kenya. *Journal of Development and Agricultural Economics*, *3*(15), 679-688.

Plasencia, J. (2004). Aflatoxins in maize: a Mexican perspective. *Journal of Toxicology: Toxin Reviews*, *23*(2-3), 155-177. http://dx.doi.org/10.1081/TXR-200027809

Probst, C., Bandyopadhyay, R., Price, L. E., & Cotty, P. J. (2011). Identification of atoxigenic *Aspergillus flavus* isolates to reduce aflatoxin contamination of maize in Kenya. *Plant Disease*, *95*(2), 212-218. http://dx.doi.org/10.1094/PDIS-06-10-0438

Probst, C., Njapau, H., & Cotty, P. J. (2007). Outbreak of an acute aflatoxicosis in Kenya in 2004: identification of the causal agent. *Applied and Environmental Microbiology*, *73*(8), 2762-2764.

Reiter, E., Zentek, J., & Razzazi, E. (2009). Review on sample preparation strategies and methods used for the analysis of aflatoxins in food and feed. *Molecular Nutrition & Food Research*, *53*(4), 508-524. http://dx.doi.org/10.1002/mnfr.200800145

Riddell, R.W. (1950). Permanent stained mycological preparations obtained by slide culture. *Mycologia*, *42*(2), 265-270.

Scheidegger, K. A., & Payne, G. A. (2003). Unlocking the secrets behind secondary metabolism: a review of *Aspergillus flavus* from pathogenicity to functional genomics. *Journal of Toxicology: Toxin Reviews*, *22*(2-3), 423-459. http://dx.doi.org/10.1081/TXR-120024100

Shabani, I., Kimanya, M. E., Gichuhi, P. N., Bonsi, C., & Bovell-Benjamin, A. C. (2015). Maize storage and consumption practices of farmers in Handeni District, Tanzania: Corollaries for mycotoxin contamination. *Open Journal of Preventive Medicine*, *5*(08), 330. http://dx.doi.org/10.4236/ojpm.2015.58037

Strosnider, H., Azziz-Baumgartner, E., Banziger, M., Bhat, R.V., Breiman, R., Brune, M., & Wilson, D. (2006). Workgroup report: public health strategies for reducing aflatoxin exposure in developing countries. *Environmental Health Perspectives, 114*, 1989-1903.

VICAM (2013). Afla-V Instructions for corn. Water Corporation. VICAM LP, Waters.

Viebrantz, P. C., Radunz, L. L., & Dionello, R. G. (2016). Mortality of insects and quality of maize grains in hermetic and non-hermetic storage. *Revista Brasileira de Engenharia Agrícolae Ambiental*, *20*(5), 487-492. http://dx.doi.org/10.1590/1807-1929/agriambi.v20n5p487-492

Wagacha, J.M., & Muthomi, J.W. (2008). Mycotoxin problem in Africa: Current status, implications to food safety and health and possible management strategies. *International Journal of Food Microbiology, 124*, 1-12. http://dx.doi.org/10.1016/j.ijfoodmicro.2008.01.008

Wagacha, J. M., Mutegi, C. K., Christie, M. E., Karanja, L. W., & Kimani, J. (2013). Changes in fungal population and aflatoxin levels and assessment of major aflatoxin types in stored peanuts (*Arachis hypogaea* Linnaeus). *Journal of Food Research*, *2*(5), 10-23. http://dx.doi.org/10.5539/jfr.v2n5p10

14

Microbiological Safety Levels of South Sudanese Bank Notes in Circulation at University of Juba Food Restaurants

Amegovu K. Andrew[1]

[1] Department of Food Science & Technology, College of Applied and Industrial Sciences, University of Juba. P. O. BOX 83, Juba, South Sudan

Correspondence: Amegovu K. Andrew, Department of Food Science & Technology, College of Applied and Industrial Sciences, University of Juba. P. O. BOX 83, Juba, South Sudan. E-mail: kiri_andrew@yahoo.com

Abstract

Food borne infections arise from either a host of bacteria, viruses and parasites originating in food or pathogens introduced through cross contamination. This study assessed the potential microbiological cross contamination risk posed by South Sudanese Pounds in circulation at University of Juba food restaurants by examining the level of microorganisms on banknotes. Bacterial contamination on the South Sudanese Pounds in circulation at University of Juba were determined using currencies collected from five different food serving points coded A,B,C, D and E respectively. From each food serving points, five samples of banknotes 5, 10and 25 South Sudanese Pounds denominations were randomly selected and their surface bacterial content enumerated. High and varying proportions of Total Coli forms (TC), Escherichia coli (*E. Coli*) and Staphylococcus aureus (*S. aureus*) were detected. Findings revealed a significant correlation between microbial levels and the denominations of the bank notes, with the smallest having the highest levels of microorganisms per square centimeter. However, there was no specific pattern in contamination levels between banknotes obtained from the different food points. Another factor that influenced the level of contamination was period the banknotes took in circulation with the older notes having higher levels of microorganisms. High levels of microorganisms on banknotes coupled with unhygienic food handling practices predisposes consumers to health risks. Strategies to reduce the risk of transmission of pathogens from the South Sudanese Pounds with specific emphasis on awareness programs and improvement in food hygiene & handling practices through physical contact between food and money in restaurants at University of Juba were mentioned in order to reduce risk of food borne illness or otherwise potentially lethal outbreak of food borne diseases.

Keywords: South Sudanese pound note (ssp), contamination, bacteria, University of Juba

List of acronyms

> *E. coli*: Escherichia coli
>
> EPA: Environmental Protection Agency
>
> MO: Microorganism
>
> PC: Plate Count
>
> S. aureus: Staphylococcus aureus
>
> SS: South Sudan
>
> SSP: South Sudanese Pounds
>
> TC: Total Coliform
>
> USA: United States of America

1. Introduction

World over the frequency in occurrence of food borne infections has become a public health issue of major concern. In the United States alone, the economy's cost associated with outbreaks of five major food borne pathogens was estimated in the region of US$ 6.9 billion per year (Hall et al., 2008). In addition to acute gastroenteritis, food borne infections may result into deaths or chronic disability, for example *E. coli O157:H7* infection was found to be the leading cause of hemolytic uremic syndrome which results into acute kidney failure in children (Boyce et al., 1995).

Several factors contribute to emergence of food borne diseases ranging from human demographics and behavior, technology in the food industry, microbial adaptation but most importantly ineffectiveness of public health measures (Institute of Medicine, 1992). Foodborne disease outbreaks are also more likely to originate from commercial food service premises with restaurants and hotels accounting for more than 75% incidence in many countries (Little & Gillespie, 2008; Altekruse, 1996). Accordingly, the main channels for pathogen entry were identified as cross contamination, unhygienic food storage facilities and infected food handlers (Little & Gillespie, 2008; Altekruse, 1996).

In general, the level of consumer awareness about food safety issues is dependent upon the number of food borne disease outbreaks reported worldwide (Jevšnik et al., 2008). Studies revealed that 79 percent of sampled consumers had been alerted about food safety issues including stories related to *E-Coli:015H7* bacteria, salmonella, food handling and preparation among others through media, and more than 50% of the consumers surveyed were likely to respond to negative stories concerning safe drinking water, bacteria in food and food preparation

Research has also shown that several attributes of paper currency make it an ideal breeding ground for microorganisms. First, the paper bills offer a large surface area for organisms and organic debris to collect. Secondly, folds and/or deliberate depressions or projections specifically engineered into the bills' as anti-counterfeit designs serve as settling sites for both organisms and debris, which allow the microorganisms to live longer in banknotes also (Creswell, Munsell, Fultz, & Zirbel, 1997) and also weave their way through the population for many years before they come to rest (Alemu, 2014; Girma et al., 2014).

The potency of currencies as means for transmission of potential pathogenic microorganisms was first suggested by Abrams and Waterman 1972. Particularly because money is frequently transferred from one person to another, if contaminated, it could play a significant role in spreading communicable diseases. However, there are limited studies on currency contamination by MOs, there is indeed limited knowledge especially among local populations in South Sudan about the risk of banknotes carrying potentially harmful pathogens.

1.1 Statement of the Problem

The SSP is widely used in South Sudan for exchange for goods and services. These banknotes are handled using bare hands, stored under unhygienic conditions and frequently dropped in dirty places. It is also believed that due to cultural and financial market constraints, people keep money in clothes like socks, underwear and bras. Poor waste disposal further exposes the SSP to microorganisms including pathogenic ones when money drops in places contaminated with urinal and fecal matter which can easily be transferred on to foods.

The problem of improper handling of banknotes is compounded with unhygienic food environments. In the public restaurants for example, preparation and serving of food are done by the person receiving money and often without washing hands or use of knives and cutting boards. Similarly, consumers make payments as they dine or eat without washing their hands which could potentially lead to contamination and therefore cause food borne illness. Due to the rising number of students, there has been a demand led emergence of food eating points, which inherently increases the health risks associated with improper handling of banknotes in food restaurants if not given the due attention it deserves.

Although studies in Sub-Saharan Africa (SSA) have investigated the contamination on some currencies, these have been mainly been limited to the Nigerian Naira, South African Rand and Ghanaian Cedi (Kawo et al., 2009; Yazah et al., 2012; Ngwai et al., 2011; Ehwarieme, 2012; Tagoe & Adams, 2011; Igumbor et al., 2012). Assessing the potential risk of the SSP as a medium for contamination of food remains to be investigated. Findings of this study will increase awareness about the potential health risk associated with unhygienic food servings and currency handling in restaurants contributing to the effective control and prevention of foodborne diseases.

1.2 Study Objectives

This study examined microbiologically safety of the South Sudanese Pounds used in University restaurants by specifically;

    a.   Determining level of Total Coliform on South Sudanese banknotes;

    b.   Determining the level of Escherichia coli on South Sudanese banknotes; and

    c.   Determining the level of Staphylococcus aureus on South Sudanese notes.

1.3 Hypothesis

The level of Total Coliforms, Escherichia coli and Staphylococcus aureus on the South Sudanese banknotes used at University of Juba restaurants are above acceptable limits

2. Materials and Methods

South Sudan currency, SSP was used for purposes of this study. After random selection of banknotes from identified points, horizontal methods were applied to obtain MOs from surfaces by use of contact plates and swabs. Subsequent enumeration of MOs were done following the ISO 18593 guidelines.

2.1 Sampling Procedures

Five food serving points A, B, C, D and E at University of Juba were randomly identified. From each point, five (5) random samples of 5SSP, 10SSP and 25SSP denominations were selected for enumeration of their bacterial content. The sampled banknotes of 5, 10, and 25SSP were swabbed with sterile swabs as follows, the tip of sterile swabs were initially moistened by immersing in a tube containing sterile quarter strength ringer's solution and pressed against container walls to remove excess liquid while rotating the swabs between the thumb and finger a demarcated area of 100cm^2 was swabbed in two directions at right angles to each other. Either side of the banknotes were swabbed and treated separately and the swab sticks were aseptically broken into bottles of 100ml sterile quarter strength ringer's solution, diluted and mixed. Serial dilutions were made from the initial suspension to form the samples which were analyzed. MOs were enumerated using procedures **2.2** for Coliforms, *E. coli* and **2.3**for *S. aureus* as described below. The number of microorganisms per swab was calculated from the mean number of colonies grown on petri dishes.

2.2 Enumeration of Coliforms and Escherichia coli

Following ISO 4832:2006 guidelines for the enumeration of Coliforms and *E. coli*, the colony count techniques was applied as below;

2.2.1 Reagents

  a) Violet Red Bile Lactose Agar
  b) Brilliant Green Lactose Bile Broth
  c) Peptone water
  d) Kovac's reagent

2.2.2 Procedure

Media was prepared by weighing the required amount of media powder and mixing thoroughly with distilled water. The mixture was heated until boiling while occasionally stirring. It was allowed to boil for 2 minutes and then cooled immediately in a water bath at 44 °C – 47 °C.

Serial dilutions were prepared by transferring 1ml of the sample or the appropriate dilutions to the center of each dish using a sterile pipette. About 10ml of medium was poured on each petri dishes, mixed with the inoculums and the mixture allowed to solidify. After complete solidification, about 5ml of the Violet Red Bile Lactose was poured onto the surface of the inoculated medium and allowed to solidify as before. A control plate with about 20ml of the medium was prepared concurrently to check for sterility. After complete solidification, the dishes were inverted and incubated at 37°C for 24 ± 2 h for Coliform s and 41.5 °C for 24 ± 2 h for *E. coli*.

2.2.3 Enumeration of Total Coliform and Escherichia coli

The purplish red colonies were considered as typical colonies of Coliforms and Escherichia coli. After counting using colony counting equipment, counts were taken from plates with colony numbers ranging from 30 to 300. In all cases spreading colonies and clusters were considered as single colonies.

2.2.4 Confirmation for coliforms

Brilliant Green Lactose bile broth was prepared by dissolving the medium powder in distilled water and boiling over a Bunsen flame. The medium was dispensed in quantities of 10ml in test tubes containing Durham tubes which were initially sterilized in an autoclave at 121 °C for 15 min. Care was taken to ensure that the Durham tubes did not contain air bubbles after sterilization. Five presumptive Coliform colonies were inoculated into tubes of the Brilliant Green Lactose bile broth and incubated at 37 °C for 24 ± 2 h. Tubes that showed positive results were counted and number of coliform colonies calculated.

2.2.5 Confirmation for Escherichia coli

Peptone water was prepared and filled in tubes autoclaved at 121°C. Five presumptive E.coli colonies were inoculated into tubes of the peptone water and incubated at 37 °C for 24 ± 2 h. 3 drops of Kovac's reagent were dropped into the tubes and after a few minutes, red rings were observed which confirmed presence of *E.coli*. Tubes that showed positive results were counted and number of *E.coli* colonies calculated.

2.3Enumeration of Staphylococcus Aureus

Following the ISO 6888-2:2003 method for the enumeration of *S. aureus*, the surface spread technique was applied as follows;

2.3.1 Media and Reagents

- a) Baird Parker Agar (BPA)
- b) Tellurite egg yolk emulsion
- c) Brain-heart infusion broth
- d) Rabbit Plasma

2.3.2Procedure

Required amount of media powder was weighed, mixed thoroughly with distilled water and sterilized by autoclaving at 121°C for 15 min. The mixture was immediately cooled in the water bath at 44 - 47° C. 50ml of egg yolk emulsion with tellurite was added to every 950ml of the molten BPA. About 20ml of the molten agar was aseptically poured on the petri dishes and left to set at room temperature. After complete solidification the plates inverted to avoid condensed water from dripping back onto the solidified agar

0.1ml of chosen dilutions was aseptically transferred onto the center of the solidified agar plate. The inoculum was evenly spread using a sterile spreader as quickly as possible on the agar surface, the plates inverted   and incubated for **24 ± 2 h** then re- incubated for a further **24 ± 2 h** at **37 °C**. We observed characteristic black /grey colonies surrounded by opaque clear zones and made counts from the plates that contained less than 300 colonies.

2.3.3Enumeration of Staphylococcus aureuscolonies

Counts were taken from plates with colony numbers ranging from 30 to 300 using colony counting equipment. As in the first case, all spreading colonies and clusters were considered as single colonies.

2.3.4 Confirmation of Staphylococcus aureus

Selected colonies were transferred to a tube of brain-heart infusion broth, incubated at 37°C for 24h.Aseptically 0.1ml of culture was added to 0.3 mL of rabbit plasma and incubated at 37°C.The clotting was examined after 4h, and negative results were reexamined after an additional 24h.The test was considered positive if the clot exceeded half of the volume and from the tubes with positive results, number of *S. ureus* colonies were calculated.

3. Results

3.1 Occurrence of Microorganisms on Banknotes

On average, all the banknotes that were obtained from the various food serving points were contaminated with microorganisms. Three different microorganism species were isolated, with the most common being TC (81.4%) and *S. aureus* (18.6%) (Table 1).The 5SSP currency notes had the highest levels of MO contamination (50.7%) as compared to the 10SSP (16.5%) and 25SSP (32.8%).

Table 1. Relative occurrence of MOs on banknotes selected from 5 food serving points in Juba University café

Microorganism	Currency denominations (SSP)Variation of Microorganisms			
	5	10	25	Total (%)
Total Coliform	398,494	113,430	183,768	695,692 (81.4)
Escherichia coli	52	142	72	266 (0.0)
Staphylococcus aureus	34,591	27,644	96,584	158,818 (18.6)
Total	433,137(50.7)	141,215(16.5)	280,423 (32.8)	854,776 (100)

3.2 Occurrence of MOs at Sampling Points

Table 2. Relative occurrence of MOs in relation to sampling points

Microorganism	Sampling points					Variation of MOs
	A	B	C	D	E	Total (%)
Total Coliform	804,667	4,263	128,200	200,073	22,283	1,159,487(81.4)
Escherichia coli	37	67	110	174	57	444 (0.0)
Staphylococcus aureus	238	49,967	136,690	15,686	62,117	264,697 (18.6)
Total	804,941 (56.5)	54,297 (3.8)	265,000 (18.6)	215,933 (15.2)	84,457 (5.9)	1,424,628 (100)

Serving point A recorded the highest levels of aggregated microorganism contamination (56.5%) amongst all the serving points from which the denominations were sampled. This was followed by serving point C (18.6%), serving point D (15.2%), serving point E (5.9%) and serving point B (3.8%) as shown in Table 2. A close analysis showed no specific pattern of MO contamination differences among the identified serving points.

Although no specific currency contamination level trends are observed at the different sampling points, relative distribution of MOs indicated higher concentration of TC, *E. coli* and *S. aureus* on the 5SSP, 10SSP and 25SSp respectively (Figures 1, 2, 3 & 4)

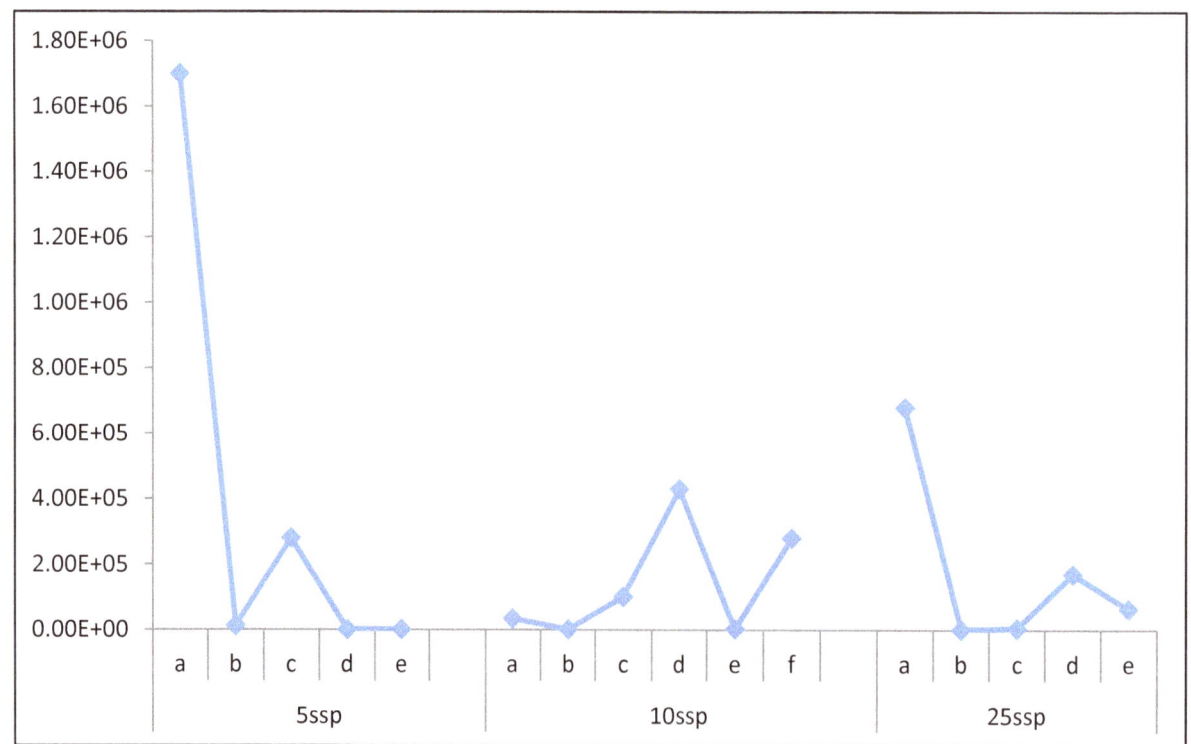

Figure 1. Variation of Total Coliform (TC) in different denominations from different serving points

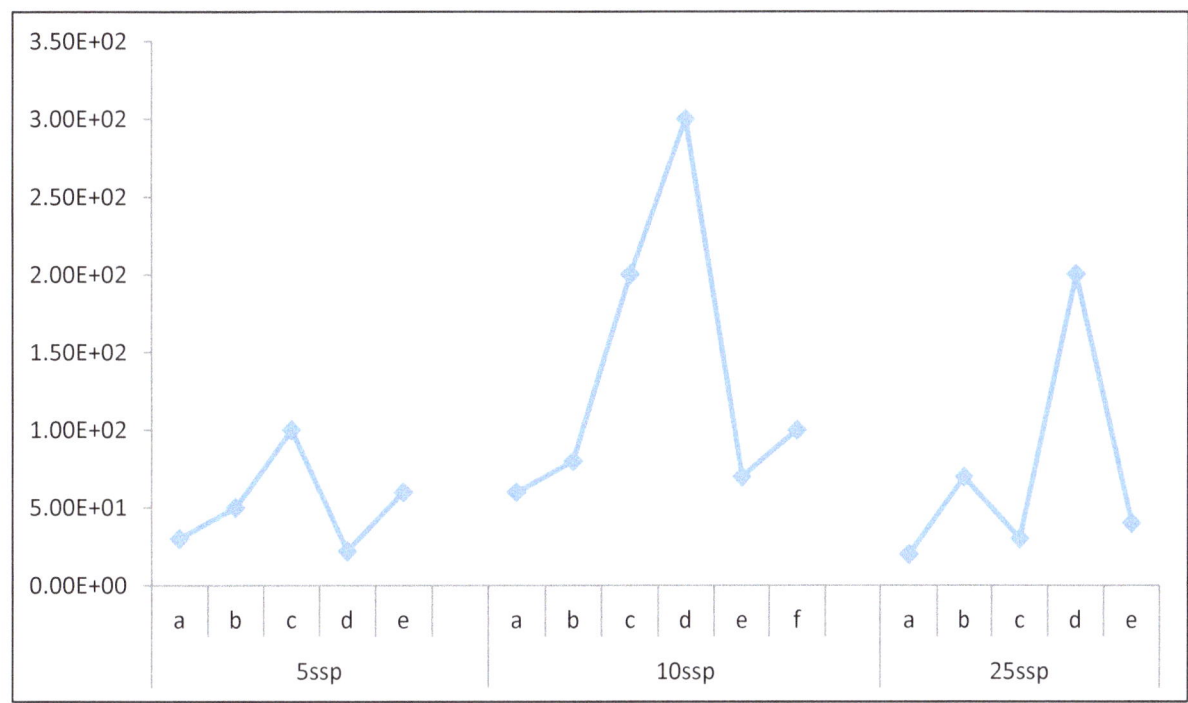

Figure 2. Variation of Escherichia coliin different denominations from different serving points

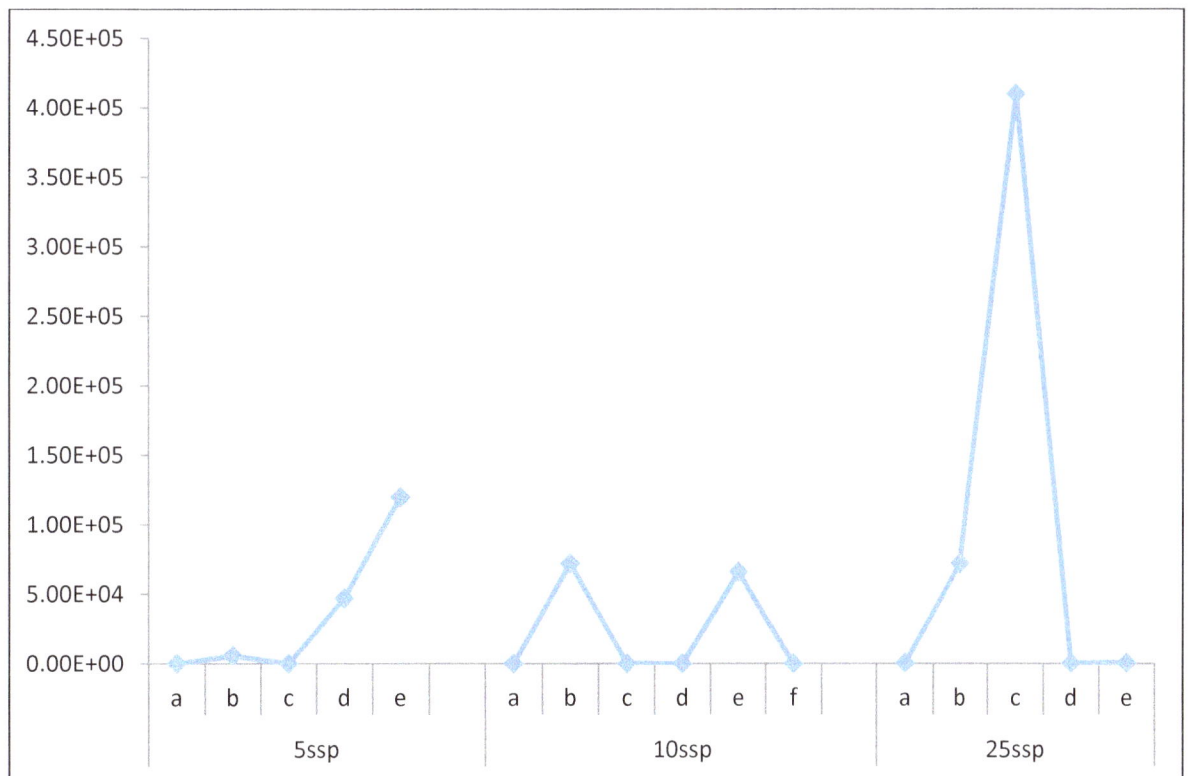

Figure 3. Variation of Staphylococcus aureusin different denominations from different serving points

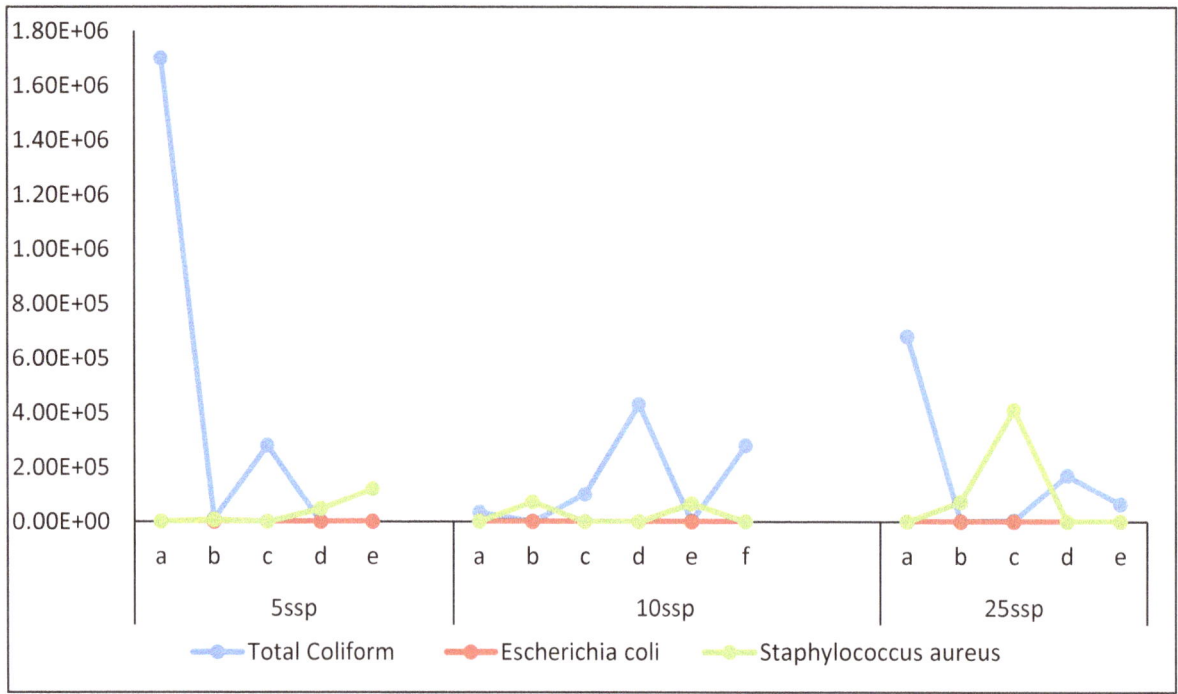

Figure 4. Variation of TC, *E. coli* and *S. aureus* in different denominations from different serving points

4. Discussions

In this study five random samples of 5, 10 and 25 SSP notes respectively obtained from five serving points coded A, B, , D and E in Juba University restaurants were analyzed for microbial load. All the sampled banknotes were found to be contaminated with microorganisms in varying proportions. Similar (100%) currency contamination rates were previously reported in Ghana and Pakistan (Tagoe et al., 2011; Sabahat & Humaira, 2011). The presence of MOs on all banknotes demonstrates the critical role played by money in transmitting pathogens, hence poses a dimensional public health threat. This finding supports reports from other parts of the world that currency notes are usually contaminated by microorganisms that can cause a wide range of diseases (El-Dars & Hassan, 2005; Khinet et al., 1989; Siddique, 2003; Pope et al., 2002) including tuberculosis (Basavarajappa et al., 2005). Several factors contribute to currency contamination namely; storage of money in clothes, body surfaces or dirty surfaces, improper washing of hands, use of saliva to wet fingers when counting currency, coughing and sneezing (Akoachere et al., 2014)

In general the extent of contamination corresponded with currency denominations. The lowest currency in circulation (5SSP) notes exhibited a highest levels of contamination (56.5%). Such findings were also reported from studies in Bangladesh, Ghana and Pakistan (Tagoe et al., 2011; Sabahat & Humaira, 2011; Ahmed et al., 2010). The relatively high level of contamination of the 5SSP denominations could be attributed to the fact that it passes through many hands as the most frequently used banknote which predisposes it to continued contamination. This was consistent with studies in Cameroon which suggested that the tendency to mishandle money is also higher in lower denominations (Akoachere et al., 2014). Apart from denomination, the physical condition of the currencies also influenced the level of contamination. Old, tattered and dirty notes were more contaminated in line with previous studies in which soiled notes especially those held together with bits of sticky tape were particularly dangerous (Siddique, 2003).

At the level of food service points there was no specific pattern of MO distribution between different sampling points. This could be because the banknotes don't necessarily get contaminated from the restaurants but other points of exchange like neighboring small vendors, bars and butcheries. Even then currencies could have moved from one part of the country to another during processes of trade, travel and recirculation.

Upon isolation of colonies, presence of TC, *E. coli* and *S. aureus* were confirmed on all bank notes albeit varied proportions. The highest levels of MOs were TC, *S. aureus* and *E. coli* in descending order.

TC presented significant prevalence in all the currency notes with the highest contamination levels registered with the 5SSP currency notes. The coliform group consist of a collection of different types of bacteria usually present in the environment and feces of warm blooded animals. Although a small amount of coliforms may not cause illnesses, high presence of coliforms indicates possible presence of additional harmful pathogens which requires further scrutiny (Connecticut Department of Public Health, June 2010). Coliformsmay be an indication of fecal contamination from poor hygiene and inappropriate waste disposal. According to the US Environmental Protection Agency (EPA) presence of TC is a health concern and for water systems, TC levels exceeding 5% were regarded as acute risk (Connecticut Department of Public Health, June 2010).

Investigation also confirmed presence of *E. coli* on banknotes. This finding concurs with reports that currency notes act as reservoirs for enteric pathogens (Goktas & Oktay, 1992; Xu et al., 2005). Presence of *E. coli* indicates recent fecal contamination, dirty environments and inadequate personal hygiene of money handlers. *E. coli* is opportunistic and has a shelf life of up to 11days on currencies during which it can be easily transferred from one medium to another. If ingested, *E. coli* may result into bloody diarrhea or acute kidney failure in children moreover person to person transmission is well documented (Boyce et al., 1995).

5. Recommendations and Conclusions

5.1 Recommendations

This study has shown that SSPs are contaminated with potentially pathogenic microorganisms and all the people handling this currency are invariably exposed to infections. Based on the results and outcomes of our laboratory diagnosis, the following suggestions are proposed to improve food safety and quality assurance:

Its strongly recommend provision of health education. The importance of basic personal hygiene though practices like frequent and thorough hand washing, avoiding use of saliva during counting money, compulsory separation of cashier from other restaurant jobs and cleanliness of power houses like kitchens, bars and restaurants and shops that serve as food outlets are strongly recommended to reduce the risk of infection.

Furthermore the study also recommend formation of market and community health education committees to oversee the day today market and business outputs/ inputs like market cleanliness, awareness campaigns, and act as advocates for behavior change and communication. This would also improve local initiative and ownership of public health improvement efforts.

Adoption and integration of a food safety framework and strategy into National Health Policy to aid food safety assurance and quality control. This should involve critical scrutiny and health certification for restaurants to operate. Impromptu inspection of restaurants and reprimands for food outlets that do not comply with food safety standards or have high frequency of public health problems be initiated. Periodic disinfection of the South Sudanese currency for 24 hours before being re-circulated to minimize cross contamination from use of dirty banknotes is also encouraged.

5.2 Conclusions

This study has shown that the SSPs circulating in University of Juba restaurants have high levels of contamination which could potentially facilitate spread of pathogens. Although contamination could have originated from other currency users like butcheries and street vendors, restaurants present a higher health risk since MOs can easily get into food at any point before or during consumption. This hazard is aggravated by poor food handling practices. Given that the most affected banknotes were the lower denominations (5SSP), we postulate that people belonging to weaker social and economic status may be at a higher risk of infections arising from cross contamination. Infants, children, the elderly or sick may be particularly vulnerable.

Due to comparative cost advantage and relative ease of implementation, we recommend emphasis on improvement of hygiene. Knowledge, appropriate behavior & attitude towards hygiene during and after handling money could be fostered by increasing awareness through health education, campaigns and advocacy. Prevention rather than treatment is targeted and improvement in hygiene presents the best means to reduce food contamination (Todd et al., 2010). At the central bank level, removal of mutilated and worn-out banknotes and periodic sanitization of the SSP before recirculation is advised.

References

Abrams, B. L., & Waterman, N. G. (1972). Dirty money. *J Am Med Assoc, 219*, 1202-1212. http://dx.doi.org/10.1001/jama.1972.03190350038011

Ahmed, M. S. U., Parveen, S., Nasreen, T., & Feroza, B. (2010). Evaluation of microbial contamination of Bangladesh paper currency notes (Taka) in circulation. *Adv. Biol. Res., 4*, 266-271.

Akoachere, J. F. T. K., Gaelle, N., Dilonga, H. M., &Nkuo-Akenji, T. K. (2014). Public health implications of contamination of Franc CFA (XAF) circulating in Buea (Cameroon) with drug resistant pathogens. *BMC research notes, 7*(1), 1-13.http://dx.doi.org/10.1186/1756-0500-7-16

Alemu, A. (2014). Microbial contamination of currency notes and coins in circulation: a potential public health hazard. *Biomedicine and Biotechnology, 2*(3), 46-53.

Altekruse, S. F., Street, D. A., Fein, S. B., & Levy, A.S. (1996). Consumer knowledge of foodborne microbial hazards and food handling practices. *Journal of Food Protection, 59*, 287-94.

Basavarajappa, K. G., Rao, P. N., & Suresh, K. (2005). Study of bacterial, fungal, and parasitic contaminaiton of currency notes in circulation. *Indian journal of pathology & microbiology, 48*(2), 278-279.

Boyce, T. G., Swerdlow, D. L., & Griffin, P. M. (1995). Escherichia coli O157:H7 and the hemolytic-uremic syndrome.*NEnglJ Med., 333*, 364-8.http://dx.doi.org/10.1056/NEJM199508103330608

Connecticut Department of Public Health. (June 2010). Connecticut Association of Directors of Health, Inc. *Connecticut   Environmental   Health   Association.*   Retrieved   from http://www.ct.gov/dph/lib/dph/drinking_water/pdf/Presence_of_Total_Coliform_at_Food_Service.pdf

Creswell, Munsel, Fultz &Zirbel-Public Relations Company. (September, 1997). Des Moines, Iowa 515-246-3500.

Ehwarieme, D. A. (2012). R-plasmids amongst *Escherichia coli 0157:H7* isolated from Nigerian currency notes. *Afr J Microbiology Res, 6*, 1966-1969.

El-Din El-Dars, F. M., & Hassan, W. M. (2005). A preliminary bacterial study of Egyptian paper money. *International   Journal   of   Environmental   Health   Research,   15*(3),   235-240. http://dx.doi.org/10.1080/09603120500105976

Girma, G., Ketema, T., & Bacha, K. (2014). Microbial load and safety of paper currencies from some food vendors   in   Jimma   Town,   Southwest   Ethiopia.   *BMC   research   notes,   7*(1),   843. http://dx.doi.org/10.1186/1756-0500-7-843

Hall, G., Vally, H., & Kirk, M. (2008). Foodborne Illness: Overview, *International Encyclopedia of Public Health, 638-653.* Food safety International. Retrieved from (http://promedmail.org/direct.php?id=3168434) http://citizen.co.za/118074/diarrhoea-outbreak-caused-salmonella/(http://teamzimbabwe.org/food-poisoning -scare-hits-nust/)

Institute of Medicine. (1992). *Emerging infections: microbial threats to health in the United States.* Washington (DC): National Academy Press.

Jevšnik, M., Hlebec, V., &Raspor, P. (2008). Consumers' awareness of food safety from shopping to eating. *Food Control, 19*(8), 737-745.http://dx.doi.org/10.1016/j.foodcont.2007.07.017

Kawo, A., Adam, M., Abdullahi, B., & Sani, N. (2009). Prevalence and public health implications of the microbial load of abused naira notes. *Bayero Journal of Pure and Applied Sciences, 2*(1), 52-57.

Little, C. L., & Gillespie, I. A. (2008). Prepared salads and public health. *Journal of Applied Microbiology, 105*(6), 1729-1743. http://dx.doi.org/10.1111/j.1365-2672.2008.03801.x

Ngwai, Y. B., Ezenwa, F. C., &Ngadda, N. (n.d.). Contamination of Nigerian currency notes by Escherichia coli in Nasarawa State.

Pope, T. M., Ender, P. T., Woelk, W. K., Koroscil, M. A. & Koroscil, T. M. (2002). Bacterial contaminaiton of paper currency. *South Med J, 95*, 1408-1410. http://dx.doi.org/10.1097/00007611-200212000-00011

Sabahat, S., &Humaira, R. (2011). Evaluation of bacterial contamination of Pakistani paper currency notes (Rupee) in circulation in Karachi. *Eur J BiolSci, 3*, 94-98.

Sharma, S., &Sumbali, G. (2014). Contaminated Money in Circulation: A Review. *International Journal of Recent Scientific Research, 5*, 1533-1540.

Tagoe, D. N. A., Adams, L., & Kangah, V. G. (2011): Antibiotic resistant bacterial contamination of the Ghanaian currency note: a potential health problem. *J Microbiol Biotech Res, 1*, 37-44.

Todd, E. C., Greig, J. D., Michaels, B. S., Bartleson, C. A., Smith, D., & Holah, J. (2010). Outbreaks where food workers have been implicated in the spread of foodborne disease. Part 11. Use of antiseptics and sanitizers in community settings and issues of hand hygiene compliance in health care and food industries. *Journal of Food Protection, 73*(12), 2306-2320.

Yazah, A. J., Yusuf, J., & Agbo, A. J. (2012). Bacterial contamination of Nigerian currency notes and associated risk factors. *Res J Med Sci, 6*, 1-6.http://dx.doi.org/10.3923/rjmsci.2012.1.6

15

On The Possibility to Trace Frozen Curd in Buffalo Mozzarella Cheese

Nadia Manzo[1], Loredana Biondi[2], Donatella Nava[2], Federico Capuano[2], Fabiana Pizzolongo[1], Alberto Fiore[3] & Raffaele Romano[1]

[1]Dept. of Agricultural Sciences, University of Naples Federico II, Portici (NA), Italy

[2]Istituto Zooprofilattico Sperimentale Del Mezzogiorno, Portici (NA), Italy

[3]School of Science, Engineering & Technology Division of Food & Drink, Abertay University, Dundee, UK

Correspondence: Raffaele Romano, University of Naples Federico II, Portici (NA), Italy. E-mail: rafroman@unina.it

Abstract

The manufacturing of Buffalo Mozzarella PDO (Protected Designation of Origin) cheese requires the exclusive use of fresh buffalo milk, which must be transformed into cheese within 60 hours after milking. The limited availability of buffalo milk and simultaneous increase in Mozzarella demand during the summer cause producers to use frozen intermediates (milk and/or curd) in the cheese-making process. These practices are not allowed. Few data are available in the literature about the effects of freezing on buffalo milk and curd. Recent studies demonstrated that the use of frozen buffalo milk can be detected in mozzarella cheese based on the increase in casein fragment γ4-CN. This work aims to verify the possibility of tracing the presence of frozen curd in Buffalo Mozzarella PDO cheese. The electrophoresis technique was used to reveal the presence of γ4-CN. Equivalent concentrations of this fragment were found in fresh and frozen curd that were stored for 9 months. Our results suggest that γ4-CN cannot be used to discriminate fresh PDO Mozzarella and Mozzarella cheese produced from frozen curd. A second objective of the work was to evaluate the effects of freezing on curd lipids. In particular, the fatty acid and mono-diglyceride profiles were evaluated. Significant differences were found in the amounts of 1,2-Dipalmitin and 1,3-Diolein between fresh curd and curd that was stored for 9 months at freezing temperatures. Although some significant differences were found in the mono-diglyceride profiles, no objective marker that can distinguish between fresh and frozen products is currently available.

Keywords: Buffalo mozzarella cheese, curd, diglycerides, freezing, γ_4-CN, PDO

1. Introduction

Buffalo Mozzarella PDO (Protected Designation of Origin) is a stretched curd cheese produced exclusively by using fresh buffalo milk, which must be processed within 60 hours after milking (EC Regulation 103/2008). The final product must satisfy certain criteria, such as a white porcelain colour, fat/dry matter $\geq$ 52% (w/w) and maximum moisture content of 65% (w/w).

Geographical Indications are the authentic expression of the values and the history of a territory. These certifications are a guarantee of quality for both producers and consumers, in terms of safety, genuineness and freshness.

Mozzarella cheese is a widespread product: in the year 2014, 38,000 tons of Mozzarella PDO cheese were produced (data obtained from the Consortium for the Protection of Buffalo Mozzarella PDO cheese). During the summer there is limited availability of buffalo milk and a concurrent increase in Mozzarella demand. To overcome this inconvenience, producers resort not to contemplated practices. In particular, they freeze milk or curd (mainly curd) in the winter and use them to manufacture Mozzarella PDO when there is a lack of fresh milk. Moreover, some producers buy frozen curd from foreign countries at low costs and mix it with the local curd to produce PDO cheese. Therefore, food control authorities require an analytical method to discriminate between fresh Mozzarella PDO cheese and Mozzarella produced from frozen intermediates, which are not produced according to the PDO manufacturing process.

Some authors examined the effects on proteolysis, lipolysis and sensory characteristics of the frozen curd addition in cheeses obtained from bovine, goat and sheep species (Alonso, Picon, Gaya & Nuñez, 2013; Picon, Alonso, Gaya, & Nuñez, 2013; Pazzola et al., 2013; Van Hekken, Tunick & Park, 2005; Todaro, Scatassa, Alicata, Mazza & Caracappa, 2011; Zhang, Mustafa, Ng-Kwai-Hang & Zhao, 2006). In particular, according to Zhang et al. (2006), good quality sheep cheese can be produced from ovine milk that is frozen at −15 and −25 °C for 6 months without affecting the cheese yield or composition. Pazzola et al. (2013) suggest that considerable decreasing renneting properties for sheep milk after a long-freezing storage can be obtained, which implies that the freezing of milk should be limited to periods shorter than 5 months. Todaro et al., (2011) reported different effects of freezing according to the species, for example goat milk is more affected by the length of the freezing period and the typology of defrosting than ovine and bovine milk. Van Hekken, Tunick & Park (2005) found no significant proteolysis during the refrigerated ageing of soft cheese.

Little data is available in the literature about the effects of freezing buffalo milk or curd. Therefore, finding a method that enables the identification of frozen curd use in the Buffalo Mozzarella PDO cheese-making process is an important goal to protect this high-value product.

Di Luccia et al. (2009) identified a casein fragment (γ4-CN), which is derived from the action of plasmin on β-casein (Ismail & Nielsen, 2010), as a marker of frozen milk in Buffalo Mozzarella PDO cheese. This fragment was detected as a faint band in fresh milk, and an increase of its intensity was observed when the milk freezing time was prolonged (-20°C for 12 months). Moreover, recently, a fragment derived from αS1-casein (αS1-I) was also suggested as a possible indicator of frozen curd in Buffalo Mozzarella cheese (Petrella et al., 2015), even if the thermal and time histories of the analysed mozzarella cheeses remained unknown in this study.

Mainly, commercial fraud regards use of frozen curd rather than frozen milk, as curd storage is cheaper and easier, requiring smaller spaces and lower energy costs. In this study, the effectiveness of γ4-CN to identify frozen curd in Buffalo Mozzarella PDO cheese was evaluated. Then the effects of freezing on the lipid component, which is a possible index of oxidation phenomena, were studied to verify possible modifications that occur during the buffalo curd storage.

2. Methods

2.1 Sampling

The following samples were collected during Buffalo Mozzarella cheese-making process in three different dairy plants in province of Caserta, Salerno and Napoli (Campania region, Italy):

- Raw milk (RM);
- Thermized milk (TM);
- Premature curd (PC), with pH 6.2-6.3;
- Mature curd (MC), with pH 5.0-4.8;
- Mozzarella cheese (M).

Three different samplings were performed in each dairy plant to obtain a total of 9 RM, 9TM, 9PC, 9MC and 9M samples.

The quality and hygiene parameters were tested to examine the sample's conformity to the current legislation. In particular, a microbiological analysis was performed to study and count the main pathogenic, spoilage and pro-technological bacteria. Then, chemical determinations, such as pH, Aw and somatic cells were performed (data not shown).

Each sample was divided into 10 aliquots: the first was used to analyse the fresh product, and the other 9 aliquots were frozen at − 20°C (freezing time: 4 hours) for a total period of 9 months. Monthly, the samples were thawed, and caseins and fat were extracted for the analysis of casein and lipids.

To simulate, in the laboratory, the actual condition of dairy plants, a fast freezing was obtained by cutting curd and mozzarella samples to small sizes, nearly 100 g, and freezing them at -25°C. The thawing process was conducted by maintaining the frozen samples at 4°C over-night, and subsequently in a water bath at 40°C. After thawing, they were used for chemical analysis.

The freezing and thawing time-temperature profiles were registered using a data logger (Ebro), which was placed in the core of the sample. Temperature was measured at regular intervals of 30 sec.

2.2 Quality And Hygiene Parameters

The pH analysis was performed with a FIVE Easy[TM] pH-metre .

The water activity was measured by The Official Methods ISO 21807:200 at 25°C ± 1°C, with Aqualab 4TE instrument.

The somatic cells were revealed through a fluorescence microscope Nucleo Counter SCC-109 – ASTORI according to the Internal Method. This method involves adding a DNA intercalator solution (500 µl) to 500 µl of milk. The number of somatic cells was automatically calculated by a software program. The value of this parameter in accordance to Italian law (Regulation 853/2004) for buffalo milk is ≤ 400,000* for ml (*rolling geometric average, which was calculated over a period of three months with at least one sample per month from bulk milk, unless otherwise specified by the appropriate authorities).

The total bacterial count was in accordance with UNI EN ISO 4833-1:2013. The legal limit is ≤ 1500,000 CFU/ml or g at 30°C (rolling geometric average, which was calculated over a period of two months with at least two samplings per month), in accordance with Regulation n°853/2004.

Microbiological analysis showed all parameters within legal limits (data not shown)

2.3 Casein Extraction and γ4-CN Analysis

Casein was separated from fresh and frozen curd by isoelectric precipitation at pH 4.6 by addition of acetic acid.

The casein analysis was performed using the Official Method N. L74/25 of 20.03.1992, SDS-PAGE (Sodium Dodecyl Sulphate - PolyAcrylamide Gel Electrophoresis), and a subsequent densitometric analysis (GS-800 Bio-Rad).

2.4 Lipid Extraction

Lipids were extracted using the modified Schimith-Bondzynsky-Ratzlaff method (Official Method of Cheese Analysis (D.M.1986)).

2.5 Fatty Acid Profile

The fatty acid profile was analysed by using gas chromatography coupled with an FID detector (Agilent Technologies 6850 SeriesII). The GC was equipped with a capillary column 50% CyanopropilMethil Silicone, 100 m 0.25 mm ID 0.20µm. The oven program temperature was: 140°C x 5 min, increase 4°C/min until reaching 175°C for 20 min, then increase 3°C/min until reaching 240°C and maintained for 1 min. The injector program temperature was: 60°C x 0.1 min, with an increase of 500°C/min until reaching 260°C and maintained for 5 min. Helium was used as the carrier gas (flow rate 1.8 ml/min). The detector temperature was set at 260°C.

A fat solution of 5% in hexane was injected, after transesterification with 2N KOH. Identification was performed with a 37 component FAME mix standard from Supelco, St.Louis, Mo, USA.

2.6 Lipolitic Index

To avoid the interference of triglycerides (TG) with mono and di-glycerides (MDG), the extracted fat from the fresh and frozen samples was separated using the SPE column, according to a modified procedure described in Caboni, M.F., Menotta, S., & Lercker, G. (1996)..

The MDG fraction was analysed by gas chromatography coupled with mass spectrometry (GC MS), using a capillary column DB-5, 25m x 0.247 mm I.D., film thickness 0.25 µm. Betulin was used as an internal standard. The samples were analysed as trimethylsilyl (TMS) derivatives. As reported by other authors, NIST and Wiley libraries do not provide the mass spectra for all products of glycerol esterification. In particular, there is no single mass spectrum of diglycerides with different substitutes (Isidorov et al., 2007).

For the quantification, we assumed that all monoglycerides had an identical response factor as monoolein, and all diglycerides had an identical response factor as 1,2 diolein, which were used as external standards.

2.7 Statistical Analysis

All experiments were performed in triplicate and results are the average values of three determinations. The data were analysed by ANOVA XL-STAT.

3. Results and Discussion

Milk and curd used in the cheese-making process showed quality and hygiene parameters within the limits provided by the current legislation.

The raw and thermized milk showed a pH of 6.6-6.8.

The somatic cell values were within the limits indicated by Reg. EC 853/2004; values ranged from 167,000 to 281,000.

The total bacterial count (TBC) ranged between 6.3 and 4.5 log CFU/ml g in RM and TM.

3.1 Evaluation of the Γ4-CN Effectiveness in Frozen Curd

The SDS-PAGE electrophoretic profiles were obtained for premature curd, mature curd and mozzarella cheese in their fresh and frozen states. The casein profiles of premature and mature curd are shown in figure 1 and 2.

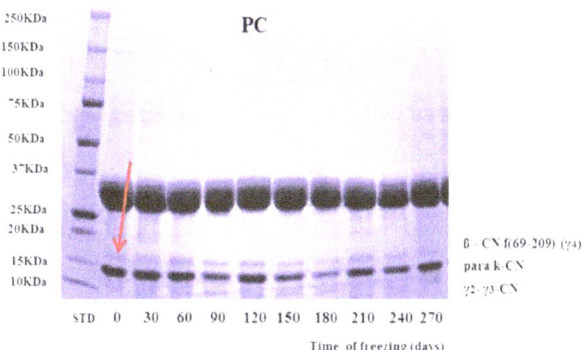

Figure 1. Polyacrylamide gel Electrophoresis (Sds-Page) Of premature curd (PC) at different freezing times

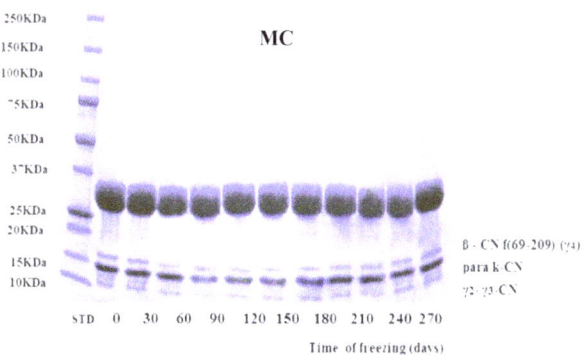

Figure 2. Polyacrylamide gel electrophoresis (SDS-PAGE) of mature curd (MC) at different freezing times

The first lane of the gel shows the standard protein (STD) with molecular weights from 10 kDa to 250 kDa. In the second lane the electrophoretic profile of fresh sample is shown. The frozen profiles, from 30 to 270 days, can be observed in the following lanes.

The profiles of premature curd, mature curd and mozzarella (data not shown) had four electrophoretic bands: γ4-CN, γ2-CN, γ3-CN and para-k-CN. The first three fragments are derived from the hydrolysis of β-casein by the action of plasmin (Ismail & Nielsen, 2010).

Our results show that the fragment γ4-CN is a band of identical intensity in all analysed samples (both fresh and frozen at different freezing times). The densitometric analysis confirmed that no significant differences (p>0.05) were found between fresh and frozen samples.

Di Luccia *et al.* 2009 reported an increase in γ4-CN intensity in refrigerated and frozen buffalo milk.

This discordant result can be explained by the proteolytic action of plasmin on β-casein, which appears more or less intense depending on many factors, such as the process conditions, animal health, phase and number of lactations (Burbrink & Hayes, 2006). Our samples showed quality and hygiene parameters within the limits provided by the current legislation.

3.2 Fatty Acid Profile

Fatty acids from C4 to C22 were detected in all samples. In figure 3, the fresh and frozen (9 months) fatty acid profiles of mature curd overlaid. Concentration of saturated fatty acids (approximately 70%), monounsaturated fatty acids (approximately 26%) and polyunsaturated fatty acids (3%) were found. Only the main fatty acids are reported in the figure.

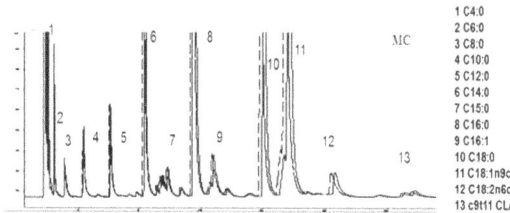

Figure 3. Fatty acids profile of mature curd (MC) in fresh samples (continuous line) and samples that were frozen for 270 days (dot line)

As observed and confirmed by the statistical analysis, no significant differences ($p>0.05$) between the fresh and frozen fatty acid profiles were detected. The complete list of fatty acids identified in fresh and frozen curd (9 months) is reported in table 1.

Table 1. Fatty acid composition (g/100 g) of fresh (t0) and frozen mature curd (t9) (mean values ± standard deviations)

FA	t0	t9
C4:0	2.04±0.09	2.42±0.16
C6:0	1.98±0.05	2.21±0.07
C8:0	1.02±0.02	1.15±0.10
C10:0	2.03±0.01	2.26±0.18
C11:0	0.09±0.01	0.09±0.01
C12:0	2.51±0.01	2.69±0.18
C13:0	0.10±0.00	0.10±0.01
C14:0	10.47±0.10	10.97±0.54
C14:1	0.84±0.00	0.88±0.06
C15:0	1.06±0.01	1.09±0.04
C16:0	34.49±0.35	34.75±0.21
C16:1	1.23±0.01	1.24±0.02
C17:0	0.97±0.01	1.09±0.01
C17:1	0.44±0.01	0.46±0.01
C18:0	13.11±0.13	12.17±0.75
C18:1n9t	1.45±0.00	1.34±0.00
C18:1n9c	22.04±0.38	21.35±0.58
C18:2n6t	0.13±0.00	0.11±0.01
C18:2n6c	2.43±0.06	2.28±0.06
C20:0	0.29±0.01	0.24±0.04
C20:1	0.57±0.02	0.54±0.02
c9t11CLA	0.20±0.01	0.20±0.01
C18:3n3	0.02±0.00	0.02±0.01
C20:2	0.02±0.00	0.02±0.01
C20:3n3	0.09±0.00	0.04±0.00
C20:4n6	0.08±0.01	0.04±0.01
C20:5n3	0.06±0.00	0.04±0.01
C22:6n3	0.05±0.00	0.03±0.01
C22:0	0.15±0.00	0.12±0.02
C22:2	0.03±0.00	0.02±0.01

($p>0.05$)

Our results of fatty acid profiles are consistent with those of Zhang, Mustafa, Ng-Kwai-Hang & Zhao (2006) for sheep milk, where no effect on the milk or cheese fatty acid concentrations was found after the freezing storage.

3.3 Lipolitic Index

Mono-diglycerides (MDG) were examined as an index of lipolitic activity becouse the lipases can attack the globules of milk fat, release free fatty acids and form mono- and diglycerides from triglycerides (TG) (Metha, 2015). According to Bareth, Strohmar & Kitzelmann, (2003) if the sample contains fat, the diglycerides detection can be interfered by the presence of triglycerides and fatty acids from C4 to C14 and diglycerides cannot be cromatographically separated. To avoid these possible interferences, TG and phospholipids (P) were separated by solid phase extraction (SPE). No significant difference between fresh and frozen mature curd was detected. The MDG qualitative profiles for fresh and frozen curd after a period of 9 months were identical. After a freezing period of 9 months, an increase of 5.2% and 8.5% was revealed for 1,2-Dipalmitin and 1,3-Diolein, respectively. However, the obtained results cannot be considered markers of freezing storage.

The quantification results of MDG, TG and P in mature curd showed no significant differences between fresh and frozen samples. The results are shown in table 2.

Table 2. % Composition in TG, MDG and P of fat extracted by fresh (t0) and frozen (t9) mature curd (MC). Mean values ± standard deviation

	TG	MDG	P
MC- t0	95.9±0.6	4.0±0.6	0.2±0.0
MC- t9	95.8±0.5	3.9±0.6	0.3±0.1

No statistical significant difference in the same column (p<0.05)

In figure 4, the MDG profiles of fresh and frozen mature curd are reported.

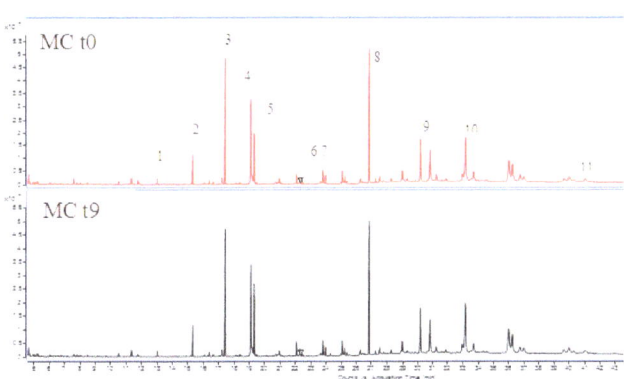

Figure 4. Fresh (t0) and frozen (9 months) mature curd MDG profile

Note 1. Decanoic acid, TMS derivative; 2:Myristic acid, TMS derivative; 3: Palmitic acid, TMS derivative; 4:Oleic acid, TMS derivative; 5:Stearic acid, TMS derivative; 6:1-Monoolein, 2TMS derivative; 7: 1-Monopalmitin, 2TMS derivative; 8: Cholesterol, TMS derivative; 9: Internal standard (betulin); 10:1,2-Dipalmitin, TMS derivative; 11: 1,3-Dipalmitin, TMS derivative

Among the identified molecules, significant differences were detected for 1,2-Dipalmitin and 1,3-Diolein between fresh and the frozen curd. In table 3, the detected amounts are reported.

Table 3. 1, 2-Dipalmitin and 1,3-Diolein content in fresh and frozen curd (9 months). The results are reported as % of total MDG (average values ± standard deviations)

	fresh	frozen
1,2-Dipalmitin	3.61[a]±0.04	3.80[b]±0.06
1,3-Diolein	1.17[a]±0.01	1.27[b]±0.01

a-b: Different letters in the same row correspond to statistical significant differences (p<0.05)

4. Conclusion

The obtained casein detection results confirmed the non-effectiveness of γ4-CN to discriminate mozzarella cheese produced by frozen curd. When curd was maintained at freezing temperatures for 9 months, no significant difference was found in the γ4-CN content compared to the fresh samples.

Therefore, the identified marker in the scientific literature (higher in frozen milk than in fresh milk) is not effective when frozen curd is used. In our study, no increase in γ4-CN was observed during the freezing time, thus false negative cases can be created when a high content of γ4-CN is detected in Mozzarella cheese.

Regarding MDG, an increase in 1,2-Dipalmitin and 1,3-Diolein was found. However, the obtained results suggest that it is currently not possible to discriminate PDO Mozzarella cheese produced from frozen curd because of the lack of an objective marker.

The effects of freezing on the rheological properties of Buffalo Mozzarella PDO cheese produced from frozen curd will be evaluated.

Acknowledgments

This research has been supported by the Ministry of Healt (IZSME 02/13 RC)

References

Alonso, R., Picon, A., Gaya, P., & Nuñez, M. (2013). Proteolysis, lipolysis, volatile compounds and sensory characteristics of Hispánico cheeses made using frozen curd from raw and pasteurized ewe milk. *J. Dairy Res., 80*, 51-57. http://dx.doi.org/10.1017/S0022029912000738

Bareth, A., Strohmar, W., & Kitzelmann, E. (2003). Gas cromatographic determination of mono-and diglycerides in milk and milk products. *Eur Food Res Technol, 216*, 365-368. http://dx.doi.org/10.1007/s00217-002-0650-7

Burbrink, C. N., & Hayes, K. D. (2006). Effect of thermal treatment on the activation of bovine plasminogen. *Int. Dairy J., 16*, 580-585. http://dx.doi.org/10.1016/j.idairyj.2005.08.017

Caboni, M. F., Menotta, S., & Lercker, G. (1996). Separation and analysis of phospholipids in different foods with a Light-Scattering Detector. *J. Am. Oil Chem. Soc., 73*, 1561-1566. http://dx.doi.org/10.1007/bf02523525

Di Luccia, A., Picariello, G., Trani, A., Alviti, G., Loizzo, P., Faccia, M., & Addeo, F. (2009). Occurrence of β-casein fragments in cold-stored and curdled river buffalo (Bubalus bubalis L.) milk. *J. Dairy Sci., 92*, 1319-1329. http://dx.doi.org/10.3168/jds.2008-1220

Isidorov, V.A., Rusak, M., Szczepaniak, L., & Witkowski, S. (2007). Gas chromatographic retention indices of trimethylsilyl derivatives of mono- and diglycerides on capillary columns with non-polar stationary phases. *Journal of Chromatography A., 1166*, 207-211. http://dx.doi.org/10.1016/j.chroma.2007.07.047

Ismail, B., & Nielsen, S. S. (2010). Plasmin protease in milk: current knowledge and relevance to dairy industry. *J. Dairy Sci., 93*, 4999-5009.

Metha, B. M. (2015). Chemical Composition of Milk and Milk Products. Handbook of Food Chemistry. http://dx.doi.org/10.1007/978-3-642-36605-5_31

Petrella, G., Pati, S., Gagliardi, R., Rizzuti, A., Mastrorilli, P., La Gatta, B., & Di Luccia, A. (2009). Study of proteolysis in river buffalo mozzarella cheese using a proteomics approach. *Journal of Dairy Science, 98*, 7560-7572. http://dx.doi.org/10.3168/jds.2015-9732

Pazzola, M., Dettori, M. L., Piras, G., Pira, E., Manca, F., Puggioni, O., Noce, A., & Vacca, G. M. (2013). The Effect of Long-term Freezing on Renneting Properties of Sarda Sheep Milk. *Agriculturae Conspectus Scientificus, 78*, 275-279 http://hrcak.srce.hr/106922

Picon, A., Alonso, R., Gaya, P., & Nuñez, M. (2013). High-pressure treatment and freezing of raw goat milk curd for cheese manufacture: effects on cheese characteristics. *Food Bioprocess Technol, 6*, 2820-2830. http://dx.doi.org/10.1007/s11947-012-0923-5

Reg. CE 103/2008, recante approvazione delle modifiche non secondarie del disciplinare di una denominazione registrata nel registro delle denominazioni d'origine protette e delle indicazioni geografiche protette - Mozzarella di Bufala Campana (DOP)

Todaro, M., Scatassa, M. L., Alicata, M. L., Mazza, F., & Caracappa, S. (2011). Latte bovino, ovino e caprino congelato: variazione dei parametri fisici, chimici e tecnologici. *Scienza e Tecnica Lattiero-Casearia, 62*,

275-281. http://hdl.handle.net/10447/61639

Van Hekken, D. L., Tunick, M. H., & Park, Y. W. (2005) Effect of frozen storage on the proteolytic and rheological properties of soft caprine milk cheese. *J. Dairy Sci., 88*, 1966-1972. http://dx.doi.org/10.3168/jds.S0022-0302(05)72872-1

Zhang, R. H., Mustafa, A. F., Ng-Kwai-Hang, K. F., & Zhao, X. (2006). Effects of freezing on composition and fatty acid profiles of sheep milk and cheese. *Small Ruminant Res., 64*, 203-210. http://dx.doi.org/10.1016/j.smallrumres.2005.04.025

16

Physical Properties of Gluten Free Sugar Cookies Containing Teff and Functional Oat Products

George. E. Inglett[1], Diejun Chen[1] & Sean. X. Liu[1]

[1] Functional Food Research Unit, National Center for Agricultural Utilization Research, USDA-ARS. 1815 N University Street, Peoria, IL 61604, USA

Correspondence: George. E. Inglett, Functional Food Research Unit, National Center for Agricultural Utilization Research, USDA-ARS.1815 N University Street, Peoria, IL 61604, USA.
E-mail: George.Inglett@ars.usda.gov

Abstract

Teff-oat composites were developed using gluten free teff flour containing essential amino acids with oat products containing β-glucan known for lowering blood cholesterol and improving texture. The teff-oat composites were used in sugar cookies for improving nutritional and physical properties. Teff and its composites had higher water holding capacities compared to wheat flour. The pasting properties were not significantly influenced by 20% oat product replacements in teff-oat composites. The pasting viscosities of teff-OBC and teff-WOF 4:1 composites were similar to teff flour, but they were all higher than wheat flour. The elastic properties of teff-OBC (oat bran concentrate) and teff-WOF (whole oat flour) doughs were slightly higher than teff dough. Differences were also found in geometrical and textural properties of the doughs and cookies. Overall, the teff-oat cookies were acceptable in colour, flavour, and texture.

Keywords: cookies, oats, texture, functional food, nutrition, elasticity/viscosity

1. Introduction

The important ancient grain Teff (*Eragrostis tef*) is finding resurgence in the modern age. Teff has an excellent nutritional profile with significant levels of minerals including calcium, iron, magnesium, phosphorus, potassium, and zinc. Also, teff is rich in vitamins, such as thiamin (B1), riboflavin, B2), vitamin A and K (USDA Nutrient Database, 2014). Furthermore, teff is high in proteins with an excellent amino acid composition including all 8 essential amino acids for humans that is superior in lysine than wheat or barley along with its high carbohydrates and fiber contents (Taha et al., 2012).

In Egypt, red teff grains are highly recommended for improving of osteoporosis and bone healing. Chemical studies of the red teff seeds reported the isolation of seven compounds from its ethanol extract, namely β-sitosterol , β-amyrin-3-*O*-(2′-acetyl)-glucoside , β-sitosterol-3-*O*-β-D-glucoside , naringenin , naringenin-4′-methoxy-7-*O*-α-L-rhamnoside, eriodictyol-30,7-dimethoxy-4′-*O*-β-D-glucoside and isorhamnetin-3-*O*-rhamnoglucoside. This was the first report on the isolation of compounds β-amyrin-3-*O*-(2′-acetyl)-glucoside and eriodictyol-30,7-dimethoxy-4′-*O*-β-D-glucoside in nature (Taha et al., 2012). A proximate analysis revealed the high nutritive value of the seeds: carbohydrates (57.27%), protein (20.9%), essential amino acids (8.15%) with major leucine and lysine (1.71 and 1.35%, respectively), vitamin B1 (1.56 mg/100 g), and potassium and calcium (32.4 and 9.63%, respectively). The seeds yielded 22% w/w of oil containing 72.46% of unsaturated fatty acids in which oleic acid was predominant (32.41%) following by linolenic acid (23.83%). The ethanolic extract and fixed oil of the seeds exhibited anti-hyperlipedaemic and antihyperglycaemic activities. Oral administration of the fixed oil for 10 days resulted in a rise in serum calcium levels in rats (Taha et al., 2012).

Oat products, such as whole oat flour (WOF) and oat bran concentrate (OBC), contain β-glucan that has beneficial health effects on reducing serum cholesterol and postprandial serum glucose levels (Klopfenstein,

1988). Also, oat products have high viscosities and water holding capacities. In addition, the phenolic and other antioxidant compounds in oats provide health benefits (Madhujith & Shahidi, 2007). Nutrim, oat hydrocolloid with 15% β-glucan, was prepared by steam jet-cooking OBC, sieving, and drum-drying (Inglett, 2000 & 2011). Nutrim containing β-glucan has numerous functional food applications to reduce fat content and calories in a variety of foods (Lee et al., 2004); control the rheology and texture of food products (Rosell et al., 2001); modify starch gelatinization and retrogradation (Rojas et al., 1999; Lee et al., 2005); and also provide freezing/thawing stability (Lee et al., 2002). Rheological and other physical evaluation of jet-cooked oat bran were studied in low calorie cookies by replacing 20% of the shortening with oat β-glucan hydrocolloids. The cookies containing oat hydrocolloid (20% β-glucan), exhibited reduced spreading characteristics and increased elastic properties compared with the control (Lee & Inglett, 2006).

The suitability of teff flour in bread, layer cakes, cookies, and biscuits has been studies (Coleman et al., 2013). However, none of the prior studies included both teff and oat products. Although there are many cookie recipes using teff flour, none were scientifically studied and reported. Thus, oat products containing β-glucan and teff with its distinctive protein profile and gluten free uniqueness were used to produce teff-oat composites, and used in cookies for this research.

In this exploratory study, the teff-oat composites (4:1) were used and evaluated in sugar cookies. The objectives of this study were to evaluate the physical properties of teff-oat composites along with their formulated dough and cookies; to determine if the textures of dough and cookie will be changed by replacing teff with 20 % of oat products; and to determine if the teff-oat cookies are acceptable in the texture, colour, and flavor compared to wheat flour. The physical properties of teff-oat composites and cookies could provide useful information for functional food applications.

2. Materials and Methods

2.1 Ingredients

Teff flour was purchased from Bob's red mill, Milwaukie, OR, USA. OBC (oat bran concentrate) was supplied by Quaker Oats, Chicago, IL, USA. Nutrim (β-glucan hydrocolloid, 15 g/100g) was provided by VDF FutureCeuticals (Momence, IL, USA). Organic whole oat flour colloidal fine (WOF) was provided by Grain Millers (Eugene, OR, USA).

Ingredients were used for cookie: sugar (C&H sugar company, Crockett, CA, USA), brown sugar (C&H sugar company, Crockett, CA, USA), nonfat dry milk (Carnation, Nestlé, Vevey Switzerland), sodium bicarbonate (Arm & Hammer, Church and Dwight, Co., Inc., Princeton, NJ, USA), shortening (Crisco, the J.M. Smucker company, Orrville, OH, USA), ammonium bicarbonate (Calumet, Kraft, Northfield, IL, USA).

2.2 Preparation of Teff-Oat Composites

Teff flour was mixed with Nutrim, OBC, or WOF (4:1, w/w) using a mixer (KitchenAid, St Joseph, MI, USA) for 2 min, respectively. The mixtures were passed through a 20 mesh sieve followed by additional mixing in a mixer for 1 min to obtain the desired consistency.

2.3 Water Holding Capacity of Composites

The water holding capacity (WHC) of the samples was determined by a previous procedure with minor modifications (Ade-Omowaye et al., 2003). Each sample (2 g, dry weight) was mixed with 25 g of distilled water and vigorously mixed for 1 min to a homogenous suspension using a Vortex stirrer, held for 2 h, and centrifuged at 1,590 g for 10 min. Each treatment was replicated twice. Water-holding capacity was calculated by the difference between the weight of water added and decanted on dry basis (g of water absorbed /100 g of dry sample).

2.4 Pasting Property Measurement of Composites

The pasting properties of teff, oat composites, and wheat flour were evaluated using a Rapid Visco Analyzer (RVA-4, Perten Scientific, Springfield, IL). Samples (2.24 g, dry basis) were made up to a total weight of 28 g with distilled water in a RVA canister (8% solids, w/w). The viscosity of the suspensions was monitored during the following heating and cooling stages. Suspensions were equilibrated at 50°C for 1 min, heated to 95°C at a rate of 6.0°C/min, maintained at 95°C for 5 min, and cooled to 50°C at rate of 6.0°C/min, and held at 50°C for 2 min. For all test measurements, a constant paddle rotating speed (160 rpm) was used throughout the entire analysis except for 920 rpm in the first 10s to disperse sample. Each sample was analyzed in duplicate. The results were expressed in Rapid Visco Analyser units (RVU, 1 RVU = 12 centipoises).

2.5 Cookie Formulation and Preparation

Cookie flour formulations are teff flour; teff -Nutrim composites 4:1; teff-OBC composites 4:1; teff-WOF composites 4:1; and wheat flour. Cookies were prepared following AACC method 10-52 for sugar cookie as described by an earlier study with modifications (AACC International, 2000; Lee & Inglett, 2006). Sugar (72 g), brown sugar (22.5 g), Nonfat dry milk (2.3 g), salt (2.8g), and sodium bicarbonate (2.3 g) were mixed using a whisk in a bowl. The mixture was placed on top of shortening (100 g), and blended with a paddle beater in a mixing bowl using a KitchenAid mixer (St Joseph, MI, USA) at speed 2 for 3 min, scraping down every minute. The mixture of water (49.5 g) containing ammonium bicarbonate (1.1 g) was added and mixed for 1 min at speed 1. After scraping down, they were mixed for another minute at speed 2. Teff, or teff-oat composites (4:1), or wheat flour (225 g) were added while mixing at speed 1, and continued mixing for 2 min at speed 2 with scraping every 30 s. Dough was flattened by using a rolling pin on a board with 7 mm gauge strips, and cut by a cookie cutter of 6 cm diameter. Cookies were baked at 205 °C in a convention oven (XAF-113 LineChef Stefania, Cadco, Ltd. Winsted, CT, USA) for 10 min, and cooled. The cookies were stored in a sealed plastic bag before measurements were taken.

2.6 Rheological Properties of Dough

A sample was loaded on a 2 cm diameter X-hatch parallel stainless plate with a 2 mm gap using a rheometer (AR 2000, TA Instruments, New Castle, DE, USA). The outer edge of the plate was sealed with a thin layer of mineral oil (Sigma Chemical Co., St Louis, MO, USA) to prevent dehydration during the test. All rheological measurements were carried out at 25°C using a water circulation system within ± 0.1°C. A strain sweep experiment was conducted initially to determine the limits of linear viscoelasticity; then a frequency sweep test was carried out to obtain storage modulus (G') and loss modulus (G") at frequencies ranging from 0.1 to 100 rad s^{-1}. A strain of 0.5 within the linear viscoelastic range was used for the dynamic experiments. All rheological measurements for samples were performed in duplicate.

2.7 Water Loss, Moisture Content, and Water Activity

Water losses during baking were measured by the weight of the difference before and after baking. After grinding three cookies with a pestle in a mortar, moisture content and water activity were measured. Moisture contents of cookies were determined by drying 5 g of sample at 105 °C to a constant weight (about 3-4 h). The water activity was measured using an Aqua Lab water activity meter (Decagon Devices Inc., Pullman, WA, USA).

2.8 Geometrical Properties

Six cookies were placed next to each other and the total diameter was measured. They were rotated by 90° and measured three more times. The average of four measurements was divided by six to calculate the average diameter of a cookie. To measure the height, six cookies were stacked, measured, restacked in different order, and measured again. The average cookie height was the mean of three readings divided by six. The spread ratio was calculated by dividing the diameter by the height. Nine measurements for each replicate were taken.

2.9 Colour

The colour of the cookies was measured with a Hunter Lab spectrocolorimeter (Labscan XE, Hunter Associates Laboratory Inc., Reston, VA, USA). The colorimeter was calibrated using a standard white plate. The colour values L^*, a^*, b^* were measured with a C illuminant and a 10° standard observer. The dimension L^* means lightness with 100 for white and 0 for black. The value a^* indicates redness when positive and greenness when negative. The value b^* indicates yellowness when positive.

2.10 Texture Analysis

Dough hardness was measured using a TA-XT2 Texture Analyzer equipped with 5 kg load cell in compression mode by penetrating with a flat probe of 5 mm diameter. The cookie dough (110 g) was gradually and evenly placed in a dough cell while compressing and flatting the surface by a plunger to avoid randomly distributed air pockets as a potential cause of variability in consistency measurements. The test was conducted at a pre-test speed of 2.00 mm s^{-1} and test speed of 3 mm s^{-1}, post-test speed of 10.0 mm s^{-1}, and distance of 20 mm.

Cookie hardness was measured using a recommended method of TA-XT2 Texture Analyzer (Texture Technology Crop., Scarsdale, New York, USA) equipped with 30 kg loading cell. The cookie hardness measurement was conducted by a cutting force using a three-point bending method with sharp-blade probe (6 cm long and 1 mm thick). The hardness of the cookies was indicated by the maximum peak force required to break the cookies. The slotted inserts were adjusted and secured on the Heavy Duty Platform to fit sample size and position centrally under the knife edge. The cookie was rested on two supporting beams spaced at a distance of 3 cm. The instrument was set to 'return to start' cycle, a pre-test speed of 1.5 mm s^{-1}, test speed of 2.0 mm s^{-1}, post-test speed of 10 mm s^{-1}, and a distance of 5.0 mm.

2.11 Sensory Evaluation

Cookies were evaluated by 26 untrained consumers. A scorecard was evaluated for attributes including surface colour, texture, flavour, and overall quality. The panelist assigned scores for each parameter using a 9-point hedonic scale.

2.12 Statistical Analysis

All data from replicated samples were analyzed with analysis of variance using Duncan's multiple comparison to determine significant differences ($P< 0.05$) between treatments (SAS Institute, 1999, Rayleigh, North Carolina, USA).

3. Results and Discussion

3.1 Nutritional Composition

Teff has endured the ages from early civilizations as an important food of a highly nutritious gluten-free grain. Teff contains the highest levels of sugar, calcium, iron, potassium, and zinc whereas oat contains the highest magnesium and phosphorus contents among products (Table 1). Also, teff contains protein (13.3%) that is similar to protein in oat (13.7%) but higher than protein in whole wheat flour (13.2%), rice (5.95%) and corn (9.42%). In addition, teff and oat are rich in vitamins. Teff contains the highest vitamin B2 while oat has the highest thiamin (B1) among all products in the Table 1. All B vitamins help the body to convert carbohydrates into energy. Vitamin B2 is important for normal vision along with other nutrients. Some early evidence shows that vitamin B2 (riboflavin) might help prevent cataracts which can lead to cloudy vision (University of Maryland Medical Center, 2013). Vitamin B1 helps the body metabolize fats and protein, and is required for healthy skin, hair, eyes, and liver. It also helps the nervous system and brain function properly (University of Maryland Medical Center, 2013). Remarkably, only teff and oat contain vitamin K among all products in Table 1. Vitamin K is a fat-soluble vitamin that the body stores it in fat tissue and liver. It is best known for its role in helping blood clot and bone health (University of Maryland Medical Center, 2013). Oat was selected in this study because of its β-glucan content that is helpful for lowering blood cholesterol and improving food texture. Moreover, oat has high quality lipids including monounsaturated and polyunsaturated fatty acid. Therefore, the nutritional value of gluten free baked products could be improved by adding the ancient grain teff and oat products to recipes compared to cookies with wheat flour.

3.2 Water Holding Capacities (WHC)

Nutrim had the highest water-holding capacity (603.0 g /100g) while wheat flour had the lowest water-holding capacity (93.8g/100g) among all the starting materials (Figure 1). Nutrim was produced by jet-cooking technology using thermal-shearing forces to promote molecular breakdown that probably contributed to increased water absorption (Inglett, 2000; Lee & Inglett, 2006). The trend of WHC for wheat flour (93.8 g/100g), WOF (158.2 g/100g), OBC (339.5 g/100g), and Nutrim (603.0 g/100g) appeared to be related to their β-glucan contents (wheat flour, 1.2 g/100g; WOF, 3.9-7.5 g/100g; OBC,12.0 g/100g; Nutrim,15.5 g/100g; Kim, Inglett, & Liu, 2008), suggesting β-glucan may be an important factor for WHC.

In general, the WHC of 4:1 composites for teff-Nutrim (135.3g/100g), teff-OBC (139.3g/100g), and teff-WOF (137.7g/100g) were all higher than wheat flour (93.8 g /100 g). Also, the actual WHC values of teff-Nutrim (135.3 g/100g), teff-OBC (139.3 g/100g), and teff-WOF (137.7 g/100g) were all much lower than theoretical WHC values that were calculated by using the actual WHC values of the starting materials (teff-Nutrim, 228.5 g/100g; teff-OBC, 175.8g/100g; teff-WOF 158.2 g/100g), respectively. Teff flour contains 13.3% protein and is rich in calcium and magnesium (Table 1). It is possible that the calcium and magnesium in teff reacted with protein causing precipitation that resulted in a reduction in WHC. It could be a similar mechanism involved in making tofu by coagulating proteins in soymilk with calcium or magnesium sulfate. The proteins coagulate when bonding occurs between the positively charged calcium ions and negatively charged anionic groups of the protein molecules resulting in protein clumping and the removal of insoluble material from solution. Nutrim was produced by jet-cooking that may possibly produce anionic groups on the surface that could allow interactions with protein and calcium in teff. Those results suggested an interaction between Nutrim and OBC with teff. Teff-oat composites could be widely used in different applications in the food industry because of improved WHC compared to wheat flour alone. These composites are notable for thickening properties, synersis control, and emulsion stabilization along with their nutrients.

Table 1. Comparison of teff composition with other cereals

Nutrient	Unit	teff	rice	corn	whole wheat	oat
		per 100 g	per 100 g	per 100 g	per 100 g	per 100 g
Proximates						
Water	g	8.82	11.89	10.37	10.74	9.37
Energy	kcal	367	366	365	340	371
Protein	g	13.3	5.95	9.42	13.21	13.7
Total lipid (fat)	g	2.38	1.42	4.74	2.5	6.87
Carbohydrate, by difference	g	73.13	80.13	74.26	71.97	68.18
Fiber, total dietary	g	8	2.4	7.3	10.7	9.4
Sugars, total	g	1.84	0.12	0.64	0.41	1.42
Minerals						
Calcium, Ca	mg	180	10	7	34	47
Iron, Fe	mg	7.63	0.35	2.71	3.6	4.64
Magnesium, Mg	mg	184	35	127	137	270
Phosphorus, P	mg	429	98	210	357	458
Potassium, K	mg	427	76	287	363	358
Sodium, Na	mg	12	-	35	2	3
Zinc, Zn	mg	3.63	0.8	2.21	2.6	3.2
Vitamins						
Thiamin (vitamin B1)	mg	0.39	0.138	0.385	0.502	0.54
Riboflavin (vitamin B2	mg	0.27	0.021	0.201	0.165	0.12
Niacin	mg	3.363	2.59	3.627	4.957	0.82
Vitamin B-6	mg	0.482	0.436	0.622	0.407	0.1
Folate, DFE	µg		4	19	44	32
Vitamin B-12	µg	-	-	-	-	-
Vitamin A, RAE	µg	-	-	11	-	-
Vitamin A, IU	IU	9	-	214	9	-
Vitamin E (alpha-tocopherol)	mg	0.08	0.11	0.49	0.71	0.7
Vitamin K (phylloquinone)	µg	1.9	-	-	-	3.2
Lipids						
Fatty acids, total saturated	g	0.449	0.386	0.667	0.43	1.11
Fatty acids, total monounsaturated	g	0.589	0.442	1.251	0.283	1.98
Fatty acids, total polyunsaturated	g	1.071	0.379	2.163	1.167	2.3

Data were selected from USDA nutrient data base. Teff : uncooked; rice: white, flour; corn: grain, yellow; whole wheat: whole wheat grain flour; oat: Quick Oats, dry.

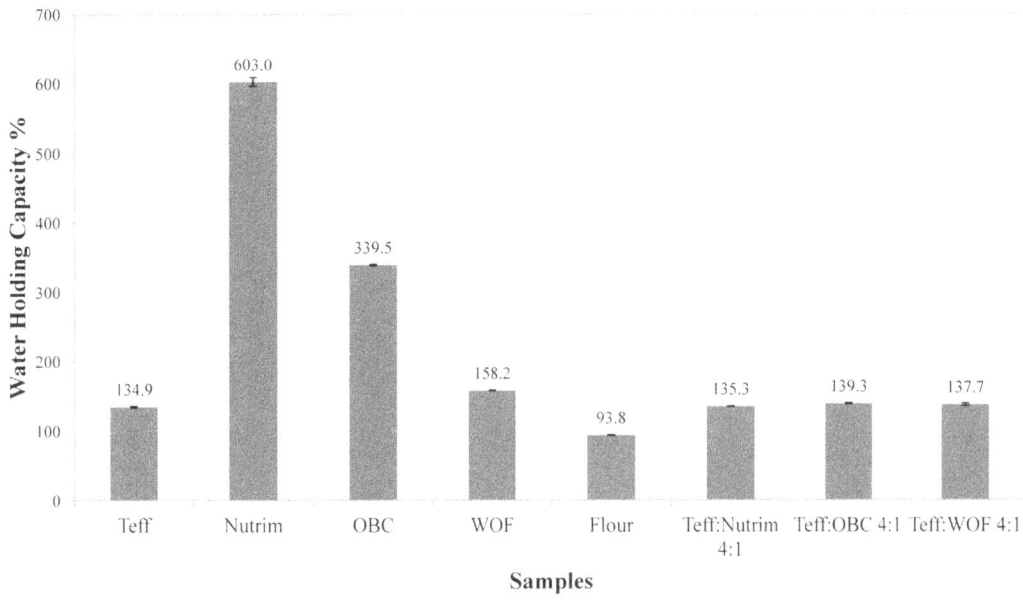

Figure 1. Water holding capacities of starting ingredients and teff-oat composites

3.3 RVA Pasting Properties of Composites

The pasting curves of all starting materials showed dissimilar patterns (Figure 2). The pasting viscosity curve of Nutrim increased sharply (~23 RVU/min) and showed a significantly high initial peak (~250 RVU) at 11 min (90°C), followed by a rapid decrease in viscosity to ~ 25 RVU during continued heating. The Nutrim viscosity slightly increased during cooling showing a considerably low final viscosity (~58 RVU) that was slightly higher than teff flour (~50 RVU). It is known that the viscosity of a completely gelatinized starch slurry decreases during heating (Guha et al., 1998). These characteristics are common for pregelatinized flour (Lai & Cheng, 2004) and typical for Nutrim since it had undergone jet-cooking during preparation where starch gelatinization occurred. The viscosity of OBC increased gradually (~7 RUV/min) to the initial peak (~100 RVU) after temperature reaching 95°C, remained almost constant viscosity during continued heating and shearing, and then increased sharply (~10 RVU/min) during cooling resulting in a considerably high final viscosity (~210 RVU). This high viscosity could be due to starch gelatinization resulting in an entanglement of molecules during cooling to form a matrix with greater stability under heating and shearing. WOF had a lower initial viscosity peak (~ 50 RVU) than Nutrim and OBC at 95°C, showing a small breakdown (peak viscosity minus the lowest point of viscosity after peak), and then slowly increased to a final viscosity (~89 RVU) that was lower than OBC but higher than the rest of the starting materials. The viscosity of teff showed a lower initial peak (~26 RVU) than all oat products but slightly higher than wheat flour (~16 RVU), and remained constant during heating and shearing, and reached a final viscosity (~50 RVU) similar to Nutrim during cooling. The wheat flour showed the lowest initial peak (~16 RVU) and final peak (~17 RVU) among all samples. The trend of initial peaks from wheat flour (~16 RVU), WOF (~48 RVU), OBC (~100 RVU), and Nutrim (~250 RVU) appeared to be related to their β-glucan contents (1.2 g/100g, 8 g/100g, 12 g/100g and 15 g/100g) as shown for water holding capacity.

In general, teff-OBC and teff-WOF 4:1 composites showed similar viscosities of initial and final peak compared to teff, indicating they would have similar viscosity properties after shearing and cooking (Figure 3). The initial peak of teff-Nutrim 4:1 (~17 RVU) was tremendously lower than Nutrim (~250 RVU). Also, the initial peak viscosity of teff-Nutrim 4:1 (~17 RVU) was lower than the teff-OBC (~22 RVU), teff-WOF 4:1 (~29 RVU) composites, and teff (~27 RVU). It suggested possible interactions between teff with Nutrim as showed by WHC. Overall, all teff-oat composites had higher final viscosities than wheat flour. It may be due to the protein of teff bonded with beta-glucan from oat products after heating. Higher initial peak viscosity was related to starch gelatinization whereas higher final peak viscosities suggested stability after heating and shearing. All teff-oat composites showed increased water holding and pasting viscosities with increasing oat contents compared to wheat flour. However, they were only significantly influenced by 80% oat products in teff-oat composites in a previous study (Inglett et al., 2015). Improvement in the textural properties of food using oat β-glucan hydrocolloids has been reported (Lee & Inglett, 2006). The RVA data were useful for providing information on food products.

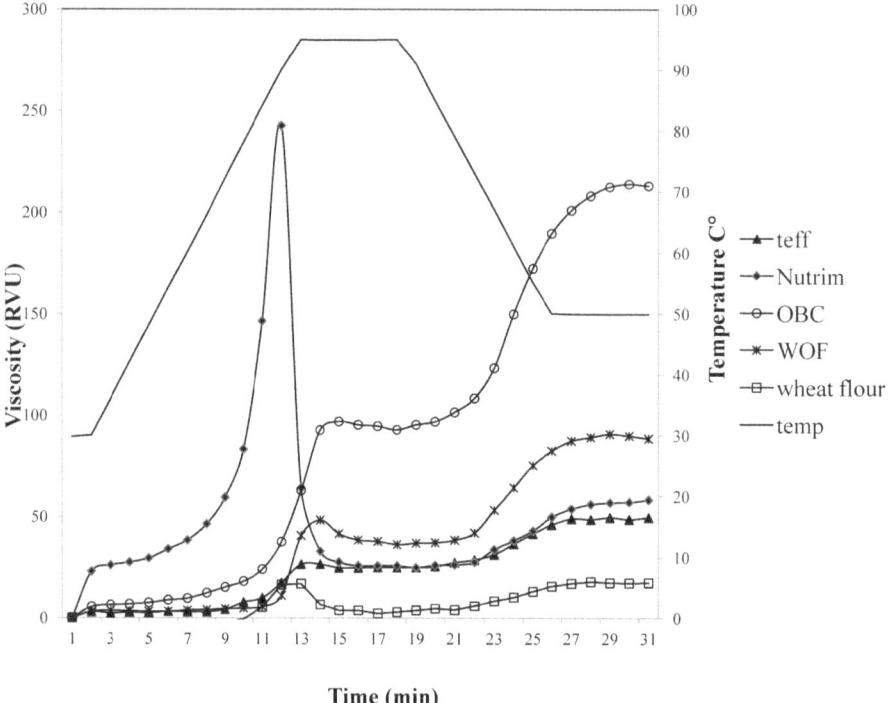

Figure 2. Pasting properties of starting ingredients

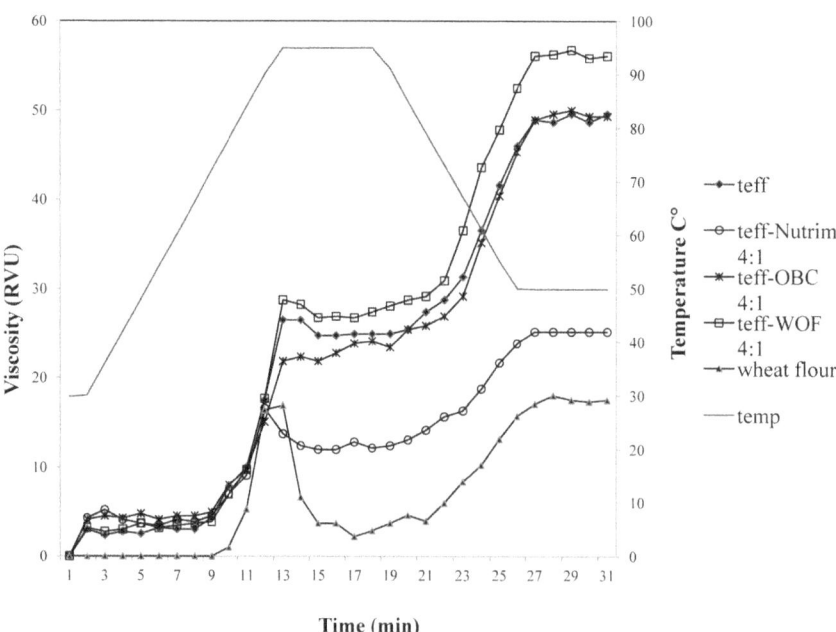

Figure 3. Pasting properties of teff-oat composites

3.4 Rheological Properties of Dough

The dynamic viscoelastic properties of product have been related to the quality of food products (Lee & Inglett, 2006). The G' elastic (storage) and G" viscous (loss) moduli against frequencies for all the doughs are displayed in Figure 4. Both moduli (G' and G") for all samples were increased with increasing frequencies, showing frequency dependence. Furthermore, elastic moduli G' were greater than viscous G" throughout the frequency range for all samples at different levels with the exclusion of dough containing teff-OBC composite. The storage modulus of the sample was higher than the loss modulus, indicating that the sample could be classified rheologically as having a higher predominance of the elastic properties versus the viscous properties (Lai & Liao, 2002). In contrast, the viscous modulus G" of dough containing teff-OBC composites was higher than the elastic modulus G' suggesting a more viscous than elastic nature was predominant in teff-OBC dough. It is probably due to the components and structures of OBC since corn bran in OBC that could not easily form elastic dough before heating. The highest elastic G' and loss G" moduli were observed for dough containing teff-WOF composites, followed by dough with teff and then teff-Nutrim. However, G' values for dough containing teff, teff-WOF, and teff-Nutrim were similar indicating related properties. The values of storage G' and loss G" moduli for wheat flour dough were considerably lower than other doughs. It suggested that dough containing teff or teff-WOF and teff-Nutrim composites could have higher elastic properties than wheat flour dough.

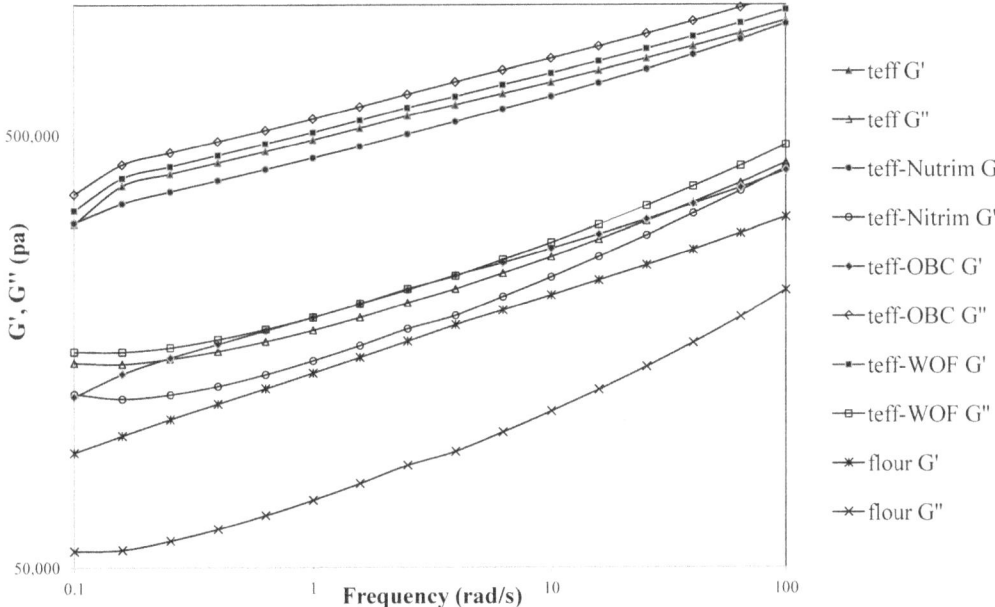

Figure 4. Dynamic viscoelastic properties of cookie doughs containing teff, teff-oat 4:1 composites, and wheat flour

Furthermore, these rheological patterns of the doughs were also confirmed by the tan δ values (loss modulus G"/storage modulus G') (Fig. 5). The values of tan δ clearly describe the ratio of energy lost to the amount of energy stored during a test cycle. The phase shift δ is defined by δ = $\tan^{-1}$(G"/G') which indicates whether a material is solid (δ= 0°), liquid (δ= 90°), or between (0° < δ< 90°). Therefore, the values of tan δ are from zero to infinity; and tan δ = 1 means G' = G", tan δ <1 represents G'>G", and tan δ >1 indicates G'<G". With the exception of teff-OBC, the tan δ values for all doughs were less than 1 representing G'>G". The tan δ values for doughs containing teff, teff-Nutrim, and teff- WOF composites were very similar across the entire frequency spectrum ranging from 0.33 to 0.49. The tan δ for wheat flour dough had a similar behavior that was 0.1 higher than the doughs containing teff, teff-Nutrim, and teff-WOF. It indicated wheat flour dough had slightly more viscous properties than the dough containing teff, teff-Nutrim, and teff-WOF. Interestingly, the tan δ values for teff-OBC dough ranged from 2.5 to 2.9. Also, the slightly decreased tan δ trend was observed with the increased frequency for teff-OBC dough. It demonstrated that loss modulus G" of teff-OBC dough was decreased with frequency showing more viscous property which may be attributed to some components and structures of OBC. These rheological data appeared to indicate that the cookie dough containing the teff and teff-Nutrim or teff-WOF had more elastic nature than wheat flour whereas teff-OBC dough had more viscous properties than wheat flour.

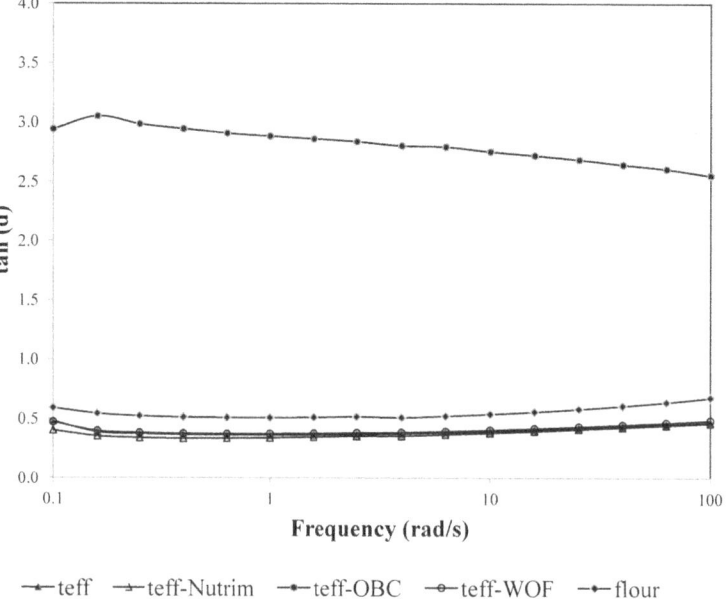

Figure 5. Values of tan (δ) versus frequencies (rad/s) for cookie doughs containing teff, teff-oat 4:1 composites, and wheat flour

3.5 Geometrical Properties of Cookies

Cookie qualities were determined by width, thickness, and cookie spread factor. It was evident from the results that the width was affected by the addition of teff flour. The largest diameter (6.57 cm) was observed for wheat flour cookies and the smallest diameter (5.92 cm) for teff-Nutrim cookies (Table 2). Baking powder released carbon dioxide gas into a batter or dough through an acid-base reaction, causing bubbles in the wet mixture to expand the volume. The smallest diameter value of cookies using teff-Nutrim may be attributed to high WHC that makes dough less expandable compared to the dough with wheat flour. The cookies using teff-OBC composites had the highest thickness (1.28 cm) whereas the cookie containing teff-Nutrim composites had the lowest thickness (1.12 cm, Table 2). The results showed that the width of the cookies using teff, teff-Nutrim, teff-OBC, and teff-WOF decreased ~6, 9, 6, 4 % while the thickness of the cookies using teff, teff-Nutrim, teff-OBC, and teff-WOF decreased ~6, 10, 0, 2% compared to cookies with wheat flour, respectively. The thickness of the cookies using teff-OBC was slightly increased compared with wheat flour cookies probably because oat bran in OBC expanded after heating and absorbing water. The differences were also found in the spread factor. Teff-Nutrim composites had the highest spread ratio (5.30) while the cookie containing teff-OBC composite had the lowest (4.83). In general, higher cookie spread factor indicates higher quality cookies. It was clear from the results that diameter, thickness, and spread factor were not greatly influenced by the use of teff flour and its oat composites compared to cookies with wheat flour.

Table 2. The geometrical properties of cookies

	Width		Thickness		Spread
	before bake	after bake	before bake	after bake	factor
	Cm	cm	cm	cm	width/thickness
Teff	6	6.18 ± 0.01 c	0.7	1.19 ± 0.01 d	5.21 ±0. 05 b
Teff: Nutrim 4:1	6	5.92 ± 0.00 d	0.7	1.12 ± 0.00 e	5.30 ± 0.00 a
Teff: OBC 4:1	6	6.18 ± 0.02 c	0.7	1.28 ± 0.01 a	4.83 ± 0.02 d
Teff :WOF 4:1	6	6.30 ± 0.02 b	0.7	1.24 ± 0.01 c	5.08 ± 0.04 c
Wheat flour	6	6.57 ± 0.01 a	0.7	1.26 ± 0.01 b	5.23 ± 0.03 b

Means ± standard deviation; n=3; means followed by the same letter within the same column are not significantly different ($P>0.05$).

3.6 Colour of Cookies

The cookies containing teff were dark in colour (*L**: 35.2) compared with the other cookies since the dimension *L** means lightness with 100 for white and 0 for black (Table 3). The value *a** indicates redness when positive and greenness when negative. All the cookies showed the redness in different degrees. The *a** value (14.09) of wheat flour cookies was the highest indicating more redness than the other cookies. All the cookies also showed the yellowness in different degree since *b** indicates yellowness when positive. All the cookies showed the yellowness in different degrees. Wheat flour cookies (*b**: 32.49) had the highest value of yellowness among the cookies. In general, ground cookie powders were lighter with less redness and more yellowness compared to the surface colour of cookies with some exceptions.

Table 3. The colour profile of cookies

	Colour (surface)		
	L*	*a**	*b**
Teff	35.20± 0.97 d	8.08 ±0.15 c	15.98 ± 0.33 d
Teff: Nutrim 4:1	36.19 ± 0.80 cd	7.90 ± 0.52 c	17.72 ± 0.34 c
Teff: OBC 4:1	37.22 ±0.32 bc	9.18 ± 0.36 b	19.09 ± 0.25 b
Teff :WOF 4:1	37.77 ± 0.57 b	9.10 ± 0.34 b	18.97 ±0.18 b
Wheat flour	52.30± 1.58 a	14.09 ± 0.49 a	32.49 ± 0.55 a
	Colour (ground powder)		
	L*	*a**	*b**
Teff	34.05 ± 0.90 e	8.84 ±0.19 b	19.91 ± 0.61 e
Teff: Nutrim 4:1	45.62 ± 0.49 b	7.29 ± 0.20 d	24.85 ± 0..47 b
Teff: OBC 4:1	38.41 ±0.12 d	8.96 ± 0.09 b	22.36 ± 0.74 d
Teff :WOF 4:1	43.10 ± 0.67 c	8.12 ± 0.31 b	23.35 ±0.44 c
Wheat flour	56.64 ± 0.23 a	10.20 ± 0.11 a	30.39 ± 0.22 a

Means ± standard deviation; n=3; means followed by the same letter within the same column are not significantly different (*P>0.05*).

3.7 Water Loss, Moisture Content, and Water Activity

Cookies with teff flour had the lowest water loss (9.95%) that was similar to cookies with teff-Nutrim (10.05%) and teff-WOF (10.11%) composites but they were lower than cookies with teff-OBC (10.44%) and wheat flour (10.32%, Table 4). Water losses during cooking ranged from 9.95% to 10.47%. All the cookies had similar moisture contents from 3.55% to 3.84%. Moisture is an important aspect of food stability and shelf-life. The growth and metabolic activities of bacteria, molds, and yeasts are retarded and eventually inhibited as the water activities (aw) of foods are decreased. Water activity of the cookies ranged from 0.246 to 0.298 (Table 4). They were much lower than the limit of *water activity* for spoilage *by bacteria,* yeasts and molds, approximately 0.90, 0.85-0.88, and 0.80, respectively. The rate of chemical reactions in food decreases much more slowly with reduced moisture content. Also, the enzymatic activity in foods may be significant at water activities as low as 0.30 (Smith, 2007). Additionally, water activity level affects lipid oxidation rates. Lipid oxidation is typically lowest when almond water activity (a_w) was ~0.25 to 0.35 (~3–4% moisture content), and increased above or below that water activity range (Huang, 2014). Overall, the differences were not intensive in water losses, moistures, and water activities among all cookies.

Table 4. The water loss during baking, moisture, and water activity

Products	Water loss during baking %	moisture %	Water activity (a_w)
Teff	9.95 ± 0.03 d	3.69 ±0.005 c	0.275±0.005 b
Teff: Nutrim 4:1	10.08 ± 0.04 cd	3.99 ±0.03 a	0.298±0.003 a
Teff: OBC 4:1	10.47 ± 0.05 a	3.86± 0.03 b	0.279 ±0.014 b
Teff :WOF 4:1	10.12 ± 0.01 c	3.70 ±0.05 c	0.268 ± 0.008 b
Wheat flour	10.35 ± 0.03 b	3.55 ±0.05 d	0.246 ± 0.006 c

Means ± standard deviation; n=3; means followed by the same letter within the same column are not significantly different (*P>0.05*).

3.8 Texture of Dough and Cookies

The texture evaluations of dough and cookies were presented in Table 5. The cookie dough containing teff-Nutrim had the highest texture value (6.28 N), followed by cookie dough containing teff-OBC (4.51 N) tested by a penetrating force using a 5 mm diameter probe. The cookie dough using teff flour had the lowest hardness value (2.55 N). The dough hardness from teff-WOF composites (3.73 N) was higher than both doughs with wheat (3.14 N) and with teff flour (2.55 N). The cookie dough hardness appeared to be related to WHC. The highest dough hardness value was from cookies containing teff-Nutrim composites with the highest WHC. In contrast, the cookies containing teff-Nutrim required the least cutting force (28.44 N) to break the cookies. The cookies containing wheat flour required the maximum cutting force (94.34 N) to break the cookies. It indicated that dough texture properties were greatly changed after heating. Hoseney and Rogers reported that hardness of the cookies was caused by the interaction of proteins and starch by hydrogen bonding (Hoseney & Rogers, 1994).

Table 5. The texture of cookies and dough

Products	Dough hardness Penetrating Force (N)	Cookie hardness Cutting force (N)
Teff	2.55 ± 0.10 e	64.33 ± 2.45 b
Teff: Nutrim 4:1	6.28 ± 0.10 a	28.44 ± 0.39 d
Teff: OBC 4:1	4.51 ± 0.20 b	57.27 ± 5.20 c
Teff :WOF 4:1	3.73 ±0.10 c	55.90 ± 3.92 c
Wheat flour	3.14 ±0.10 d	94.34 ± 4.41 a

Means ± standard deviation; n=3; means followed by the same letter within the same column are not significantly different (*P>0.05*).

3.9 Sensory Evaluation

The score for colour, texture, flavour, and overall ranged from 6.2 to 7.4, 6.4 to 7.2, 6.0 to 6.8, and 6.2 to 7.2, respectively based on the panelist assigned scores for each parameter using a 9-point hedonic scale (Table 6). The statistical results indicated that no significant differences were found in texture, flavour, and overall quality among cookies (Table 6). These results may be due to the fact that only a small portion (20%) was replaced by oat products, and also both teff and oat apparently had no distinct difference in flavour compared to wheat flour. In addition, the same formulation was used for all cookies. However, the scores of colour preference for wheat flour cookies were higher than cookies with teff and teff-oat composites. It may be due to the difference in colour between teff and wheat flour. Teff flour has brown colour while wheat flour has the light tan colour. If cocoa powder is used in the cookie formula for chocolate flavored cookies, it will make the colour of teff-oat cookies more attractive. Overall, sensory evaluation suggested that the cookies using teff-oat composites were acceptable in all respects.

Table 6. Sensory evaluation

Products	Colour	Texture	Flavour	Overall
Teff	7.17 ± 0.57 b	7.33 ± 0.56 a	7.17 ± 0.54 a	7.33 ± 0.53 a
Teff: Nutrim 4:1	6.83 ± 1.00 b	7.17 ± 1.15 a	6.83 ± 1.03 a	7.00 ± 1.15 a
Teff: OBC 4:1	7.83 ± 1.73 b	7.67 ± 1.08 a	7.50 ±1.31 a	7.67 ± 1.09 a
Teff :WOF 4:1	7.00 ± 0.58 b	7.50 ± 0.51 a	7.00 ± 0.53 a	7.17 ± 0.55 a
Wheat flour	8.33 ± 1.13 a	7.83 ± 0.57 a	7.67 ± 0.54 a	7.83 ± 0.58 a

Means ± standard deviation; n=26; means followed by the same letter within the same column are not significantly different. The number 1 referred to dislike extremely; 2 to dislike very much; 3 to dislike moderately; 4 to dislike slightly; 5 to neither like nor dislike; 6 to like slightly; 7 to like moderately; 8 to like very much; and 9 to like extremely.

4. Conclusion

These composites improved nutritional value, water holding capacity, and the pasting properties along with gluten free quality. In general, teff-oat composites are very suitable for preparing cookies that were acceptable in colour, flavour and texture qualities compared to the cookies with wheat flour. OBC and WOF probably are more suitable to use for cookies by cost. Information on the physical properties of these innovative cookies provides useful information for new functional foods.

References

AACC International. (2000). *Approved Methods of the American Association of Cereal Chemists* (10th ed.) Methods I0-50D and 10-52. AACC International, St. Paul, MN, U.S.A.

Ade-Omowaye, B. I. O., Taiwo, K. A., Eshtiaghi, N. M., Angersbach, A., & Knorr, D. (2003). Comparative Evaluation of the Effects of Pulsed Electric Field and Freezing on Cell Membrane Permeabilisation and Mass Transfer during Dehydration of Red Bell Peppers. *Innovative Food Science & Emerging Technologies, 4*, 177-188. http://dx.doi.org/10.1016/S1466-8564(03)00020-1

Coleman, J., Abaye, A. O., Barbeau, W., & Thomason, W. (2013). The suitability of teff flour in bread, layer cakes, cookies and biscuits. *International Journal of Food Sciences and Nutrition, 64*, 877-881. http://dx.doi.org/10.3109/09637486.2013.800845

Guha, M., Zakiuddin ALI, S., & Bhattacharya, S. (1998). Effect of barrel temperature and screw speed on rapid viscoanalyser pasting behaviour of rice extrudate. *International Journal of Food Science and Technology, 3*, 259-266. http://dx.doi.org/10.1046/j.1365-2621.1998.00189.x

Hoseney, R. C., & Rogers, D. E. (1994). Mechanism of sugar functionality in cookies: the science of cookie and cracker production. In H. Faridi (ed.), *American Association of Cereal Chemists* (Ist ed., pp. 203-225). St. Paul, MN.

Huang, G. (2014). *Almond shelf life factor*s. Technical summary. Almond Board of California.

Inglett, G. E. (2000). Soluble hydrocolloid food additives and method of making. U. S. Patent Number 6,060,519.

Inglett, G. E. (2011). Low-Carbohydrate Digestible Hydrocolloidal Fiber Compositions. U. S. Patent Number 7,943,766B2.

Inglett, G. E., Chen, D., & Liu. X. S. (2015). Functional properties of teff and oat composites. *Food and Nutrition Science, 6*, 1591-1602. http://dx.doi.org/10.4236/fns.2015.67065.

Kim, S., Inglett, G. E., & Liu, S. X. (2008). Content and molecular weight distribution of oat β-glucan in oatrim, nutrim, and C-trim products. *Cereal Chemistry, 85*, 701-705. http://dx.doi.org/10.1094/CCHEM-85-5-0701

Klopfenstein, C. F. (1988). The role of cereal beta-glucans in nutrition and health. *Cereal Food World, 33*, 865-869.

Lai, H. M., & Cheng, H. H. (2004). Properties of pregelatinized rice flour made by hot air or gum puffing. *International Journal of Food Science and Technology, 39*, 201-212. http://dx.doi.org/10.1046/j.0950-5423.2003.00761.x

Lai, L. S., & Liao, C. L. (2002). Steady and dynamic shear rheological properties of starch and decolorized Hsian-tsao leaf gum composite systems. *Cereal Chemistry, 79,* 58-63. http://dx.doi.org/10.1094/CCHEM.2002.79.1.58

Lee, S., & Inglett, G. E. (2006). Rheological and physical evaluation of jet-cooked oat bran in low calorie cookies. *International Journal of Food Science and Technology, 41,* 553-559. http://dx.doi.org/10.1111/j.1365-2621.2006.01161.x

Lee, M. H., Baek, M. H., Cha, D. S., Park, H. J., & Lim, S. T. (2002). Freeze-thaw stabilization of sweet potato starch gel by polysaccharide gums. *Food Hydrocolloids, 16,* 345-352. http://dx.doi.org/10.1016/S0268-005X(01)00107-2

Lee, S., Inglett, G. E., Carriere, C. J. (2004). Effect of nutrim oat bran and flaxseed on rheological properties of cakes. *Cereal Chemistry, 81,* 637-642. http://dx.doi.org/10.1094/CCHEM.2004.81.5.637

Lee, S., Warner, K., & Inglett, G. E. (2005). Rheological properties and baking performance of new oat β-glucan-rich hydrocolloids. *Journal of Agricultural and Food Chemistry, 53,* 9805-9809. http://dx.doi.org/10.1021/jf051368o

Madhujith, T., & Shahidi, F. (2007). Antioxidative and antiproliferative properties of selected barley (Hordeum vulgare L.) cultivars and their potential for inhibition of low-density lipoprotein (LDL) cholesterol oxidation. *Journal of Agricutlural and Food Chemistry, 55,* 5018-5024. http://dx.doi.org/10.1021/jf070072a

Rojas, J. A., Rosell, C. M., & Benedito de Barber, C. (1999). Pasting properties of different wheat flour-hydrocolloid systems. *Food Hydrocolloids, 13,* 27-33. http://dx.doi.org/10.1016/S0268-005X(98)00066-6

Rosell, C. M., Rojas, J. A., & Benedito de Barber, C. (2001). Influence of hydrocolloids on dough rheology and bread quality. *Food Hydrocolloids, 15,* 75-81. http://dx.doi.org/10.1016/S0268-005X(00)00054-0

SAS Institute INC. (1999). The SAS® system for Windows®, version 8e. Cary, NC.

Smith, P. G. (2007). *Applications of Fluidization to Food Processing Introduction* (pp. 116-117). Wiley-Blackwell.

Taha S. E., Shahira M. E., & Amani, A. S. (2012). Chemical and biological study of the seeds of Eragrostis tef (Zucc.) Trotter, *Natural Product Research: Formerly Natural Product Letters, 26,* 619-629. http://dx.doi.org/10.1080/14786419.2010.538924.

University of Maryland Medical Center. (2013). Retrieved from http://umm.edu/health/medical/altmed/supplement/vitamin-k#ixzz33t70aIeT

USDA Nutrient Database. (2014). Retrieved from http://ndb.nal.usda.gov/ndb/search/list

17

Concentrations of Some Trace Elements in Vegetables Sold at Maun Market, Botswana

Keagile Bati[1], Oarabile Mogobe[2] & Wellington R. L. Masamba[2]

[1]Department of Biological Sciences, University of Botswana. Botswana

[2]Okavango Research Institute, University of Botswana. Botswana

Correspondence: Oarabile Mogobe, Okavango Research Institute, Private Bag 285, Maun, Botswana. E-mail: omogobe@ori.ub.bw

Abstract

Contamination of vegetables with toxic metals is one of the most important contributing factors to ill health throughout the world, more so because vegetables are considered essential for human health and their consumption is highly recommended by health authorities. The aim of this study was to determine the concentrations of selected essential elements (Fe, Cu, Mn, Mo, Zn) and toxic elements (As, Cd, Cr, Pb) in common vegetables sold for human consumption in supermarkets and open market of Maun village, Botswana. Five vegetables (cabbage, rape, tomatoes, onions and potatoes) were purchased from different selling points, washed with de-ionised water, cut into small pieces and digested with aqua regia on a block digester, following the US. EPA method 200 - 7 and analysed for metal content using Inductively Coupled Plasma - Atomic Emission Spectroscopy (ICP-AES). The results showed that concentrations of essential and toxic metals varied with the type of vegetable and also with the market category (supermarket or street vendor). The highest concentration of essential elements was obtained from cabbage with a Zn concentration of 135.4mg/Kg and the lowest was from onion with a Mo concentration of 1.35mg/Kg. For toxic elements the highest concentration was obtained from rape vegetable with a Pb concentration of 4.73mg/Kg and the lowest from the same vegetable with Cr concentration below the detection limit. Also observed was that leafy vegetables, especially cabbage, had the highest concentrations of most trace metals. It was concluded that vegetables sold in Maun had sufficient levels of essential elements but also some had high concentrations of toxic metals. We thus recommend consumption of vegetables from the studied markets with reduced frequency to avoid metal poisoning.

Keywords: trace metals, vegetables, aqua regia

1. Introduction

Food quality and safety has become a global concern due to contamination of food products with toxins such as pesticides, heavy metals, mycotoxins and other microbiological contaminants (WHO, 2003; Kumar *et.al*, 2012). Agricultural produce, free of chemical contaminants is therefore one of the most important aspects of food safety since consumption of food products contaminated with metals may pose health risks to people. The sources of metal contaminants in food such as vegetables occurs through anthropogenic activities like mining, metal and chemical industries, motor vehicle emissions as well as agricultural practices like use of inorganic fertilizers (Das *et.al*, 2008).

The contamination of vegetables may also occur from irrigating with sewage and industrial water. These metals accumulate in agricultural soils, taken up by vegetable crops and get transferred to humans through consumption of vegetables. Examples of these trace elements include iron (Fe), copper (Cu), cobalt (Co), arsenic (As), mercury (Hg), gold (Au), silver (Ag) and others (Kumar *et.al*, 2012). Some of these elements (Fe, Cu, Ni, Zn) are common in our environment and diet and are actually necessary for good health. However, when they occur in high concentrations, they may cause acute or chronic toxicity (Das *et.al*, 2008) . Some metals like Pb, Hg, and Cr are toxic even at low concentrations and have been associated with many health issues like cancers, biochemical disorders and diseases of the nervous system (Orisakwe *et al.*, 2012).

Trace elements such as Cadmium (Cd), Arsenic (As) and Chromium (Cr) have been listed by the World Health

Organization (WHO) as carcinogens while mercury (Hg) and Lead (Pb) are associated with the developmental disorders in children (Sharma *et.al,* 2008; Singh *et al.* 2011). Studies carried out in various countries, investigating the impact of daily intake of vegetables contaminated with toxic metals, showed a strong correlation with prevalence of different types of cancers and other health issues like bone and the nervous system disorders, (Hough *et al.,* 2004; Kumar, *et.al,* 2012; Türkdoğan *et.al,* 2003). The aim of this study was to assess the concentrations of some trace elements (Fe, Cu, Pb, Cr, Cd, As, Mn, Mo, and Zn) in selected vegetables purchased from Maun central market area, in Botswana and assess their safety and potential contribution to good health.

2. Materials and Methods

2.1 Study Area

The study was conducted in Maun, a major town of Ngamiland District in the north west of Botswana (*figure 1*). Maun has a population of 60, 263 (CSO, 2011). The town is located in a tourism destination area, where most tourists pass through and spend some time before proceeding to the famous Okavango Delta. Although there are no chemical or metal industries within the area, use of inorganic fertilizers in commercial arable farming is common. Also, with its relatively high population, exposure to metal contaminants in vegetables needs investigation in order to protect the population from metal poisoning. Vegetables in the markets here are basically from two main sources; local suppliers and imports from South Africa.

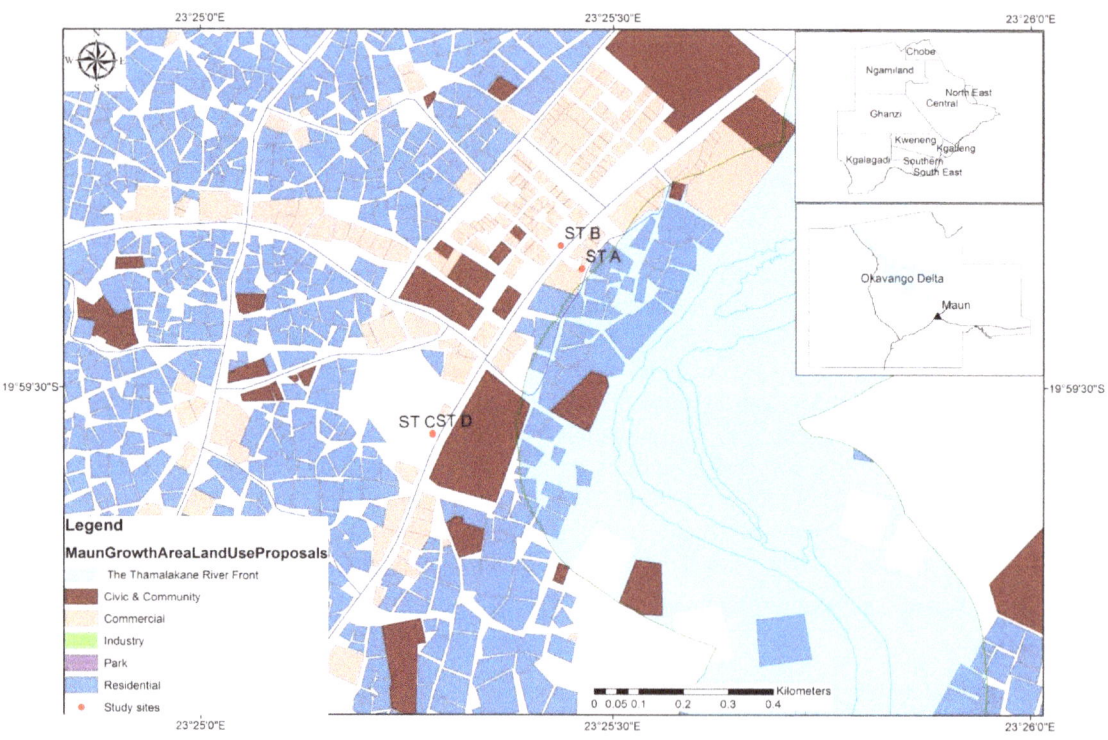

Figure 1. Location of the study area (Maun)

2.2 Chemicals and Reagents

All chemicals and reagents used were of analytical grade and purchased from Sigma - Aldrich company (USA). 37% hydrochloric acid (HCL) and 65% nitric acid (HNO₃) were used for digestion of the samples. Calibration standards were prepared from single element (1000 ppm) standard solutions of each element analysed. Internal quality control samples were used for quality assurance.

2.3 Instrument and Apparatus

Inductively coupled plasma optical emission spectrophotometer (ICP-OES), model Perkin Elmer, Optima 2100 with auto-sampler of model S10 was used for analyzing the trace metals of interest. All glassware were soaked overnight in 0.1 M nitric acid and then rinsed with deionized water to reduce chances of contamination and any other interference.

2.4 Sample Collection and Treatment

Vegetables were purchased from street vendors and supermarkets across the main market area of Maun. Names of vegetables purchased are shown in table 1. They were separated and placed into large labelled polyethylene sampling bags and transported to the Environmental laboratory at Okavango Research Institute for analysis. Vegetables were first washed with fresh running water to remove dirt, soil sediments and other surface contaminants as in Shah *et al*, 2013. The edible parts (table 1) were then further washed with deionized water, cut into small pieces, and then dried overnight in oven at 40 $^{\circ}$C. The dried samples were crushed and ground to powder using a clean mortar and pestle. The ground samples were put in well labelled polyethylene bags and kept at 4 $^{\circ}$C awaiting further processing (Mapanda *et al.*, 2005).

Table 1. Sampled vegetables and their characteristics

English Name	Scientific Name	Family	Part used	n
Cabbage	*Brassica oleracea var. capitata*	*Brassiaceace*	Leaf	12
Rape	*Brassica napus*	*Brassiaceace*	Leaf	27
Tomatoes	*Solanum lycopersicum*	*Solanaceace*	Fruit	27
Onion	*Allium Cepa*	*Alliaceace*	Bulb	27
Potatoes	*Solanum tuberosum*	*Solanaceace*	Tuber	12

2.5 Sample Digestion and Analysis

Extraction of metals from the samples was done by *aqua regia* following a method described by Taghipour and Mosaferi, (2013) with minor modifications. About 1.0 ± 0.01 g of the powdered samples were put into labelled digestion tubes to which aqua regia in ratio 3: 1 (12mL of conc. HCl + 4 mL of conc. HNO_3) was added. The contents were then mixed by swirling and then covered with digestion tube glass covers and allowed to stand for at least 12 hours in a fume cupboard. The digestion tubes containing the sample solutions were placed on a digestion block under fume-hood and heated at 95 $^{\circ}$C for 30 minutes. The samples were then cooled and filtered using a Whatman 47 mm filter paper. The filtrates were topped to 100 mL in volumetric flasks using 0.1 M HNO_3. An internal control sample (P124) was digested together with the samples. Prior to analysis calibration standard solutions of 1, 3 and 5 ppm for the metals were prepared from 1000 ppm stock solutions by pipetting the required volume into a 100 mL volumetric flask and diluting to the mark using 0.1 M Nitric acid (HNO_3). All samples were then analysed in triplicates for presence of trace metals using ICP-OES based on recommended wavelengths.

2.6 Statistical Analysis

Statistical analyses were conducted using Sigma Plot software, version 11. Mean and standard error were computed and expressed in mg/Kg dry weight of vegetables.

3. Results and Discussion

The results obtained from this study are presented in Tables 2 and 3. The sellers communicated that most vegetables sold by street vendors were grown in small gardens by small scale farmers along the local Thamalakane River. Some of the vegetables sold by these street vendors were actually harvested from their backyard gardens under the government poverty eradication scheme. On the other hand, supermarkets purchased their vegetables from large scale farmers in the country and imports from Zambia, Zimbabwe and South Africa. All potatoes sold by supermarkets were from South Africa. It was observed that distribution of metals in vegetables was not uniform as some vegetables recorded high concentrations of certain metals. This can be attributed to different crops having differing mineral uptake and accumulation and also different agricultural practices. It is also apparent from the results given on Table 2 that vegetables from supermarkets had higher concentrations of essential and toxic elements compared to those from street vendors. These differences were not statistically significant (P>0.05) and they may be due to the fact that commercial farmers tend to opt for intensive production practices, that is use of fertilizers, pesticides and wastewater for irrigation. All these products contain high levels of nutrients and minerals.

Table 2. Concentrations of essential elements (mg/Kg) in vegetables from street vendors (STV) and supermarkets (SP)

Vegetable	Seller	Zn	Fe	Cu	Mn	Mo
Brassica napus	STV	27.8 ± 1.85	76.8 ± 3.81	2.73 ± 0.27	70.2 ± 4.35	8.41 ± 4.00
	SP	47.3 ± 4.99	94.5 ± 9.99	5.63 ± 0.37	51.1 ± 4.84	2.83 ± 0.29
Solanum lycopersicum	STV	27.9 ± 3.22	55.7 ± 8.01	9.56 ± 0.79	17.9 ± 1.74	8.68 ± 4.00
	SP	45.5 ± 5.72	119.6 ± 32.8	14.7 ± 6.09	19.1 ± 1.40	3.19 ± 0.23
Allium cepa	STV	23.8 ± 2.20	40.7 ± 3.28	8.79 ± 0.57	18.4 ± 2.24	1.35 ± 0.08
	SP	40.1 ± 6.23	59.7 ± 3.94	3.19 ±0.33	19.8 ± 0.70	1.73 ± 0.13
Brassica oleracea var. capitata	SP	135.4 ± 12.1	33.1 ± 4.91	8.06 ± 0.71	38.6 ± 2.09	5.79 ± 0.94
Solanum tuberosum	SP	89.8 ± 5.98	54.6 ± 3.40	2.57 ± 0.12	17.2 ± 0.45	1.79 ± 0.08

Values are mean ± standard error, BDL: below detection level (< 5 µg/L)

Table 3. Concentrations of toxic metals (mg/Kg) in vegetables from street vendors (STV) and supermarkets (SP)

Vegetable	Seller	Cr	Cd	Pb	As
Brassica napus	STV	BDL	0.31 ± 0.02	0.88 ± 0.07	4.82 ± 2.96
	SP	2.02 ± 0.62	0.53 ± 0.05	4.73 ± 0.50	1.83 ± 0.07
Solanum lycopersicum	STV	0.07 ± 0.04	0.38 ± 0.03	1.42 ± 0.05	1.21 ± 0.31
	SP	0.95 ± 0.12	0.33 ± 0.02	0.88 ± 0.05	0.97 ± 0.29
Allium cepa	STV	0.003 ± 0.00	0.33 ± 0.02	1.62 ± 0.13	2.91 ± 0.10
	SP	1.01 ± 0.02	0.36 ± 0.04	1.39 ± 0.06	2.95 ± 0.89
Brassica oleracea var. capitata	SP	0.36 ± 0.06	0.80 ± 0.05	4.65 ± 0.34	4.50 ± 0.10
Solanum tuberosum	SP	0.98 ± 0.02	0.61 ± 0.05	3.62 ± .028	1.52 ± 0.16

Values are mean ± standard error, BDL: below detection level (< 5 µg/L).

Table 4. Recommended dietary allowances (RDA) and Adequate Intake (AI) for trace elements studied based on WHO (1989) and FDA (2001) standards

Element	RDA
Zinc	12 - 15 mg/day (RDA)
Iron	8 - 18 mg/day (RDA)
Copper	0.2 - 1.3 mg/day (RDA)
Manganese	1.8 - 2.3 mg/day (RDA)
Molybdenum	2 - 50 µg/day (AI)
Lead	<250 µg/day (AI)
Chromium	20 - 25 µg/day (AI)
Cadmium	0.5 to 0.8 µg/L (AI)
Arsenic	50 µg/L (water) (AI)

3.1 Distribution of Metals in Different Vegetables

3.1.1 Zinc

The concentration of zinc in vegetables collected from street vendors was found to range from 13.24 ± 0.68 mg/Kg in onions to 65.80 ± 6.86 mg/Kg in tomatoes. From supermarkets, zinc concentrations were in the range 17.74 ± 0.15 mg/Kg to 185.12 ± 5.08 mg/Kg in onions and cabbages respectively. In all samples, some vegetables recorded zinc concentration above the permissible limit of zinc in plants, which is 50 mg/Kg (WHO, 2007). Other studies (Taghipour & Mosaferi 2013, Harmanescu et al. 2011 and Arora et al. 2008) obtained zinc concentration ranges of 10 - 50 mg/Kg, 22.7 - 138.7 mg/Kg and 21 - 46 mg/Kg in Iran, Romania and India respectively in the same type of vegetables and these results are comparable to values obtained in this study. Therefore, based on the determined concentrations of zinc, consumption of 100 g of most vegetables analyzed would provide more than 100% of the daily required amount of zinc for good health shown on Table 4, hence they are nutritious especially for pregnant and lactating mothers who need at least 14 mg Zn/day. Zinc is known to be a cofactor of many enzymes involved in metabolic pathways and important in cellular growth hence an important trace element for human life. However, overconsumption of the vegetables would be unsafe due to the tendency of zinc interfering with copper metabolism (Umar et al., 2014). The high concentrations of zinc in vegetables may be associated with rampant use of zinc fertilizers and metal based pesticides (Singh et al., 2010).

3.1.2 Iron

This study found out that the concentration of iron in vegetables bought from street vendors was in the range 7.98 - 122.7 mg/Kg and 17.7 - 177.8 mg/Kg for those purchased from supermarkets. The lowest concentration of iron recorded was 7.98 mg/Kg in tomatoes obtained from street vendors while the highest concentration of iron was 177.8 mg/Kg in rape obtained from supermarket stores. The permissible concentration limit of iron in plants is 20 mg/Kg (WHO, 2007) therefore some of the investigated vegetables have concentrations above the permissible limit. A study by Umar et al. (2014) found the mean concentration of iron in onions as 139.383 ± 13 mg/Kg which is higher than the result of this study. Concentration of iron in cabbage was in the range 16 - 60 mg/Kg. Iron is important in the human diet as it facilitates the oxidation of fats, proteins and carbohydrates to control body weight which is an important factor in diseases like diabetes. It also plays a pivotal role in the transportation of oxygen around the body. The recommended dietary allowance of Fe is 8 - 18 mg/day (FDA, 2001) shown on table 4, therefore; the studied vegetables are able to sustain the daily requirement of Fe in the body for consumption of a 100g portion of vegetables. The vegetables are also suitable for pregnant and lactating mothers who require at least 27 mgFe/day (WHO, 2003).

3.1.3 Copper

The concentration of copper in vegetables collected from the street vendors were in the range 1.14 - 14.8 mg/kg of which the lowest concentration (1.14 mg/Kg) was recorded in rape while the highest concentration of 14.8 mg/Kg recorded in tomatoes. On the other hand, concentrations of copper in vegetables obtained from supermarkets ranged between 1.28 - 61.93 mg/Kg with the least concentration in tomatoes and the highest still in tomatoes. Concentrations of copper were found in the range of 0.75 - 2.00 mg/Kg in tomatoes grown in Glen Valley farms in Botswana (Dikinya & Areola, 2010). Harmanescu et.al, (2011) found mean concentrations of 1.77 ± 0.23, 1.37 ± 0.14 mg/Kg of copper in cabbage and onions. The permissible limit of copper in plants is 10 mg/Kg (WHO, 2007) and vegetables sold by street vendors were within the limit, except for *solanum lycopersicum* (14.7 ± 6.09) sold by supermarkets. However, some vegetables, like some cabbages and rape sold by supermarkets were above the permissible limit. The recommended dietary allowance of copper by the Food and Drug Administration (FDA) is 0.7 - 0.9 mg/daily for a normal adult body (FDA, 2001). Therefore, based on the mean concentrations of copper in all analyzed vegetables a 100g portion of vegetables is able to supply the daily required amount of copper to the human body without need for use of supplements. Adequate amounts of copper are needed for a healthy body as it is very vital in the functioning of the liver and kidneys (Arif et al, 2011).

3.1.4 Manganese

Manganese is an important trace element in the human body. It acts as a catalyst and cofactor in many enzymatic processes involved in the synthesis of fatty acids and glycoproteins which coat body cells and protect against invading pathogens (Umar et al., 2014). Results of this study found out that in vegetable samples collected from street vendors, the concentration of manganese was in the range 9.05 - 95.7 mg/Kg with high concentrations in rape. However, for supermarkets the concentrations were in the range 11.1 - 49.9 mg/Kg. Other studies (Harmanescu et al., 2011) obtained concentrations of 4.07 ± 0.5, 10.47 ± 1.5 mg/Kg for onions and cabbages respectively. The recommended dietary allowance for manganese is 1.800 - 2.300 mg/day (FDA, 2001) and the obtained concentrations shows that the analyzed vegetables can individually sustain the required daily intake on just a 100g portion. Intake of lower than the required amount of manganese can result in manganese deficiency which causes defective muscular coordination and retarded nerve development and function (Liu et al., 2013).

3.1.5 Molybdenum

Analytical results of this study shows that the concentrations of molybdenum in vegetable samples collected from street vendors were in the range 0.31 - 54.6 mg/Kg while vegetable samples collected from supermarkets had concentrations in the range 1.02 - 13.7 mg/Kg. The highest concentration of molybdenum in vegetables obtained from street vendors was 54.6 and 52.1 mg/Kg were recorded in rape and tomatoes respectively. The recommended dietary allowance for molybdenum is 34 - 45 µg/day (FDA, 2001). Therefore based on the mean concentrations of each vegetable, it can be noted that consumption of 100 g portion each vegetable would provide more than 100% of the required daily intake. Molybdenum act as a cofactor in certain essential enzymes that play a vital role in carbohydrate metabolism, utilization of iron, sulfite detoxification, and uric acid formation (Balch and Balch, 1990). It has also been found out that molybdenum works with riboflavin (vitamin B2) to incorporate iron into hemoglobin hence playing a pivotal role in production of red blood cells (Powers,

2003). Deficiency of molybdenum has been linked to sexual impotence in older males, mouth and gum diseases (Balch and Balch, 1990). However, according to the Institute of Medicine, Food and Nutrition Board, (2001) excessive amount of molybdenum in the body may result in anemia and low white blood cell counts due to lack of copper as it interferes with copper metabolism.

3.1.6 Lead

Lead is the most recognized toxic environmental pollutant. This has resulted in more studies being carried out to determine its source and effects on humans, hence formulation of stringent regulations like ban of leaded petrol worldwide (Wang, 2005). Toxic levels of lead have been associated with encephalopathy seizures and mental retardation (Umar et al., 2014). The results of this study show that vegetable samples collected from street vendors had concentrations of lead in the range 0.37 - 2.93 mg/Kg while those from supermarkets the concentration range was 0.52 - 8.89 mg/Kg. Some of results were above the FAO/WHO maximum permissible limit that is 2 mg/Kg. Rape and potatoes sold by supermarkets, had high concentrations of lead which was above the permissible limit and therefore regarded unsafe for human consumption. Other studies (Guerra et al. 2012; Harmanescu et al. 2011; Abbas et al. 2010; Yang et al. 2011) obtained ranges; 470 - 1660 μg/Kg, 4.5 - 21.8 mg/Kg, 18 - 150 μg/Kg and 6 - 16 mg/Kg in Brazil, Nigeria, Pakistan and China respectively.

3.1.7 Chromium

The concentrations of chromium in vegetable samples collected from street vendors were in the range 0.00 - 0.48 mg/Kg. The highest concentration (0.48 mg/Kg) was obtained in tomatoes. However, many vegetables obtained from street vendors had chromium levels below the detection level. On the other hand, the concentration range of chromium in vegetable samples collected from supermarkets was 0.00 - 7.26 mg/Kg, with highest concentration of 7.26 mg/Kg obtained in rape. Other studies, (Taghipour & Mosaferi 2013; Maobe et al. 2012; Guerra et al. 2012b) obtained chromium concentrations in the range 0.32 - 10.96 mg/Kg, 58 - 118 μg/Kg and 7 - 43 μg/Kg in Iran, Kenya and Brazil, respectively. The adequate intake of chromium 20 - 25 μg/day (FDA, 2001) and therefore, some of the vegetables can provide more than the required concentration hence may result in toxic effects. At levels below 105 mg/day (US EPA, 2010), chromium is important for effective insulin activity and DNA transcription (Guerra et al. 2012b). The maximum permissible limit of chromium in food is 2300 μg/Kg (FAO/WHO, 2003) therefore all vegetables analyzed were regarded as safe for human health as the concentrations of chromium were all below the maximum permissible limit.

3.1.8 Cadmium

Cadmium is a hazardous element which has been associated with various complications like bone de-calcification, renal, membrane and DNA damages (Guerra et al. 2012). Research studies have found that vegetables contribute about 70% cadmium intake by humans, varying according to the level of consumption (Wagner, 1993 in Rivera-becerril et al. 2002). Results of this study show that the concentration of cadmium in vegetables collected from street vendors was in the range 0.18 - 0.56 mg/Kg while it was 0.19 - 1.00 mg/Kg for vegetable samples from supermarkets. Other studies (Dikinya & Areola, 2010; Sharma et al, 2009) obtained concentrations in the ranges 0.04 - 0.14 mg/L, 0.10 - 4.30 mg/Kg in Botswana and India respectively. The permissible limit of cadmium in plants recommended by FAO/WHO is 0.02 mg/Kg. The analyzed vegetables have concentrations above the permissible limit hence regarded as unsafe for human consumption

3.1.9 Arsenic

The concentrations of arsenic in vegetable samples collected from street vendors was in the range 0.24 - 45.5 mg/Kg while that of supermarkets was 0.08 - 12.0 mg/Kg. Other studies (Baig and Kazi, 2012; Ramirez-Andreotta, 2013) obtained concentrations in the ranges; 962 - 4810 μg/Kg and 10 - 1960 μg/Kg in Pakistan and Arizona in United States of America respectively. The permissible limit by FAO/WHO for arsenic is 100 μg/Kg (Abbas et al., 2010) therefore; most of the analyzed vegetables were unsafe for consumption since the concentrations were above the permissible limits. Arsenic is extremely toxic and long term exposure results in health effects like cardiovascular diseases and diabetes (Abbas et al., 2010; Zhuang et al., 2009).

4. Conclusion

The concentrations of essential trace metals (Zn, Fe, Mn, Mo, Cu) were found to be high enough to meet the recommended daily intake hence it was concluded that vegetables sold in Maun were good for human health and may contribute towards preventing nutritional mineral deficiencies in people who eat vegetables in Maun. Vegetables from supermarkets had higher concentrations of essential metals compared to vegetables from street vendors probably due to intensive use of fertilizers and wastewater for irrigation. In terms of toxic metals (Pb, Cr, Cd, As), concentrations greatly varied between vegetable type and source (vendors and supermarkets).

Vegetables from street vendors generally had lower concentrations of toxic metals compared to the ones obtained from supermarkets. This could still be attributed to intensive use of fertilizers and pesticides by the more sophisticated farmers who supply supermarkets. From the results of this study, the conclusion is that most vegetables from Maun supermarkets have higher concentrations of both essential and toxic metals (above WHO acceptable limits) and therefore must be consumed less frequently to avoid metal poisoning. Vegetables from street vendors have good levels of essential minerals and lower concentrations of toxic metals and therefore can be consumed freely for promoting good health. It is also concluded that leafy vegetables accumulate the highest concentrations of both essential and toxic metals.

Acknowledgements

The authors would like to thank the Okavango Research Institute for funding this study.

References

Abbas, M., Parveen, Z., Iqbal, M., Riazuddin, M., Iqbal, S., Ahmed, M., & Bhutto, R. (2010). Monitoring of toxic metals (cadmium, lead, arsenic and mercury) in vegetables of Sindh, Pakistan. *Kathmandu University Journal of Science, Engineering and Technology.*

Arif, I. A., Khan, H. A., Homaidan, A. A. Al, & Ahamed, A. (2011). Determination of Cu , Mn , Hg , Pb , and Zn in the Outer Tissue Washings, Outer Tissues , and Inner Tissues of Different Vegetables Using ICP-OES, *20*(4), 835-841.

Arora, M., Kiran, B., Rani, S., Rani, A., Kaur, B., & Mittal, N. (2008). Heavy metal accumulation in vegetables irrigated with water from different sources. *Food Chemistry*, *111*(4), 811-815. https://doi.org/10.1016/j.foodchem.2008.04.049

Baig, J. A., & Kazi, T. G. (2012). Translocation of arsenic contents in vegetables from growing media of contaminated areas. *Ecotoxicology and Environmental Safety*, *75*(1), 27-32. http://dx.doi.org/10.1016/j.ecoenv.2011.09.006

Balch, J. F,, & Balch, P. A., (1990). Prescription for nutritional healing: A practical A-Z reference to drug-free remedies using vitamins, minerals, herbs & food supplements. New York: *Avery Publishing.*

Central Statistics Office (CSO), 2011 Botswana Population and Housing Census, Alphabetical Index of Villages. http://www.cso.gov.bw/templates/cso/file/File/2011%20Census%20_Alphabetical%20Index%20_Populatio n%20of%20Villages.pdf (Accessed on 03/10/2016).

Das, N., Vimala, R., & Karthika, P. (2008). Biosorption of heavy metals - An overview. *Indian Journal of Biotechnology.*

Dikinya, O., & Areola, O. (2010). Comparative analysis of heavy metal concentration in secondary treated wastewater irrigated soils cultivated by different crops, *7*(2), 337-346.

EPA. (2009). National Primary Drinking Water Regulations. *Environmental Protection Agency.* https://www.epa.gov/dwregdev/drinking-water-regulations-and-contaminants.

FAO/WHO. (1999). Codex Alimentarious Commission. Draft Maximum Levels for Lead. CX/FAC 00/24. Joint FAO/WHO Food Standards Programme Codex Committee on Food Additives and Contaminants.

FAO/WHO. (2003). Food and Agriculture Organization of the United Nations World Health Organization JOINT FAO / WHO EXPERT COMMITTEE ON FOOD ADDITIVES Sixty-first meeting, (52). Retrieved from Http;//www.who.com

FDA. (2001). Determination Of Food Contaminants. *Food and Drug Administration.* http://www.fda.gov/Food/GuidanceRegulation/GuidanceDocumentsRegulatoryInformation.

Guerra, F., Trevizam, A. R., Muraoka, T., Marcante, N. C., & Caniatti-Brazaca, S. G. (2012). Heavy metals in vegetables and potential risk for human health. *Scientia Agricola.* https://doi.org/10.1590/S0103-90162012000100008

Harmanescu, M., Alda, L., Bordean, D., Gogoasa, I., & Gergen, I. (2011). Heavy metals health risk assessment for population via consumption of vegetables grown in old mining area; a case study: Banat County, Romania. *Chemistry Central Journal.* https://doi.org/10.1186/1752-153X-5-64

Hough, R. L., Breward, N., Young, S. D., Crout, N. M. J., Tye, A. M., Moir, A. M., & Thornton, I. (2004). Assessing potential risk of heavy metal exposure from consumption of home-produced vegetables by urban populations. *Environmental Health Perspectives*, *112*(2), 215-221. https://doi.org/10.1289/ehp.5589

Institute of Medicine, Food and Nutrition Board, (2001). Dietary reference intakes for vitamin A, vitamin K, arsenic, boron, chromium, copper, iodine, iron, manganese, molybdenum, nickel, silicon, vanadium, and zinc. Washington, DC: *National Academy Press*.

Kumar, A., Zaidi, A., Wani, P. A., Khan, M. S., & Kurek, E. (2012). Toxicity of Heavy Metals to Legumes and Bioremediation. *Management*, 163-178. http://dx.doi.org/10.1007/978-3-7091-0730-0

Kumar, N., Soni, H., & Kumar, R. (2010). Characterization of Heavy Metals in Vegetables Using Inductive Coupled Plasma Analyzer (ICPA). *Journal of Applied Sciences and Environmental Management*. http://dx.doi.org/10.4314/jasem.v11i3.55131

Liu, X., Song, Q., Tang, Y., Li, W., Xu, J., Wu, J., … Brookes, P. C. (2013a). Human health risk assessment of heavy metals in soil-vegetable system: A multi-medium analysis. *Science of the Total Environment*. http://dx.doi.org/10.1016/j.scitotenv.2013.06.064

Liu, X., Song, Q., Tang, Y., Li, W., Xu, J., Wu, J., … Brookes, P. C. (2013b). Human health risk assessment of heavy metals in soil-vegetable system: A multi-medium analysis. *Science of the Total Environment*. http://dx.doi.org/10.1016/j.scitotenv.2013.06.064

Maobe, M. A. G., Gatebe, E., Gitu, L., & Rotich, H. (2012). Profile of Heavy Metals in Selected Medicinal Plants Used for the Treatment of Diabetes , Malaria and Pneumonia in Kisii Region , Southwest Kenya, *6*(3), 245-251. http://dx.doi.org/10.5829/idosi.gjp.2012.6.3.65128.

Mapanda F, Mangwayana E, Nyamangara J, (2005). The effect of long-term irrigation using wastewater on heavy metal contents of soils under vegetables in Harare, Zimbabwe. *Agr Ecosyst Env., 107*(2-3), 151-165. https://doi.org/10.1016/j.agee.2004.11.005

Orisakwe, O. E., Kanayochukwu, N. J., Nwadiuto, A. C., Daniel, D., & Onyinyechi, O. (2012). Evaluation of Potential Dietary Toxicity of Heavy Metals of Vegetables. http://dx.doi.org/10.4172/2161-0525.1000136

Powers H.J, (2003). Riboflavin (vitamin B-2) and health. American Journal of Clinical Nutrition, *77*, 1352-60.

Ramirez-Andreotta, M. D., Brusseau, M. L., Artiola, J. F., & Maier, R. M. (2013). A greenhouse and field-based study to determine the accumulation of arsenic in common homegrown vegetables grown in mining-affected soils. *The Science of the Total Environment, 443*, 299-306. http://dx.doi.org/10.1016/j.scitotenv.2012.10.095

Rivera-becerril, F., Calantzis, C., Turnau, K., Caussanel, J., Belimov, A. A., Gianinazzi, S., … Gene, D. (2002). Cadmium accumulation and buffering of cadmium-induced stress by arbuscular mycorrhiza in three Pisum sativum L . genotypes. *Journal of Experimental Botany, 53*(371), 1177-1185. https://doi.org/10.1093/jexbot/53.371.1177

Sharma, R. K., Agrawal, M., & Marshall, F. M. (2009). Heavy metals in vegetables collected from production and market sites of a tropical urban area of India. *Food and Chemical Toxicology, 47*(3), 583-591. https://doi.org/10.1016/j.fct.2008.12.016

Singh, A., Sharma, R. K., Agrawal, M., & Marshall, F. M. (2010). Health risk assessment of heavy metals via dietary intake of foodstuffs from the wastewater irrigated site of a dry tropical area of India. *Food and Chemical Toxicology, 48*(2), 611-619. https://doi.org/10.1016/j.fct.2009.11.041

Singh, R., Gautam, N., Mishra, A., & Gupta, R. (2011). Heavy metals and living systems: An overview. *Indian Journal of Pharmacology, 43*(3), 246-253. https://doi.org/10.4103/0253-7613.81505

Taghipour, H., & Mosaferi, M. (2013). Heavy metals in the vegetables collected from production sites. *Health Promotion Perspectives, 3*(2), 185-93. http://dx.doi.org/10.5681/hpp.2013.022

Türkdoğan, M. K., Kilicel, F., Kara, K., Tuncer, I., & Uygan, I. (2003). Heavy metals in soil, vegetables and fruits in the endemic upper gastrointestinal cancer region of Turkey. *Environmental Toxicology and Pharmacology*. http://dx.doi.org/10.1016/S1382-6689(02)00156-4

Umar, M. A., Salihu, Z. O. O., & Pepper, C. (2014). Heavy metals content of some spices available within FCT-Abuja , Nigeria . *International Journal of Agricultural and Food Science, 4*(1), 66-74.

Wang, X., Sato, T., Xing, B., & Tao, S. (2005). Health risks of heavy metals to the general public in Tianjin, China via consumption of vegetables and fish. *Science of the Total Environment*. http://dx.doi.org/10.1016/j.scitotenv.2004.09.044

WHO. (2003). *Codex Alimentarius Commission Procedural Manual*.

http://www.fao.org/fao-who-codexalimentarius/procedures-strategies/procedural-manual/en.

WHO. (2007). Specifications for the identity and purity of food additives and their toxicologicalevaluation: some emulsifiers and stabilizers and certain other substances. Geneva: FAO and WHO.

Yang, Q. Wei, Xu, Y., Liu, S. Jiang, He, J. Feng, & Long, F. Yan. (2011). Concentration and potential health risk of heavy metals in market vegetables in Chongqing, China. *Ecotoxicology and Environmental Safety*, *74*(6), 1664-1669. https://doi.org/10.1016/j.ecoenv.2011.05.006

Zhuang, P., McBride, M. B., Xia, H., Li, N., & Li, Z. (2009). Health risk from heavy metals via consumption of food crops in the vicinity of Dabaoshan mine, South China. *Science of the Total Environment*, *407*(5), 1551-1561. https://doi.org/10.1016/j.scitotenv.2008.10.061

18

Nutrition Intakes and Nutritional Status of School Age Children in Ghana

Justina Serwaah Owusu[1,2], Esi Komeley Colecraft[2], Richmond Aryeetey[3], Joan A. Vaccaro[1], & Fatma G. Huffman[1]

[1]Department of Dietetics and Nutrition, Robert Stempel College of Public Health and Social Work, Florida International University, Miami, FL, USA

[2]Department of Nutrition and Food Science, University of Ghana, Accra, Ghana

[3]School of Public Health, University of Ghana, Accra, Ghana

Correspondence: Fatma G. Huffman, Department of Dietetics and Nutrition, Robert Stempel College of Public Health and Social Work, Florida International University, 11200 SW, 8th ST, AHC-5 Room 306, Miami, FL 33199, USA. E-mail: huffmanf@fiu.edu

Abstract

This paper compares nutrition intakes and nutritional status of school children from two public schools in neighbouring communities of Ghana with different school feeding programmes. One hundred and eighty-two caregiver and school-age child pairs were interviewed concerning socio-demographics, dietary practices, and food security in a cross-sectional design. The independent t-test was used to compare the contribution of the publicly funded Ghana School Feeding Programme and private School Feeding Programme meals to total daily nutrient intakes of the children. Predictors of nutritional status of the children were assessed using logistic regression models. The private school feeding programme contributed more energy, protein, and micronutrients as compared to the government school feeding programme. About two-thirds (67.0%) of the children were stunted, underweight, or anaemic. Child's age was a significant predictor of stunting. Undernutrition was prevalent among children from both programmes. Improved quality of diet from the feeding programmes may contribute to addressing malnutrition in these children.

Keywords: nutrient intakes, nutritional status, school-age children, school feeding programmes, Ghana

1. Introduction

Undernutrition among school-age children (SAC) is a public health problem in developing countries (Best, Neufingerl, Van Geel, van den Briel, & Osendarp, 2010). The most commonly reported nutrition problems among SAC include underweight and micronutrient deficiencies of iron, zinc, iodine, and vitamin A (Best et al., 2010). While nationwide data on the nutritional status of Ghanaian SAC are currently limited, a study in the Eastern region of the country found that 44% of the 645 rural SAC assessed were stunted, and 70% of them were anaemic (Fentiman, Hall, & Bundy, 2001). More recently, a study in Ashanti region found 52.2% of primary five pupils who were either participants or non-participants of Ghana School Feeding Programme (GSFP) were stunted and 46.5% of them underweight (Danquah, Amoah, Steiner-Asiedu, & Opare-Obisaw, 2012). In Ghana, 13% and 19% of preschool age children are underweight and stunted, respectively (Ghana Statistical Service [GSS], Ghana Health Service [GHS], and Inner City Fund [ICF] Macro, 2015). Uncorrected, undernutrition among children in preschool age tracks into school-age (Bundy, 2009).

Undernutrition negatively affects children's health and capacity to perform in school. Researchers have reported a link between malnutrition and poor school attendance, intelligence quotient, school achievement, and morbidity among SAC (Ghazi, Isa, Aljunid, Tamil, & Abdalqader, 2012; Mukudi, 2003; Omwami, Neumann, & Bwibo, 2011; Schaible & Kaufmann, 2007). Malnutrition among SAC in Africa has been linked to morbidity, hygienic practices, dietary intakes and family socioeconomic status (Herrador et al., 2014; Mesfin & Berhane, 2015; Mwaniki & Makokha, 2013; Ndukwu, Egbuonu, Ulasi, & Ebenebe, 2013). Interventions for improving nutrition and health of SAC include deworming, health, nutrition and hygiene education, provision of portable water and sanitary latrines, micronutrient supplementation as well as school feeding programmes (World Food Programme [WFP] and United Nations Children's Fund [UNICEF], 2005). Well-designed school-feeding programmes tailored to specific needs of SAC have been effective in improving nutritional outcomes such as nutrient intake, physical growth, as well as micronutrient status (Adelman, Gilligan, & Lehrer, 2008). Primarily, school-feeding programmes have been established to address hunger and its negative effect on the nutritional status and learning capacity of school-age children (WFP, 2015). Although, the provision of school meals to SAC cannot reverse all damage caused by early nutritional deficits (Bundy, 2009), many studies have reported nutritional and health benefits of school feeding programmes (Afridi, 2010; Ahmed, 2004; Arsenault et al., 2009; Musamali, 2007; Neervoort et al., 2013; Neumann, Murphy, Gewa, Grillenberger, & Bwibo, 2007; Zeba, Martin Prevel, Some, & Delisle, 2006). A systematic review of school-feeding programmes have reported improvements in weight, height in younger children (6-8 years), school attendance, mathematics performance, bone mineral density, arm muscle, concentrations of B – vitamins and behaviour among participants (Kristjansson et al., 2007). It has been recommended that school feeding meals should meet at least 30% of energy and micronutrients need of children (Bhatia, 2013).

School-feeding programmes are implemented across the globe in both developing and developed nations (Bundy, 2009). The Ghana School Feeding Programme (GSFP) which started in 2005, aims to curb hunger and malnutrition through the provision of one nutritious hot meal prepared from locally-grown foodstuffs on school days (Ghana School Feeding Programme [GSFP], 2010). Longitudinal assessment of GSFP has reported improvements in school attendance, enrolment and educational performance among children (Abotsi, 2013). There is limited evidence, however, on nutrient intakes and nutritional status of SAC enrolled in school feeding programmes in Ghana. Additionally, there are limited studies that compare nutrient intakes of SAC in public-versus private-school feeding programmes feeding programmes. This paper presents evidence from a study which assessed dietary energy and nutrient intakes from school meals and nutritional status of school-age children enrolled in two school feeding programmes.

2. Materials and Methods

2.1 Study Design and Population and Selection

This was a comparative cross-sectional study with school children attending two public primary schools in neighbouring communities in the La- Nkwatanang Madina District in the Greater Accra region. These children were in classes 1 to 6 and between the ages of 6 and 12 years. One of the schools had a publicly funded School Feeding Programme (GSFP) and the other, a private School Feeding Programme (NSFP). Study procedures were explained to children in each class. Since there were more children at the GSFP site (300) compared to the NSFP site (125), all school-age children at GSFP site (n=300) were asked to choose a concealed ballot for participation status: eligible or not eligible.

Caregivers of the children were verbally informed about the study at a Parent Teacher Association (PTA) meeting in each school. Caregivers provided signed informed consent for themselves and their children's participation in the study. Additionally, all participating children signed a child's assent form, prior to being interviewed. A total of 232 caregivers (112 from NSFP and 120 from GSFP) responded positively to permission slips sent out. Out of the total consenting respondents, complete interviews were obtained from 214 caregiver- and child- pairs (100 children from NSFP and 114 from GSFP). Eighteen children were discovered to be ineligible (older than 12yr; or did not have no data on academic performance in the past school term). Thirty-two of the school-age children did not participate in the school feeding programmes and were eliminated in the final analysis. Their socio-demographic data was compared to those (182 children) participating in the school feeding programmes and there were no significant differences.

2.2 Data Collection Procedures

Data collection took place from January, 2013 through April 2013. Data collected included dietary intake assessment, anthropometry measurements and assessment of haemoglobin levels. Two sets of questionnaires were developed and pretested in a similar community in the La- Nkwatanang Madina District in the Greater Accra region prior to the study. Caregivers provided information on their personal and household characteristics, household food security and child morbidity, and these were recorded using a questionnaire. Another questionnaire was administered to the school-age children during break times at school and solicited information on personal characteristics and dietary habits throughout the day, both at school and at home.

2.3 Dietary Intake Assessment

The 24-hour recall method was used to collect information on all foods (including school meals) and beverages, except water, consumed by the school-age children on two non-consecutive school days. Days where the children were absent from school due to ill-health or other reason, or when the feeding programme did not serve meals to any of the children were excluded from the recall days. In recalling foods consumed in the past 24-hours, the SAC were also asked to indicate the time and source of the food (whether purchased, home-made, or school-meal) for each eating event. Household measures and wooden food models were used to help children estimate quantities of foods consumed. For foods that the children reported purchasing from food vendors within the school compound or the community, the cost of food was obtained and the same quantities were purchased and weighed to estimate the quantities consumed by the children. Additionally, a sample of foods served to children by the feeding programmes were weighed daily over a five-day period to estimate the average serving size provided. Caregivers or older siblings helped in estimating the quantities of foods eaten at home for children younger than seven years old who could not estimate their food quantities.

2.4 Assessment of Anthropometry and Haemoglobin Levels

Height and weight measurements were completed for SAC using standard procedures (Centers for Disease Control and Prevention [CDC], 2015). The children's heights were measured to the nearest 0.1 cm with a Tanita HR-200 stadiometer and a Tanita BWB 800 weighing scale was used to take their weights to the nearest 0.1kg. Height and weight measurements were taken in duplicates and the average recorded. Haemoglobin levels of 54% (n=99; 48 from GSFP site and 51 from NSFP site) of the school-age children were assessed. Selection of children was done by choosing every other child from the list of participants in order of interview. However, three children who did not want to participate in haemoglobin testing were replaced from the sample. Two to three drops of blood, enough to fill a cuvette, was obtained from each child using the finger prick method by a professional phlebotomist (Inter Tribal Council of Arizona [ITCA], 2012). The haemoglobin level of each participant was measured in grams per decilitre using the Hb 201$^+$ Hemocue machine.

2.5 Statistical Analysis

SPSS version 16. 0 was used for data management and analysis. Dietary information was converted to energy and nutrients using the Research to Improve Infant Nutrition and Growth (RIING) Project Nutrient Database available at Department of Nutrition and Food science, University of Ghana. Anthropometric data of weight and height were converted to height for age Z scores (HAZ) and BMI for age Z scores (BAZ) using WHO Anthro Plus version 10.4. Haemoglobin levels of school-age children were categorized by anaemia status (anaemic vs. non-anaemic) based on age specific cut offs set by WHO (World Health Organization [WHO], 2011). Dietary diversity scores (DDS) were also calculated using the Food and Agriculture Organization (FAO) guideline (Kennedy, Ballard, & Dop, 2011). Descriptive statistics was used to summarize categorical and continuous variables. The independent *t*-test was used to compare the contribution of NSFP and GSFP meals to intakes of school-age children and total daily intakes of school-age children. Predictors of nutritional status of SAC were assessed using logistic regression models. Level of significance was set at $P < 0.05$.

3. Results

3.1 Socio-Demographic and Household Characteristics of School Age Children and Their Caregivers in the Study

The mean age of caregivers was 41 ± 11 years and most of them (68.1%) were the children's biological parents (Table 1). The majority were either Ga/Ga-Adangme or Ewe ethnicity (61.6%) and 70% had completed at least junior secondary or middle school. Caregivers were mainly traders (62.6%) and the majority were married (81.3%). Overall, the most common economic activities of the caregiver's spouses were masonry, carpentry, electricians, driving, and tailoring (vocational employment). Majority (70.2%) of caregivers reported a household income of 500 Ghana cedis (USD 256 in 2013) or less. The average household size was 6 ± 2 members comprising of a mean of 3 ± 1 adults and 4 ± 2 children (<18yrs). Caregivers of school-age children in the GSFP and NSFP had similar socio-demographic characteristics, except that the majority (76.5%) of spouses of caregivers of the SAC in the GSFP were in vocational employment compared to less than 50% of spouses of caregivers of the NSFP children (P=0.002). The mean age of the SAC was 10 ± 2 years and the majority were in classes one through three (Table 2). About 55% of the children were female and Christianity was the most common religion reported. A little over 50% of the children lived in the same community where the school was located, and were living with both parents. The majority of the children received money for school from their parents; those who were given money received 0.80 ± 0.36 Ghana cedis (USD 0.41 ± 0.18 in 2013) daily. There were no significant differences in the background characteristics of SAC in the GSFP and the NSFP.

3.2 Dietary Habits, Nutrient Intakes and Nutritional Status

During school days the children ate an average of 4 ± 1 times (inclusive of meals and snacks), per day (Table 3). More than one- half (56 ± 16%) of the children's eating events on school days were consumed at home. An average of 50.3 ± 18% of the children's eating events on school days was home-prepared and 25 ± 4% was a meal from the school feeding programmes. Figure 1 shows the food groups consumed in past 24 hours (average of 2 non-consecutive 24hour recalls). Nearly all the SAC had consumed foods from grains, roots and tubers, fish and meat, and fruits and vegetables, over the two days of recall. Less than 20% of the children consumed dark green vegetables and organ meat was the least consumed food group. The mean dietary diversity score for school-age children diet was 5 ± 1 food groups.

Compared to children in the GSFP, children in the NSFP had significantly higher intakes of energy (2413 ± 626 kcal vs. 1988 ± 627 kcal; P<0.001), protein (63 ± 17 g vs. 53 ± 19 g; P<0.001) and zinc (10 ± 3 mg vs. 9 ± 3 mg; P=0.004) (Table 4). Amounts of iron, calcium, vitamin A and vitamin C consumed were similar for the two groups. GSFP meals contributed 12% to 18.1% of the total energy and nutrient intakes of the children; whereas, NSFP meals contributed from 17% up to 30% of the total energy and nutrient intakes of the children (Table 5). With the exception of iron, where the contribution from meals by both types of programmes was similar, meals provided through the NSFP contributed significantly more energy (28 ± 10% vs. 16.2 ± 7%; P<0.001), protein (24.6 ± 9% vs. 13.3 ± 7%; P<0.001) and micronutrients, specifically zinc (P<0.001), calcium (P<0.001), vitamin C (P<0.001), and Vitamin A (P=0.042) to the children's total energy intakes compared to meals provided through the GSFP.

About 28% of the school-age children had low haemoglobin levels indicative of anaemia (Figure 2). Approximately 48% of the children were stunted and 35% had low BMI-for-age or were thin and about 1% were overweight. Sixty seven percent of the school-age children in the study had at least one nutritional deficit (including either anaemia, stunting, or thinness). A logistic regression model for stunting is shown in Table 6. Energy intake, type of feeding programme child was enrolled in, and food security status were not associated with stunting. Children under 10 years had lower odds of stunting as compared to children who were 10 years or older [OR = 0.47 (0.24, 0.91) p=0.025]. Sex was marginally significant with a trend for boys to have higher odds of stunting as compared to girls.

Table 1. Socio-demographic and Household characteristics of caregivers of Participants Government School Feeding Programme (GSFP) and NGO School Feeding Programme (NSFP)

Characteristics	Total (N = 182)		NSFP (N = 98)		GSFP (N = 84)		P
	n	%	n	%	n	%	
Mean Age (yrs)	41		40		42		0.331
SD	11		12		11		
Relationship to child							
Biological parent	124	68.1	70	71.4	54	64.3	0.303
Other relative[1]	58	31.9	28	28.6	30	35.7	
Ethnicity							
Ga /Ga– Adangme	56	30.8	32	32.7	24	28.5	0.313
Ewe	56	30.8	34	34.7	22	26.2	
Akan	40	22.0	17	17.3	23	27.4	
Northern ethnicity	30	16.5	15	15.3	15	17.9	
Formal education completed							
None	33	18.3	17	17.7	16	19.0	0.702
Primary	46	25.6	26	27.1	20	23.8	
JHS/Middle school	80	44.4	45	46.9	35	41.7	
>JHS	21	11.7	8	8.3	13	15.5	
Occupation							
Trader	114	62.6	62	63.3	52	61.9	0.525
Vocational Occupation	19	10.4	8	8.2	11	13.1	
None	15	8.3	7	7.1	8	9.5	
Other[2]	34	18.7	21	21.4	13	15.5	
Marital status							
Married	148	81.3	82	83.7	66	78.6	0.379
Single /Divorced/Widowed	34	18.7	16	16.3	18	21.4	
Spouse's (Paternal) education level*							
≤Primary	36	23.7	21	25.0	15	22.1	0.165
JHS/Middle school	75	49.3	36	42.9	39	57.3	
> JHS/Middle school	41	27.0	27	32.1	14	20.6	
Spouse's occupation*							
Vocational	91	59.9	39	46.4	52	76.5	0.002
Professional	18	11.8	13	15.3	5	7.4	
None	4	2.6	2	2.4	2	2.9	
Other[3]	39	35.7	32	37.6	9	13.2	
Monthly household income							
≤ GH ¢500 (≤USD 256)	127	70.2	63	64.9	64	76.2	0.099
> GH ¢500 (>USD 256)	54	29.8	34	35.1	20	23.8	
	Mean	SD	Mean	SD	Mean	SD	
Household size	6	2	6	2	6	2	0.390
No. of adults	3	1	3	1	3	1	0.969
No. of children (<18yrs)	4	2	4	2	4	2	0.649

NSFP: Non-Government School Feeding Programme, GSFP: Government School Feeding Programme.

[1]Other relatives include aunts, uncles, siblings, grandparents, step parents and non-relatives; [2] Other includes stone winnower, farmer and professional workers; [3]Other includes maintenance/security workers, stone winnower, trader, pensioner, susu collector and farmer.

* Paternal caregiver.

Table 2. Background characteristics of school age children participating in NSFP and GSFP

Characteristics	Total (N = 182)		NSFP (N = 98)		GSFP (N = 84)		P
	n	%	n	%	n	%	
Mean Age (yrs)	10		10		10		0.888
SD	2		2		2		
Class							
1 to 3	112	61.5	57	58.2	55	65.5	0.312
4 to 6	70	38.5	41	41.8	29	34.5	
Sex							
Female	100	54.9	51	52.0	49	58.3	0.395
Male	82	45.1	47	48.0	35	41.7	
Religion							
Christian	163	89.6	87	88.8	76	90.5	0.708
Islam	19	10.4	11	11.2	8	9.5	
Residence							
Within School's Community	99	54.4	50	51.0	49	58.3	0.323
Outside School's Community	83	45.6	48	49.0	35	41.7	
Lives with							
Both parents	98	53.8	55	56.1	43	51.2	0.456
Single parent	34	18.7	16	16.3	18	21.4	
Other relatives[1]	50	27.5	27	27.5	23	27.4	
Received money for school							
No	30	16.5	14	14.3	16	19.0	0.388
Yes	152	83.5	84	85.7	68	81.0	
	Mean	SD	Mean	SD	Mean	SD	
Money received for food on school days (GH ₵)	0.80	0.36	0.79	0.31	0.70	0.31	0.081

NSFP: Non-Government School Feeding Programme, GSFP: Government School Feeding Programme.

[1]Other relatives include aunts, uncles, siblings, grandparents and non-relatives.

Table 3. Dietary habits of school children on school days

Characteristics	Mean	SD
Frequency of eating events	4	1
Location for eating events (%)		
Home	56	16
School	38	13
Food vendor's place	6	13
Source of food (%)		
Home prepared	50.3	18
School feeding programme	25	4
Street food	18	20
School compound vendor	8	12

Table 4. Daily total intakes of energy and selected nutrients of Participants of NSFP and GSFP (2-day 24-hour recall)

	DRI	Total (N = 180)		NSFP (N = 97)		GSFP (N = 83)		P
		Mean	SD	Mean	SD	Mean	SD	
Energy (kcal)	1428 - 2341	2217	660	2413	626	1988	627	<0.001
Protein (g)	19 - 34	58	19	63	17	53	19	<0.001
Fats (g)	ND	53	30	55	26	50	36	0.302
Iron (mg)	8 -10	25	8	25	7	25	10	0.617
Zinc (mg)	5 - 8	9	3	10	3	9	3	0.004
Calcium (mg)	800 - 1300	384	173	389	162	378	186	0.687
Vitamin A (IU)	1333 -2000	947	1230	977	1035	912	1430	0.723
Vitamin C (mg)	25 - 45	87	50	93	49	81	50	0.106

NSFP: Non-Government School Feeding Programme, GSFP: Government School Feeding Programme, ND: not determined.

Table 5. Mean percent contribution of school meals to the total daily energy and nutrients of school-age children by type school feeding programme

	GSFP (N = 83)		NSFP (N = 97)		P
	Mean	SD	Mean	SD	
Energy	16.2	7	28.0	10	<0.001
Protein	13.3	7	24.6	9	<0.001
Fats	13.3	11	22.3	12	<0.001
Iron	18.1	10	20.3	8	0.109
Zinc	15.7	8	26.3	10	<0.001
Calcium	12.5	9	22.4	11	<0.001
Vitamin A	12.6	14	17.4	18	0.042
Vitamin C	16.1	13	30.2	18	<0.001

Table 6. Socio-demographic characteristics and dietary variables associated with stunting levels among children

	N	Odds Ratio	95% Confidence Interval	P
Child's age				
< 10yrs	63	0.468	0.241, 0.908	0.025
≥ 10yrs (reference)	114	1.000		
Child's gender				
Male	79	1.784	0.960, 3.314	0.067
Female (reference)	98	1.000		
Food Security status				
Food secured	123	0.773	0.398, 1.503	0.448
Food Insecure (reference)	54	1.000		
Energy (kcal)	177	1.000	0.999, 1.000	0.089
Type of feeding programme				
GSFP	82	0.823	0.424, 1.595	0.563
NSFP (reference)	95	1.000		

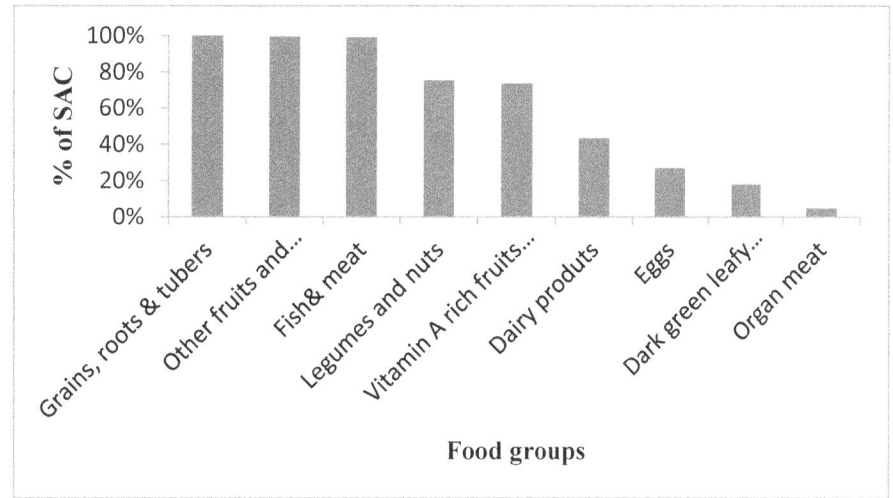

Figure 1. Food groups consumed by school-age children in the past 24 Hours (2-day 24 Hour recall)

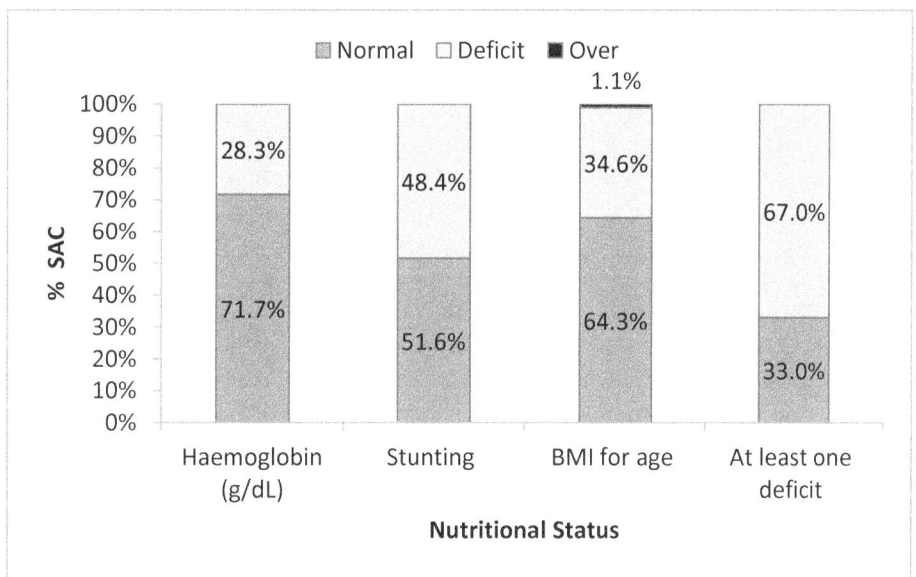

Figure 2. Prevalence of Anaemia, Stunting and Thinness and proportions of school-age children with at least one form of undernourishment

4. Discussion

4.1 Nutritional Status

The present study found a high prevalence of malnutrition among the school-age children. The prevalence of anaemia among school-age children was lower (28.3%) than the national prevalence of 71% reported in 1995 (Agble, Bader, Solal Céligny, Palma, & Dop, 2009). The prevalence of anaemia found in this study was however close to WHO global estimation on worldwide prevalence of anaemia 1993-2005. According to WHO (WHO, 2008), 25.8% of the world's populace school-age children are anaemic. High rates of anaemia among school age children may be partly due to worm infestation (Osazuwa, Ayo, & Imade, 2011). The Ghana Health Service had undertaken deworming exercise prior to the study and this may have ameliorated haemoglobin levels among the school children. Other studies in Ghana which were not on a national scale have also found anaemia prevalence of 70 % among school age children (6-11 years) in Eastern region who were either enrolled or non-enrolled in schools (Fentiman et al., 2001), 30.8% among children 6-12 years in a rural area in Volta Region (Egbi et al., 2014) and 25% among 9 months to 11 years old vegans and matched non-vegan children in Central region of Ghana (Osei-Boadi, Lartey, Marquis, & Colecraft, 2012).

Stunting was the most prevalent nutritional deficit (45.8%) among school-age children in this study. An even higher prevalence of stunting was found among rural school-aged children in Ghana (52.2%) by Danquah and colleagues (Danquah et al., 2012) among class 5 pupils attending schools with or without GSFP. In contrast, other studies have reported a lower prevalence of stunting 30.2% (Chesire, Orago, Oteba, & Echoka, 2008) in

Nairobi, Kenya and 30.7% (Mekonnen, Tadesse, & Kisi, 2013) in Northwest of Ethiopia among school-age children. The differences observed between children from Kenya (Chesire et al., 2008) and present study may be attributed to regional differences: participants in the current study were from semi-rural setting; whereas in the Kenyan study was conducted in peri-urban area. Children from the Ethiopia study were from a rural region; however, the age group (6-12 years vs. 6-14 years) may account for differences in stunting levels observed.

Stunting was more pronounced among children 10 years and older than younger children (6-9 years). Similar results have been reported in Ethiopia and Kenya among school age children (Herrador et al., 2014; Mwaniki & Makokha, 2013). Younger age associated with stunting may be due to the fact that these SAC may have had little nutritional deficit during their pre-school age. Most nutritional interventions are targeted at preschoolers (Bundy, 2009). Considering most nutrition interventions are targeted at pre-schoolers which is a stage closer to younger SAC group, it is possible these younger SAC may have benefited from such interventions.

The prevalence of thinness among the school-age children in this study was 34.6%; whereas, a lower prevalence has been observed among peri-urban Kenyan school-age children (4.5%) (Chesire et al., 2008), and among participants and non-participants of GSFP in Ashanti Region (3.4%) (Danquah et al., 2012). The higher prevalence of thinness observed in this study than elsewhere may be a result of different age groups. Although Danquah et al., (2012) study was conducted in a rural setting of Ghana, the participants were older 9-17 years, whereas; this study had 6 -12 years old children. There were more adolescents in Danquah and colleagues' (Danquah et al., 2012) investigation. Higher stunting observed in Danquah and colleagues' investigation as compared to the current study could account for lower thinness, since BMI for age is based on weight to height. Overweight and obesity was not a problem in this population of SAC. In contrast, other studies suggest that overweight and obesity may be a growing concern among SAC. In a survey of 1373 Malaysian primary school age children (9-10yrs), 16.3% of them were overweight and 6.3% were obese (Zaini, Lim, Low, & Harun, 2005).

There was a high prevalence of malnutrition with nearly 7 out of 10 of the school-age children interviewed with at least one nutritional deficit. In Pakistan, Mian and colleagues reported that 4 out of 10 of school-age children had at least one nutritional deficit. Nutritional status indicators measured in the study (Mian, Ali, Ferroni, & Underwood, 2002) were underweight, stunting and wasting and not anaemia status. Joshi and colleagues (Joshi, Gupta, Joshi, & Mahajan, 2011) found 1% of 789 Nepalese children to be either stunted or wasted. Unlike the Nepalese and Pakistan study, we included anaemia status in assessing at least one nutritional deficit. This may explain the high level of malnutrition observed in our study.

4.2 Energy and Nutrient Intakes of School-Age Children and Contribution of School Feeding Meals to Total Intakes

Most of the school-age children had adequate intakes of energy and selected nutrients except for calcium. Calcium intakes were very low among school-age children in both the GSFP and NSFP. This finding corroborates Henry and Chapman (Henry & Chapman, 2002) assessment of low calcium intakes in African populations of 300-400 mg/day which is well below the recommended daily intakes. Low intakes of food groups with calcium-rich food sources (green leafy vegetables and dairy products) among the children may explain their low levels of calcium intake. Participants in the NSFP had significantly higher intakes of energy, protein and zinc than the GSFP children. Studies that have compared dietary intakes of participants and non-participants of school feeding programmes have generally reported higher dietary intakes among participants than non-participants of school feeding programmes. Ahmed (Ahmed, 2004) reported that participants of school feeding programmes in rural areas had a higher intake of energy than non-participants. Among 320 Kenyan primary pupils, it was observed that participants had higher intake of energy (2089 ± 12.41 kcal vs. 1841 ± 15.68 kcal; $P<0.05$) and protein (58 ± 7.5 g vs. 40 ± 2.4g; $P<0.05$) than non-participants (Musamali, 2007). In the present study only children participating in one of two school feeding programmes were assessed without comparison to non-participators.

The superior contributions of the NSFP meals to the children's energy and nutrient intakes may be because of the larger portion size of meals served to the children. In Bangladesh, fortified biscuits are used as a school feeding programmes mid-morning snack and this meal only contributes to 16.4% (urban communities) and 14.8% (rural communities) of energy intakes (Ahmed, 2004). The energy contribution of the meals provided by the Bangladesh school feeding programme (Ahmed, 2004) were less than the NSFP meals which contribute to more than 25% of daily intake; albeit, they match the GSFP meals (Danquah et al., 2012). Danquah and colleagues (2012) assessed the energy and nutrient content of meals served by the GSFP in three schools and reported a mean energy content of 460.4 ± 30.1 kcal which did not supply the required 30% of daily energy and nutrients. In the present study, the energy content of the GSFP meals was significantly lower (295 ± 94 kcal) than what

Danquah et al., (2012) reported and supplied less of the recommended energy and nutrients of children (Bhatia, 2013; Food and Agriculture Organization [FAO], 2004).

Based on FAO guidelines (Burgess & Glasauer, 2004), the children in this study had adequate meal frequency. The children's eating events consisted of three meals and one snack daily. Most of the children's eating events occurred at home. Consistent with this finding, Olusanya (Olusanya, 2010) reported that a higher percentage of meals consumed by rural Nigerian primary pupils (1-3) were at home (breakfast: 89.7%, lunch: 100% and supper: 100%). In this same study of Nigerian pupils, the source of meals was mostly home prepared (breakfast: 72.4% and 100% for both lunch and supper) (Olusanya, 2010). In this present study, more than 80% of school-age children were given money for food on school days and the average amount of 70 pesewas (0.36 USD in 2013) that they received was equivalent to the cost of meals served by NSFP per child. The meals served by NSFP contributed more energy and selected nutrients than GSFP with a lower cost of 40 pesewas (0.21 USD in 2013). About 30% of the Nigerian pupils did not receive money for school on school days. Thus, it is not surprising that school-age children in this study ate more non-home prepared food than the Nigerian pupils whose major source of meals were home prepared.

Consistent with typical African diets (Oniang'o, Mutuku, & Malaba, 2003), all school-age children interviewed reported eating food belonging to grains, roots and tubers groups. Food groups that were less consumed by school-age children included organ meats, dairy products, eggs and dark green leafy vegetables. Olumakaiye (Olumakaiye, 2013), also found a similar result among 600 school-age children at South Western Nigeria school-age children involved in the study were less likely to consume food groups such as organ meats, milk and milk products, eggs, and vitamin A rich fruits and vegetables (Olumakaiye, 2013). At the time of the study, local fruits such as black velvet tamarind (locally called Yooyi), dawadawa fruit snack, and mangoes were in season and readily available and may explain the high prevalence of fruits intake among the study children in contrast to consistent documentation that children are less likely to eat fruits and vegetables as part of their diet (Lock, Pomerleau, Causer, & McKee, 2005). Approximately 75% of the children in the current study compared to less than 5% of Nigerian school-aged children reported eating fruit in the past 24 hours (Ene-Obong & Ekweagwu, 2013). It is worth noting that the fruits that the participants ate were mostly black velvet tamarind (known locally as yooyi), dawadawa fruit snack and mango. The aforementioned fruits that were eaten by the children were in season and widely available in the community and school-age children plucked them by themselves to eat. The fruits that they had to buy such as bananas, oranges and apples which were not locally available were less consumed among the participants due to the cost. Almost all (99%) the school-age children ate an animal source food in the past 48 hours compared to 2% of school-aged children (aged 5-14 years) in rural Nigeria reported by Ene-Obong and Ekweagwu (Ene-Obong & Ekweagwu, 2013). Most school-age children involved in this present study were either part of NSFP or GSFP. NSFP meals observed during the data collection period were always accompanied with fish, chicken, egg or sausage. GSFP also had anchovies added to the meals on some days so there is a likelihood of children consuming an animal source food. Another possibility is that most of the foods that school-age children reported eating based on the 24-hour recall at school were with an animal source food. School-age children at GSFP site reportedly bought fried yam with chicken or sausage mostly at break sections of school hours. Also considering the fact that most school-age children ate at least a meal at home in a day, it is likely caregivers added an animal source food.

Dietary diversity score, DDS for GSFP participants was 5.46 ± 1.00 and NSFP participants was 5.51 ± 1.15 based on 9 food groups proposed by FAO (Kennedy et al., 2011). While assessing the impact of GSFP in 4 districts in the Central region of Ghana, Martens (Martens, 2007) observed that GSFP increased the DDS of the participants of school feeding programmes by 1.0 ± 0.8 (Martens, 2007). Our study found school feeding meals included vegetables like okra and cabbage as well as legumes such as beans which may have the potential of increasing dietary diversity scores if such foods were absent in the diet of school-age children.

An important strength of the present study is that it contributes to the limited data available on the dietary intakes and nutritional status of school-aged children in Ghana. It also provides some insights into the nutritional quality of the two categories of school feeding programmes. A methodological weakness of our study is not using the same sample selection techniques at both schools. This was due to the unequal number of school age children at the two schools and hence we invited participation from all the 125 school-age children at the NSFP school and we invited a random selection of 50% of the 300 children found at the GSFP to participate in the study.

5. Conclusion

There was a high prevalence (67%) of undernutrition (either stunted anaemic or thin) among school-age children participating in NSFP and GSFP. At least 3 out of 10 school-age children were anaemic or thin (low BAZ) and about 5 out of 10 were stunted. The age of SAC was a significant predictor of stunting. Meals served by both programmes should be reviewed to ensure they meet at least 30% of recommended energy and micronutrients intake. Since school meals provide the majority of nutrition for school-aged children, the potential to mitigate malnutrition lies with school and community programmes to enhance nutrition.

Competing Interest

The authors declare no conflict of interest.

Acknowledgements

The authors thank Prof. Anna Lartey for granting permission to use RIING Nutrient Database, Dr. Seth Adu-Afarwuah, Mr. Boateng Bannerman, Phinehas Adjei, Esther Sam, Uriah Smith Karikari, Nathaniel Adu Adjei and Prince Anyitomise Kelly for their contribution to data collection and management.

Funding Information

This work was partially funded by Prembaf Ghana, an affiliate of Prem Rawat Foundation. Prembaf Ghana did not have any role in design, analysis or writing of this article.

References

Abotsi, A. K. (2013). Expectations of School Feeding Programme: Impact on School Enrolment, Attendance and Academic Performance in Elementary Ghanaian Schools *British Journal of Education, Society & Behavioural Science, 3*(1), 76-92.

Adelman, S., Gilligan, D., & Lehrer, K. (2008). How effective are food for education programs? A critical assessment of the evidence from developing countries *International Food Policy Research Institute.* http://dx.doi.org/10.2499/0896295095FPRev9.

Afridi, F. (2010). Child welfare programs and child nutrition: Evidence from a mandated school meal program in india. *Journal of Development Economics, 92*(2), 152-165. https://doi.org/10.1016/j.jdeveco.2009.02.002

Agble, R., Bader, E., Solal Céligny, A., Palma, G., & Dop, M. C. (2009). Nutrition country profile for the republic of Ghana. *FAO: Nutrition country profiles* FAO.

Ahmed, A. U. (2004). Impact of feeding children in school: Evidence from Bangladesh. *Washington, DC: International Food Policy Research Institute,*

Arsenault, J. E., Mora-Plazas, M., Forero, Y., Lopez-Arana, S., Marin, C., Baylin, A., & Villamor, E. (2009). Provision of a school snack is associated with vitamin B-12 status, linear growth, and morbidity in children from Bogota, Colombia. *The Journal of Nutrition, 139*(9), 1744-1750. http://dx.doi.org/10.3945/jn.109.108662

Best, C., Neufingerl, N., Van Geel, L., van den Briel, T., & Osendarp, S. (2010). The nutritional status of school-aged children: Why should we care? *Food & Nutrition Bulletin, 31*(3), 400-417. https://doi.org/10.1177/156482651003100303

Bhatia, R. (2013). Operational guidance on menu planning HGSF working paper series #3. Retrieved from http://hgsf-global.org/en/bank/downloads/doc_download/347-operational-guidance-on-menu-planning.;

Bundy, D. A. (2009). *Rethinking school feeding: Social safety nets, child development, and the education sector* World Bank Publications.

Burgess, A., & Glasauer, P. (2004). *Family nutrition guide* Food & Agriculture Org.

Centers for Disease Control and Prevention [CDC]. (2015). About child and teen BMI. Retrieved from http://www.cdc.gov/healthyweight/assessing/bmi/childrens_bmi/about_childrens_bmi.html

Chesire, E., Orago, A., Oteba, L., & Echoka, E. (2008). Determinants of under nutrition among school age children in a Nairobi peri-urban slum. *East African Medical Journal, 85*(10), 471-479.

Danquah, A., Amoah, A., Steiner-Asiedu, M., & Opare-Obisaw, C. (2012). Nutritional status of participating and

non-participating pupils in the Ghana school feeding programme. *Journal of Food Research, 1*(3), 263. https://doi.org/10.5539/jfr.v1n3p263

Egbi, G., Steiner-Asiedu, M., Kwesi, F. S., Ayi, I., Ofosu, W., Setorglo, J., ... Armar-Klemesu, M. (2014). Anaemia among school children older than five years in the Volta region of Ghana. *The Pan African Medical Journal, 17 Suppl 1*, 10. http://dx.doi.org/10.11694/pamj.supp.2014.17.1.3205

Ene-Obong, H., & Ekweagwu, E. (2013). Dietary habits and nutritional status of rural school age children in

Ebonyi state, Nigeria. *Nigerian Journal of Nutritional Sciences, 33*(1), 23-30.

Fentiman, A., Hall, A., & Bundy, D. (2001). Health and cultural factors associated with enrolment in basic education: A study in rural Ghana. *Social Science & Medicine, 52*(3), 429-439. https://doi.org/10.1016/S0277-9536(00)00152-0

Food and Agriculture Organization [FAO]. (2004). Dietary reference intakes. Retrieved from http://www.sochinut.cl/pdf/Recomendaciones/DRISummaryListing.pdf

Ghana School Feeding Programme [GSFP]. (2010). Snapshot of Ghana school feeding. (Presentation given to the 2010 Global Child Nutrition Forum held in Accra, Ghana.). Ghana:

Ghana Statistical Service [GSS], Ghana Health Service [GHS], and Inner City Fund [ICF] Macro. (2015). *Ghana demographic and health survey 2014*. Rockville, Maryland, USA: GSS, GHS, and ICF International.

Ghazi, H. F., Isa, Z. M., Aljunid, S., Tamil, A. M., & Abdalqader, M. A. (2012). Nutritional status, nutritional habit and breakfast intake in relation to IQ among primary school children in Baghdad city, Iraq. *Pakistan Journal of Nutrition, 11*(4), 379-382. https://doi.org/10.3923/pjn.2012.379.382

Henry, C., & Chapman, C. (2002). The nutrition handbook for food processors, *Elsevier.* https://doi.org/10.1533/9781855736658

Herrador, Z., Sordo, L., Gadisa, E., Moreno, J., Nieto, J., Benito, A., ... Custodio, E. (2014). Cross-sectional study of malnutrition and associated factors among school aged children in rural and urban settings of Fogera and Libo Kemkem districts, Ethiopia. *PloS One, 9*(9), e105880. https://doi.org/10.1371/journal.pone.0105880

Inter Tribal Council of Arizona [ITCA] Training, (2012). *Module 5: Haemoglobin Testing* retrieved from http://itcaonline.com/wp-content/uploads/2011/10/Unit-3-Hemoglobin-Testing.pdf

Joshi, H., Gupta, R., Joshi, M., & Mahajan, V. (2011). Determinants of nutritional status of school children-A cross sectional study in the western region of Nepal. *National Journal of Integrated Research in Medicine 2*(1), 10-15.

Kennedy, G., Ballard, T., & Dop, M. C. (2011). Guidelines for measuring household and individual dietary diversity *Nutrition and Consumer Protection Division, Food and Agriculture Organization of the United Nations*. Retrieved from:
http://www.fao.org/fileadmin/user_upload/wa_workshop/docs/FAO-guidelines-dietary-diversity2011.pdf

Kristjansson, B., Petticrew, M., MacDonald, B., Krasevec, J., Janzen, L., Greenhalgh, T., . . . Shea, B. (2007). School feeding for improving the physical and psychosocial health of disadvantaged students. *The Cochrane Database System Review, 24*(1). CD004676. https://doi.org/10.1002/14651858.CD004676.pub2

Lock, K., Pomerleau, J., Causer, L., & McKee, M. (2005). Low fruit and vegetable consumption. *Bulletin of the World Health Organization, 83* (2), 100-108.

Martens, T. (2007). Impact of the Ghana school feeding programme in 4 districts in central region, Ghana. *Wageningen University: Division of Human Nutrition (Thesis)*

Mekonnen, H., Tadesse, T., & Kisi, T. (2013). Malnutrition and its correlates among rural primary school children of Fogera district, northwest Ethiopia. *Journal of Nutritional Disorders and Therapy S, 12*, 2161-0509. https://doi.org/10.4172/2161-0509.s12-002

Mesfin, F., & Berhane, Y. (2015). Prevalence and associated factors of stunting among primary school children in Eastern Ethiopia.

Mian, R. M., Ali, M., Ferroni, P. A., & Underwood, P. (2002). The nutritional status of school-aged children in an urban squatter settlement in Pakistan. *Pakistan Journal of Nutrition, 1*(3), 121-123. https://doi.org/10.3923/pjn.2002.121.123

Mukudi, E. (2003). Nutrition status, education participation, and school achievement among Kenyan middle-school children. *Nutrition, 19*(7), 612-616. https://doi.org/10.1016/S0899-9007(03)00037-6

Musamali, B. (2007). Impact of school lunch programmes on nutritional status of children in Vihiga district, western Kenya. *African Journal of Food, Agriculture, Nutrition and Development, 7*(6), Retrieved from: http://www.bioline.org.br/request?nd07048

Mwaniki, E., & Makokha, A. (2013). Nutrition status and associated factors among children in public primary schools in Dagoretti, Nairobi, Kenya. *African Health Sciences, 13*(1), 38-46.

https://doi.org/10.4314/ahs.v13i1.6

Ndukwu, C., Egbuonu, I., Ulasi, T., & Ebenebe, J. (2013). Determinants of undernutrition among primary school children residing in slum areas of a Nigerian city. *Nigerian Journal of Clinical Practice, 16*(2), 178-183. https://doi.org/10.4103/1119-3077.110142

Neervoort, F., von Rosenstiel, I., Bongers, K., Demetriades, M., Shacola, M., & Wolffers, I. (2013). Effect of a school feeding programme on nutritional status and anaemia in an urban slum: A preliminary evaluation in Kenya. *Journal of Tropical Pediatrics, 59*(3), 165-174. doi:10.1093/tropej/fms070

Neumann, C. G., Murphy, S. P., Gewa, C., Grillenberger, M., & Bwibo, N. O. (2007). Meat supplementation improves growth, cognitive, and behavioral outcomes in Kenyan children. *The Journal of Nutrition, 137*(4), 1119-1123.

Olumakaiye, M. (2013). Dietary diversity as a correlate of undernutrition among school-age children in southwestern Nigeria. *Annals of Nutrition and Metabolism, , 63* 569-569.

Olusanya, J. (2010). Assessment of the food habits and school feeding programme of pupils in a rural community in odogbolu local government area of ogun state. *Nigeria Pakistan Nutrition Journal, 9*, 198-204. https://doi.org/10.3923/pjn.2010.198.204

Omwami, E. M., Neumann, C., & Bwibo, N. O. (2011). Effects of a school feeding intervention on school attendance rates among elementary schoolchildren in rural Kenya. *Nutrition, 27*(2), 188-193. https://doi.org/10.1016/j.nut.2010.01.009

Oniang'o, R., Mutuku, J., & Malaba, S. J. (2003). Contemporary African food habits and their nutritional and health implications. *Asia Pacific Journal of Clinical Nutrition, 12*(3), 331-336.

Osazuwa, F., Ayo, O. M., & Imade, P. (2011). A significant association between intestinal helminth infection and anaemia burden in children in rural communities of Edo state, Nigeria. *North American Journal of Medical Sciences, 3*(1), 30-34. doi:10.4297/najms.2011.330

Osei-Boadi, K., Lartey, A., Marquis, G., & Colecraft, E. (2012). Dietary intakes and iron status of vegetarian and non-vegetarian children in selected communities in Accra and Cape coast, Ghana. *African Journal of Food, Agriculture, Nutrition and Development, 12*(1), 5822-5842.

Schaible, U. E., & Kaufmann, S. H. (2007). Malnutrition and infection: Complex mechanisms and global impacts. *Public Library of Science Medicine, 4*(5), e115, 0806-0812. https://doi.org/10.1371/journal.pmed.0040115

World Food Programme [WFP]. (2015). WFP and school meals. Retrieved from http://www.wfp.org/school-meals/wfp-school-meals

World Food Programme [WFP] and United Nations Children's Fund [UNICEF]. (2005). *The essential package: Twelve interventions to improve the health and nutrition of school-age children.* Retrieved from: http://www.un.org/esa/socdev/poverty/PovertyForum/Documents/The%20Essential%20Package.pdf

World Health Organization [WHO]. (2008). Worldwide prevalence of anaemia 1993-2005: WHO global database on anaemia. Edited by de Benoist, B., McLean, E., Ines Egli, I. & Cogswell, M. Retrieved from: http://apps.who.int/iris/bitstream/10665/43894/1/9789241596657_eng.pdf

World Health Organization [WHO]. (2011). *Haemoglobin concentrations for the diagnosis of anaemia and assessment of severity.* (No. WHO/NMH/NHD/MNM/11.1). Geneva: World Health Organization. Retrieved from: http://www.who.int/vmnis/indicators/haemoglobin.pdf

Zaini, M. Z., Lim, C. T., Low, W. Y., & Harun, F. (2005). Factors affecting nutritional status of Malaysian primary school children. *Asia-Pacific Journal of Public Health / Asia-Pacific Academic Consortium for Public Health, 17*(2), 71-80. https://doi.org/10.1177/101053950501700203

Zeba, A. N., Martin Prevel, Y., Some, I. T., & Delisle, H. F. (2006). The positive impact of red palm oil in school meals on vitamin A status: Study in Burkina Faso. *Nutrition Journal, 5*, 17. http://dx.doi.org/1475-2891-5-17

19

Determinants and Constraints of Pulse Production and Consumption among Farming Households of Ethiopia

Alemneh Kabata[1], Carol Henry[2], Debebe Moges[3], Afework Kebebu[3], Susan Whiting[2], Nigatu Regassa[2] & Robert Tyler[4]

1Hawassa University, School of Nursing and Midwifery, Ethiopia

2College of Pharmacy and Nutrition, University of Saskatchewan, Saskatoon, S7N 5A2. Canada

3Hawassa University, School of Nutrition, Food Science and Technology, Ethiopia

4Department of Food and BioProduct, University of Saskatchewan, 9 Campus Drive, Saskatoon, S7N 5A5. Canada

Corresponding Author: Carol Henry, College of Pharmacy and Nutrition, University of Saskatchewan, Saskatoon, S7N 5A2, Canada. E-mail: cj.henry@usask.ca

Abstract

In low income countries the agricultural sector is essential to growth, poverty reduction, and food security. Pulse crops are important components of crop production in Ethiopia's smallholders agriculture, providing an economic advantage to small farm holders as an alternative source of protein and other nutrients, cash income, that seeks to address food security. This study sought to gain an understanding of determinants and constraints to production and usage of pulse crops based on data collected in 2013 from 256 households in Oromia region of Ethiopia. Determinants of production and consumption were identified using logistic regression. The result showed that Haricot bean was produced, but not widely consumed. Lentil was widely consumed but not produced. Production of haricot bean was hampered by problems related to weed control, disease, pests, yield and soil quality, a seasonal market, and a shortage of farmland. Consumption of haricot bean was low due to perceived gastrointestinal distress after eating and the culture of it being a taboo food. Logistic regression showed household head educational status and age, land size and household size statistically significantly (p-value<0.05) affected household pulse (haricot bean and lentil) consumption frequency. Agronomic, market, culture and household characteristics related determinants and constraints were identified. Also a mismatch of production and consumption was observed in the study. It is recommended that agronomic and market concerns related to production of haricot bean and other pulses be addressed and that household food preparation techniques for pulses that reduce gastrointestinal symptoms be promoted and evaluated.

Keywords: pulse production, pulse consumption, gastrointestinal distress, haricot bean, lentil

1. Introduction

Pulses have the potential to provide important benefits to small holder farmers (SHFs) as a source of protein and nutrition, and as a potential source of income as a cash crop. Pulses contain a wide range of nutrients including protein, iron and zinc which are lacking in the cereal-based Ethiopian diet (Gibson, *et al.*, 2009).

Pulses rank second among ingredients used in national dishes in Ethiopia and are an integral component of the cooking culture of Ethiopians. However, despite their importance, systematic assessment of pulses use in the Ethiopian diet has not yet been carried out at the household or individual level. A few studies suggest usage is very low (Kebebu, *et al.*, 2013; Roba, *et al.*, 2015). From a community perspective, caloric intake from consumption of pulses and oilseeds combined was reported at 9% for rural and 14% for urban communities (IFPRI, n.d). Of a total of 12.4 million hectares of farmland in Ethiopia, the majority is used for production of cereals (9.16 million hectares); a relatively small area is seeded to pulses (1.41 million hectares) (FAO, 2010).

The diverse and important roles played by pulses in farming systems and in the diets of people make them ideal crops for achieving the Sustainable Developmental Goals of reducing poverty and hunger, improving human health and nutrition, and enhancing ecosystem resilience. Moreover, some pulses (chickpeas, peas) have certain qualities that enhance soils and improve productivity (Campbell, et al., 1992 and Schwenke, et al., 1998).

Nutrition and agriculture are interlinked. In farming households' agricultural products directly inputs for family consumption. Even though community perspective pulse production and consumption discuss the gap no study tried to see agriculture and nutrition side by side and assess bottle necks of pulse production and consumption in the study area. The aim of this study was to identify barriers to production and consumption of pulse crops in the HurufaLole district of Oromia, Ethiopia: a highlands locale.

2. Method

2.1 Study population and Data Source

Oromia is a regional state in Ethiopia. Oromia occupies the largest part of the country and at present consists of 12 administrative zones and 180 woredas. East Shoa zone is one of the 12 administrative zones of the Region and it comprises 18 woredas/ districts. The estimated area of the state of Oromia is about $353,690 Km^2$ almost 32% of the country. Hurufalole, one of the kebeles (lowest government administrative division) in Oromia, is the particular site where this study was done inhabiting around 705 households. Household heads were the target population in this study.

2.2 Sampling

The study was conducted from April to May 2013 in the Hurufa Lole kebele of Oromia, Ethiopia. This rural village was selected due to its huge potential of pulse production and a project site for research on increased pulse production funded by Foreign Affairs, Trade and Development Canada (DFATD) and International Development Research Canada (IDRC). Using power analysis, a requirement for 256 households was calculated based on the estimated proportion of households in the region consuming pulses. The households were selected after mapping and house-to-house registration via systematic random sampling.

2.3 Study Design and Data Collection

A cross sectional study design incorporating quantitative and qualitative aspects was employed. Quantitative data included socio-demographic information of respondents and households, economic, pulse production and household consumption. Qualitative information was collected using focus group discussion and key informant interview.

Four focus group discussions, two with male farmers and two with female farmers, were conducted to assess barriers to the production and consumption of pulse crops. Participants (20 male and 20 female) were selected in collaboration with development agents in the area. Notes were taken by a trained research assistant while the principal investigator asked questions using a semi-structured interview guide. Probe questions were asked for clarification and to obtain details. Notes taken during the focus group interviews were transcribed from 'Affan Oromo' to English and then analyzed for common themes. The resulting narrative was used to substantiate quantitative results.

2.4 Data Analysis

Quantitative data were checked for completeness and inconsistencies, cleaned, coded and then entered into SPSS version 20. Predictors of pulse consumption patterns were identified with binary logistic regression.

Haricot bean and lentil weekly consumption frequencies are the dependent variables and socio-demographic and economic information of households and household headswere the independent variables. The independent variables were selected after assessing the existing situation and consulting previous research works. Pulse crop consumption frequency was categorized using mean consumption frequency as high and low consumers.

3. Results

3.1 Socio-Demographic and Economic Characteristics of Households

In total, 256 households were interviewed, with a response rate of 100% (Table 1). However, data from one household was not included in the analysis because of an incomplete response. The majority of the respondents were Muslims from the Oromo ethnic group. Almost all were married. All households owned agricultural land or farmland, holdings were of various sizes, and nearly all were headed by males. Mean values for the age of the household head, household size and land holding size were 37.7 ± 9.8 years, 6.9 ± 2.6 and 3.9 ± 2.1 hectares, respectively. Similar proportions of household heads had 0, 1-5 or ≥ 6 years of formal education. More than 80% of households had monthly incomes ≤ 2500 Birr (equivalent to 120 USD) i.e 1USD=20 Birr. All households used water from a public tap with a mean distance of 1.8 ± 1.0 km from the water source.

Table 1. Socio-demographic and economic characteristics of respondents and households (n=255).

Socio-demographic and economic characteristics	Frequency	Percent
Respondents' Religion		
Muslim	215	84.3
Orthodox	36	14.1
Other	4	1.6
Respondents' Marital status		
Married	239	93.7
Widowed	16	6.3
Household head		
Male	239	93.7
Female	16	6.3
Household Head Age		
20-30	70	27.5
31-40	103	40.4
$\geq$41	82	32.1
Household head educational status		
None	91	35.8
1st-5th Grade	82	32.1
$\geq$6th Grade	82	32.1
Household Income		
700-1500 Birr	96	37.6
1501-2500 Birr	112	44.0
$\geq$2501 Birr	47	18.4
Household Size		
1-6	120	47.1
$\geq$7	135	52.9
Land size		
0.75-4.00 Hectare	169	66.3
$\geq$4.5 Hectare	86	33.7

3.2 Pulse Production

Only haricot bean was produced to a significant extent in the project area. Haricot bean was produced by 62.7% of households. A large majority of respondents (95.6%) rated growing of haricot bean as "less productive" as compared to other cereals they produced. The previous year's mean farmland used for haricot bean and maize was 0.68±0.71 hectares and 2.47±1.03 hectares, respectively. Seed used by haricot bean producing households was stored from previous production (67.9%) or bought from the market (48.1%).The remainder of the households (37.3%) did not grow haricot bean for the following reasons – shortage of farmland (94.7%), lower productivity of the crop (75.8%), not used as a family food (32.6%) and unable to obtain financing (31.6%), depicted in Figure 1.

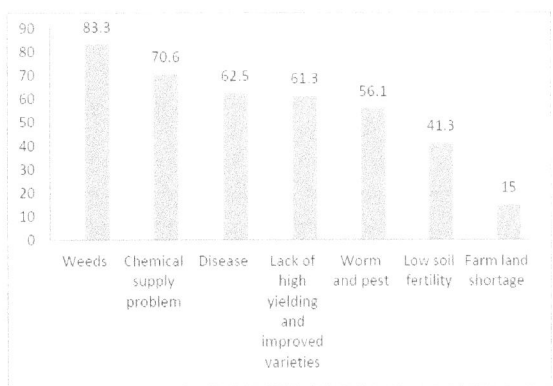

Figure 1. Percentage of households surveyed in HurufaLole, Oromia, Ethiopia, facing various constraints to production of haricot bean

Some male farmers participating in the focus group discussion voiced constraints to haricot bean production in addition to those identified in the quantitative data (figure 1). These included the need to use uncultivated land and the seasonality of the market. Focus group participants also revealed that haricot bean is produced because of its early maturity and funds are obtained from its sale is used to employ labourers to harvest maize. One participant said "Production of haricot been is challenged by weeds, and by a worm which stays in the soil during the daytime and then destroys plants at night." A middle-aged farmer stated "For me, weeds might not be a challenge if the seed is sown in rows, a new practice, because this eases weeding. I practiced this last year. What concerns me is the seasonality of the haricot bean market. No one will look for it in the market after October. The other problem is a seasonal disease, we know it locally as 'waag', which makes the leaves white and spreads throughout the farm." Another older participant in the focus group discussion compared maize and haricot bean by saying "Maize is the king of the poor! It gives more products (up to 30 quintal/per hectare), is one of the important foodstuffs, can be stored for a long period and is marketable throughout the year. Haricot bean needs uncultivated land every year which will compromise our yearly maize production, yields much less product compared to maize (not more than 10-12 quintals/hectare), its market is seasonal, and it will be eaten by pests if stored for a long period."

Focus group discussions with male farmers indicated that the area was unsuitable for the production of broad bean, chickpea, pea or lentil. An older farmer stated "Broad bean and pea need moist weather conditions and strong soil, but as you can see the area is hot and the soil is loose. Regarding chickpea, some years back I had seen a farmer trying it and it did not grow up." A young farmer said "Two years back I tried broad bean, pea and chickpea on a small proportion of my farmland. Chickpea did not grow up, pea grew up, had good flowering but grew taller than my expectation, and the same was true for broad bean. Lastly, rather than seed I got green worm inside the seed coat of pea and the production that I got from broad bean was not more than what I had sown. "Another farmer confirmed the earlier discussion by saying "Some years back I also tried to grow broad bean, pea and chickpea. I remember the chickpea did not grow. Plants of broad been were affected by disease which discoloured the leaves, and the yield was too low and the seed was destroyed by a worm. The pea grew up, had good flowering, but it become too tall and fell down after it started setting seed."

3.3 Pulse Consumption

Pea, chickpea and broad bean were not available or consumed in the study area, whereas lentil was widely consumed in the form of a stew or sauce (100% of households) or mixed with vegetables (7.8%). Haricot bean was consumed boiled (98.5% of households) or as a sauce (25.5%) or stew (20.0%).

Both female and male focus groups expressed various concerns related to cultural practices and taboos regarding the preparation and consumption of haricot bean. The following were identified as barriers to haricot bean consumption. One elderly man said "It is known that consumption of haricot bean causes abdominal discomfort like distension, flatulence and diarrhoea, and I think it is because of this that using haricot bean as a foodstuff is considered a sign of over shouldered poverty." Another elderly man supported this idea saying "The locality is hot and we are used to consuming lunch outdoors, but this is not true if we are going to have food prepared from haricot bean due to fear of what passersby will say." A middle-aged member of the female group confirmed this by saying "Consumption of haricot bean is a taboo! Male household heads in particular are not happy to consume it. In my house I use boiled red haricot bean as a snack when my kids come home from school. "Another female participant said "We females are responsible for preparing food and serving the whole family. I feel guilty to prepare and go with food from haricot bean while my husband is farming. If I do, it is certain that everybody sharing the food will be pointing at me and discussing what I prepared and served for him." A relatively young woman opposed this point of view and focused on preparation saying "Even though there is a cultural problem with haricot bean, the main concern of male household heads is abdominal discomfort and what matters is how we prepare the food. As I see it, the problem is because of the seed coat, so we should discard the water used to boil the haricot bean after a few minutes and then boil it again as the seed coat will have come off. "From the above statements, it appears that while gender played a role, the age of the participants also influenced their views about consumption of haricot beans, with females expressing a greater willingness to incorporate them into the diet.

Table 2 suggests a gap between the production and consumption of haricot bean, as it was produced by a majority of households and consumed by only one-fourth of the households. On the other hand, lentil was not produced but was consumed by all of the households surveyed.

Table 2. Pulse (lentil and haricot bean) consumption patterns of households in Hurufalole, Oromia, Ethiopia

Consumption Pattern	Percent
Lentil	
≤ 2 days/Week	21.6
≥ 3days/Week	78.4
Kidney/Haricot Bean	
≤ 2 days/week	52.3
≥ 3 days/week	47.7

Farm size and the educational status of the head of the household had a statistically significant (p-value<0.05) association with households' consumption frequency of haricot bean, i.e. households with land size of 0.75-4.0 hectares consumed haricot bean less frequently than did households with land size of ≥4.5 hectare, and households where the head had no formal education consumed haricot bean less frequently than did households where the head of the household had or more years of formal education.

Table 3. Logistic regression of socio-demographic variables with haricot bean consumption by households in HurufaLole, Oromia, Ethiopia

Socio-demographic and economic characteristics	Haricot bean			
	Haricot bean consumption status			
	Low consumer	High consumer		
	No.(%)	No.(%)	COR(95%CI)	AOR(95%CI)
Household Head				
Male	31(47.7)	29(44.6)	1.4(.22 - 9.01)	1.02(.12-8.69)
Female	3(4.6)	2(3.08)	1	1
Household Head Age				
20-30	6(9.2)	3(4.6)	1	1
31-40	17(26.2)	14(21.5)	0.39(.08 - 1.94)	0.22(.03-1.39)
≥41	11(16.9)	14(21.5)	0.65(0.22-1.86)	0.40(.10-1.58)
Head Education				
0	13(20)	5(7.7)	0.19(.46-.79)	0.13(.03-.67)*
Grade 1-5	15(23.1)	14(21.5)	0.47(.14-1.58)	0.28(.67-1.2)
Grade ≥6	6(9.2)	12(18.5)	1	1
Household Income				
700-1500 Birr	16(24.6)	13(20)	1	1
1501 - 2500 Birr	13(20)	12(18.5)	0.68(.17-2.73)	0.55(.09-3.19)
≥2501	5(7.7)	6(9.2)	0.77(.19- 3.19)	0.48(.09-2.53)
Household Size				
1-6	11(16.9)	14(21.5)	1.72(.63-4.72)	3.11(.8-12.01)
≥7	23(35.4)	17(26.2)	1	1
Land size				
0.75 - 4.0 Hectare	21(32.3)	11(16.9)	0.34(.12-.935)	0.30(.93-.94)*
≥4.5 Hectare	13(20)	20(30.8)	1	1

*p-value is significant at < 0.05. COR=Crude Odds Ratio, AOR=Adjusted Odds Ratio

Age of household head and family size had a statistically significant (P<0.05) association with households' consumption frequency of lentil (Table 4). For example, households with a head of age 31-40 years consumed lentil less frequently than households with a head of age ≥41, and households with a family size of 1-6 consumed lentil more frequently than did those with a family size of ≥7.

Table 4. Logistic regression of socio-demographic variables with lentil consumption by households in HurufaLole, Oromia, Ethiopia, (n=255)

Socio-demographic and economic characteristics	Lentil			
	Lentil Consumption Status			
	Low Consumer	High Consumer		
	No.(%)	No.(%)	COR(95%CI)	AOR(95%CI)
Household Head				
Male	25(9.8)	187(73.3)	0.83(.23-3.02)	0.91(.24-3.52)
Female	3(1.17)	13(5.09)	1	1
Household Head Age				
20-30	12(4.7)	58(22.76)	0.75(.31-1.82)	0.52(.19-1.37)
31-40	32(12.55)	71(27.84)	0.34(.16-.74)	0.29(.13-.66)*
≥41	11(4.31)	71(27.84)	1	1
Head Education				
0	18(7.06)	73(28.63)	1.31(.64-2.69)	1.26(0.56-2.78)
1-5 Grade	17(6.67)	65(25.5)	1.23(.59-2.57)	1.25(.58-2.72)
≥6 Grade	20(7.84)	62(24.3)	1	1
Household Income				
700-1500 Birr	21(8.2)	75(29.4)	0.85(.35-2.03)	0.68(.26-1.80)
1501 - 2500 Birr	25(9.8)	87(34.12)	0.82(.35-1.93)	0.73(.29-1.82)
≥2501	9(3.5)	38(14.9)	1	1
Household Size				
1-6	18(7.06)	102(40)	2.14(1.14-4.01)	2.59(1.27-5.32)*
≥7	37(14.5)	98(38.4)	1	1
Land size				
0.75 - 4.00 Hr	33(12.9)	136(53.3)	1.42(.77-2.62)	1.42(.73-2.78)
≥4.5 Hr	22(8.63)	34(13.3)	1	1

*P.V is significant at < 0.05. COR=Crude Odds Ratio, AOR=Adjusted Odds Ratio

4. Discussion

Several studies conducted in both developed and developing countries have reported on the factors affecting consumption of different foodstuffs, including consumption of pulses by households. Household total monthly income directly affected consumption of food items such as meat, rice, milk, fruit and pulses (Begum, *et al.*, 2010). Culture, tradition and food consumption patterns are influenced by each other. An understanding of cultural influences on eating habits is essential for the health promoter who wants to provide realistic educational interventions designed to modify dietary practices (Kaufman-Kurzrock, 1989; Airhihenbuwa, 1995).

Even though increased production of pulses would be expected to increase farm household income and contribute to greater household food security, production and productivity of pulses in Ethiopia are low due to low input usage, especially chemical fertilizers capable of increasing yields in field trials by 10 to 80 percent, limited availability of seed, limited familiarity with the variety of existing pulse types, and limited usage of modern agronomic practices. Marketing is another reason for the low production. Poor linkage between producers and the market has been observed to hinder producers' willingness to produce more pulses (IFPRI, 2010).In this study, both the quantitative and qualitative results showed that production of haricot bean was challenged and constrained by weeds, problems with chemical supply, disease, lack of high yielding and improved varieties, worms and pests, low soil fertility, a seasonal market and a shortage of farm land. A study from India also ranked lack of improved(high yielding and disease resistant) varieties, pest infestations and disease, not using pesticides or chemical fertilizers, lack of knowledge about improved pulse production practices, and high cost of inputs as leading biophysical and socio-economic constraints to pulse production (Roy, 2008).

Haricot bean was produced by a majority of the households in the study area, mainly for the sake of earning money. Another recent study also found that production of haricot bean was limited to about one-fourth of the farm land and for the sake of earning money to harvest maize (Gete, *et al.*, 2013). Haricot bean is considered the main cash crop and protein source for farmers in many lowland and mid-altitude zones of Ethiopia. The country's export earnings from haricot bean exceed that of other pulses such as lentil, faba bean and chickpea

(Negash, 2007).

Consumption of haricot bean was reported to be low. Several reasons were given, and these included abdominal distension, flatulence and diarrhoea following consumption, and a culture of considering pulses as a taboo food. Carbohydrates constitute the main fraction of grain legumes, accounting for up to 55-65% of the dry matter. Of these, starch and non-starch polysaccharides (dietary fiber) are the major constituents, with smaller but significant amounts of oligosaccharides (Bravo *et al.,* 1998). Oligosaccharides, which are common in legume seeds, are thought to be the major producers of flatus when consumed in significant quantities (Reddy and Salunkhe, 1980). A study conducted to evaluate the combined effect of soaking, dehulling, cooking and fermentation with *Rhizopusoligosporus* on the oligosaccharides, trypsin inhibitor, phytic acid and tannins of different pulse varieties indicated that soaking, dehulling, washing and then cooking decreased raffinose by 50% and sucrose and stachyose by more than 55-60% (Egounletya and Aworh, 2003).It is not clear, however, whether changing the oligosaccharide component will alter the health benefits of legumes. The sensation of increased gas does seem to modulate with more frequent pulse consumption (Livesey, 2001). After a few weeks of daily pulse consumption, consumers perceived that flatulence occurrence returned to normal levels. Only a small percentage of individuals were bothered by increased flatulence regardless of the length of time they had consumed pulses (Jenkins and Kendall, 2000).

Although some consumers recognize that beans and other pulses are an important protein source or meat substitute, many avoid eating them because they believe that pulse consumption will cause excessive intestinal gas or flatulence. Many may not be aware that like other vegetables, pulses are rich sources of fiber, vitamins and minerals (Fleming *et al.*, 1985). An increasing body of research supports the benefits of a plant-based diet, and pulses specifically, in the reduction of chronic disease risks and in improving food security (Christopher E., *et al* 2012; Hosseinpour-Niazi S., *et al* 2012; Mattei, *et al.*, 2011 and Celleno , *et al.*, 2007). Weekly consumption frequencies of pulses (haricot bean and lentil) were affected by socio-demographic and economic characteristics such as the age and educational status of the household head, household size and farm land size. Even though statistical associations were not calculated, in agreement to finding of this study descriptive survey done in Canada showed differences in pulse consumption frequency related to household size, educational status and income, with pulse consumption less frequent in single-person households, and increasing with increasing educational level from high school to university (Ipsos, 2010).

5. Conclusion

Increased agricultural productivity can improve nutritional intake of households through improving purchasing capacity and consumption from one's own production. Few studies have looked at the two variables, production and consumption.

Findings from this study showed that three varieties (broad bean, chickpea and pea) were tried by different famers some years back and in their trial they identified the unsuitability of the area, worm, disease and unproductiveness as barriers. Only haricot bean was produced covering too limited farm land as compared to other cereal (maize) grown and constrained by weed, lack of chemical supply, disease, lack of high yielding and improved varieties, worm and pest, low soil fertility, seasonal marketability and not being member of food stuffs. Despite the production consumption of haricot bean was too limited because of abdominal discomforts after and culture of considering it as a taboo food especially by male household heads. Household consumption frequencies of lentil and haricot bean were affected by land size, household head education and age, and household size.

Agronomic and market concerns related to production of haricot bean and other pulse crop be addressed and that household food preparation techniques for pulses that potentially reduce gastrointestinal symptoms be promoted and evaluated. Besides looking for production, availability and access to foods influence of culture, tradition, age and gender on utilization/eating habit of a community has to be well understood before nutrition intervention to modify dietary practices.

Acknowledgments

We are grateful to the International Development Research Center (IDRC), and Global Affairs the Department of Foreign Affairs and Development Canada for providing financial support for the project through the Canadian International Food Security Research Fund (CIFSRF) and the partnership team from the University of Saskatchewan and Hawassa University.

Conflict of Interest

The authors declare no conflict of interest.

References

Airhihenbuwa, C. O. (1995). Health & Culture: Beyond the Western Paradigm. Thousand Oaks.Sage Publications, Inc., London.

Begum, S. M., Khan, M. F., Begum, N., & Shah, I. U (2010). Socio economic factors affecting food consumption patternin rural area of district Nowshera, Pakistan. *Sarhad Journal of Agricultural, 26*(4), 649-653.

Bravo, L., Siddhurahu, P., & Saura-Calixto, F. (1998). Effect of various processing methods on the in vitro starch digestibility and resistant starch content of Indian pulses. *Journal of Agriculture and Food Chemistry, 46*, 4667-4674. https://doi.org/10.1021/jf980251f

Campbell, C. A. Zentner, R. P., Selles, F., Biederbeck V. O., & Leyshon, A. J. (1992). Comparative effects of grain lentil-wheat and monoculture wheat on crop production, N economy and N fertility in a Brown Chernozem. *Canadian Journal of Plant Science, 72*(4), 1091-1107. https://doi.org/10.4141/cjps92-135

Celleno, L., Tolaini, M. V., D'Amore, A., Perricone, N. V., & Preuss, H. G. (2007). A Dietary supplement containing standardized Phaseolus vulgaris extract influences body composition of overweight men and women. *International Journal of Medical Sciences, 24*, 4(1), 45-52. https://doi.org/10.7150/ijms.4.45

Christopher, E.S, Rebecca C. M, Bohdan L. L., & Anderson, G. H. (2012). The effect of yellow pea protein and fibre on short-term food intake, subjective appetite and glycaemic response in healthy young men. *British Journal of Nutrition, 108*, S74-S80. https://doi.org/10.1017/S0007114512000700

Egounlety, M., & Aworh, O. C. (2003). Effect of soaking, dehulling, cooking and fermentation with Rhizopusoligosporus on the oligosaccharides, trypsin inhibitor, phytic acid and tannins of soybean (Glycine max Merr.), cowpea (Vignaunguiculata L. Walp) and groundbean (Macrotylomageocarpa Harms). *Journal of Food Engineering, 56*, 249-254. https://doi.org/10.1016/S0260-8774(02)00262-5

FAO (2010). Special Report Fao/Wfp Crop and Food Security Assessment Mission to Ethiopia. FAO/WFP.

Flemming, SE, O'Donnell, AU, & Perman, JA (1985). Influence of frequent and long-term bean consumption on colonic function and fermentation. *American Journal of Clinical Nutrition, 41*, 909-918.

Schwenke, G. D., Peoples, M. B., Turner, G. L., & Herridge, D. F. (1998). Does nitrogen fixation of commercial, dryland chickpea and faba bean crops in north-west New South Wales maintain or enhance soil nitrogen? *Australian Journal of Experimental Agriculture, 38*(1), 61-70. https://doi.org/10.1071/EA97078

Gete, T., Nigatu, R., Henry, C. J., & Elabor-Idemudia, P. (2013). Smallholder Farmers Pulse Production and Marketing in four Districts of Ethiopia: Examining Access and Control of key resources from a gender lens. *Humboldt International Journal of Gender, Agriculture and Development, 1*(1), 90-109.

Gibson, R. S., Abebe, Y., Hambidge, M. K. Abride, I., Teshome, A., & Stoecker, B. J. (2009). Inadequate feeding practices and impaired growth among children from subsistence farming households in Sidama, Southern Ethiopia. *Journal of Maternal and Child Nutrition, 5*, 260-275. https://doi.org/10.1111/j.1740-8709.2008.00179.x

Hosseinpour-Niazi, S., Mirmiran, P., Amiri, Z., Hosseini-Esfahani, F., Shakeri, N., & Azizi, F. (2012). Dry Beans and Weight Management, Satiety, Glucose Control, and/or Diabetes. *Archives of Iranian medicine. 15*, 538-44.

International Food Policy Research Institute (IFPRI). (2010). Pulses Value Chain in Ethiopia: constraints and opportunities for enhancing exports.

Ipsos Reid, Alberta Agriculture and Rural Development and Alberta Pulse Growers Commission and Pulse Canada. (2010). Factors Influencing Pulse Consumption in Canada. Final Report

Jenkkins, J., & Kendall, C. W. (2000). Resistant starch.*Current Opinion in Gastroenterology, 16*, 178-83. https://doi.org/10.1097/00001574-200003000-00014

Kaufman-Kurzrock, D. L. (1989). Cultural aspects of nutrition.*Topics in Clinical Nutrition, 4*, 1-6. https://doi.org/10.1097/00008486-198904000-00003

Kebebu, A., Whiting, S. J., Dahl, W. J., Henry, C. J., & Kebede, A. (2013). Formulation and acceptability testing of a complementary food with added broad bean (V*iciafaba*) in southern Ethiopia. *African Journal of Food Agriculture Nutrition and Development, 13*(3).

Livesey, G. (2001). Tolerance of low digestible carbohydrates: a general view. *British Journalof Nutrition, 85*,

S7-S16. https://doi.org/10.1079/BJN2000257

Mattei, J., Hu, F. B., & Campos, H. (2011). A higher ratio of beans to white rice is associated with lower cardiometabolic risk factors in Costa Rican adults. *American Journal of Clinical Nutrition, 94,* 869-76. https://doi.org/10.3945/ajcn.111.013219

Negash, R. (2007). Determinants of adoption of improved haricot bean production package in Alaba special Woreda, Southern Ethiopia. Available at http://cgspace.cgiar.org/handle/10568/682]

Reddy, N. R., & Salunkhe, D. (1980). Changes in oligosaccharides during germination and cooking of black gram and fermentation of black gram rice blends. *Cereal Chemistry, 57,* 356-360.

Roba, A. C., Gebremichael, K., Zello, G. A., Whiting, S. J., & Henry, C. J. (2015). Nutritional status and dietary intakes of rural adolescent girls in southern Ethiopia. *Ecology of Food and Nutrition, 54*(3), 240-54. https://doi.org/10.1080/03670244.2014.974593

Burman, R., Singh, S. K., Singh, L., & Singh, A. K. (2008). Extension Strategies for increasing Pulses Production for Evergreen Revolution. *Indian Research. Journal of Extension and Education, 8*(1). 79-82

20

Sensory, Physicochemical and Microbiological Characteristics of Venison Jerky Cured with NaCl and KCl

Wannee Tangkham[1] & Frederick LeMieux[1]

[1] Department of Agricultural Sciences, McNeese State University, Lake Charles, LA 70609, USA

Correspondence: Wannee Tangkham, Department of Agricultural Sciences, McNeese State University, Lake Charles, LA 70609, USA. E-mail: wtangkham@mcneese.edu

Abstract

Traditionally, jerky is produced from sliced whole muscle marinated in a high sodium chloride (NaCl) concentration and dried. Because a high salt diet has been linked to hypertension, salt substitutes are often recommended as a healthier alternative. However, potassium chloride (KCl), a popular salt substitute may impart an undesired bitterness and metallic aftertaste. The objective of this study was to evaluate specific attributes of venison jerky prepared in three different (NaCl/KCl) salt solutions. Through sensory testing, each preparation was evaluated for consumer product acceptance and purchase intent. Additionally, the venison jerky was assayed for physicochemical characteristics and microbial counts. Using a 9-point hedonic scale, sixty-eight consumers evaluated the jerky for acceptability of flavor, texture, taste, saltiness, bitterness and overall liking. Physicochemical characteristics were evaluated for moisture content, pH, color and TBAR. Jerky was assayed for microbial counts via aerobic plate count, *Escherichia coli*, *Staphylococcus aureus* and *Campylobacter* spp. Results show that jerky prepared with 100% KCl received the most desirable score (8.75), compared to jerky prepared with 100% NaCl (6.28), and jerky prepared with 50% NaCl + 50% KCl (6.13). Acceptability and purchase intent questionnaires indicate jerky prepared with 100% KCl ranked the highest at 86.8% and 70.6%, respectively. Jerky prepared with 100% KCl had the lowest moisture content, TBAR, and a* values (P<0.05). No *E. coli*, *S. aureus* and *Campylobacter* spp. were detected over the 28 day period. Our study suggests that jerky prepared with KCl represents a low sodium alternative to traditional jerky.

Keywords: venison, jerky, sensory testing, salt substitute, microbial counts

1. Introduction

Jerky is a popular snack item and is classified by the U.S. Department of Agriculture (USDA) as a heat-treated and shelf-stable ready-to-eat meat product (USDA-FSIS, 2004). Whole or molded from chopped or ground meat, jerky may be cut into strips or stuffed into narrow casings. Jerky is made from a diversity of meat types and additives including sodium nitrite, table salt (sodium chloride) and spices (Ingram, 1973; Quintion et al., 1997; Gailani & Fung, 1986; Konieczny et al., 2007; Lim et al., 2014). Salt enhances the flavor of jerky (Gillette, 1985), acts as a preservative and provides sensations termed mouthfeel (Pszczola, 2006). However, overconsumption of table salt might increase blood pressure which has been linked to other health concerns, such as hypertension and stroke. Strokes result in 130,000 deaths in the United States (CDC, 2015) and 6 million deaths worldwide (WHF, 2015) annually. In the United States, strokes cost $34 billion annually in health care services, medications and lost productivity (CDC, 2015). Additionally, high blood pressure might be associated with in heart and kidney disease and congestive heart failure. Faced with these health issues, many consumers are concerned with limiting their salt intake. Their concerns have likewise influenced food processing manufacturers.

It has been reported (IOM, 2015) that the tolerable upper intake level for table salt is 5.8 g (or 2.3 g of sodium) per day. Healthy 19 to 50 year old adults can safely consume 3.8 g of salt or 1.5 g of sodium per day (Palar & Sturm, 2009). It is estimated that reducing table salt by 3 g per day will annually reduce the number of Americans with high blood pressure by 11 million, coronary heart disease by 120,000 and stroke by 66,000. Additionally, this amount of salt reduction would save up to 392,000 quality-adjusted life-years and $10 to $24 billion in health care costs (Palar & Sturm, 2009).

Using alternatives, such as potassium chloride (KCl), is one of the most common ways to reduce the sodium content in processed foods. According to the US Dietary Guideline (US-HHS, 2005), a potassium-rich diet reduces the effects of table salt on blood pressure and an intake of 4.7 g potassium/day is recommended. KCl has physicochemical properties closely resembling NaCl, but some consumers might experience an unpleasant aftertaste. However, it was reported (Mickelsen et al., 1977) that a mixture containing equal amounts of NaCl and KCl tasted similar to that of pure NaCl.

Eating quality has long been recognized as a factor for repeat customer purchasing (Grunert, 2004). Venison meat has become more popular in recent years partly due to its low intramuscular fat content (Hoffman & Wiklund, 2006) and higher n-3 polyunsaturated fatty acids (Bures et al., 2015) which is considered healthier than other red meats (Cordain et al., 2002; Giordano et al., 2010). In addition, venison meat is highly desirable for sensory properties such as aroma, taste, juiceness, and tenderness (Daszkiewicz et al., 2009).

To date, no studies have addressed the sensory, physicochemical and microbiological characteristics of venison jerky cured with a mixture of NaCl and KCl. The objective of this study is to evaluate specific attributes of venison jerky prepared in three different (NaCl/KCl) salt solutions.

2. Method

2.1 Preparation of Venison Jerky Samples

Venison semimembranosus muscles were obtained locally in Lake Charles, Louisiana. Muscles were cut into 5.0×10.0×0.5 cm slices. All subcutaneous and intermuscular fat and visible connective tissue were removed from the muscles. The muscle samples were refrigerated at 3°C. Samples were subjected to three treatments and cured using the following salt solutions: 100% KCl, 50% KCl + 50% NaCl and 100% NaCl. Other ingredients included: 5.91% Worcestershire sauce (Lea & Perrins Inc., Pittsburgh, PA), 4.22% soy sauce (Kikkoman Foods, Inc., Walworth, WI), 2.53% liquid smoke (The Colgin companies, Mint Way, Dallas, TX), 0.46% garlic powder (Bolner's Flesta Products Inc., San Antonio, Texas), 0.46% onion powder (Kroger Co Cincinnati, Ohio) and 0.23% black pepper (Kroger Co Cincinnati, Ohio). The samples were allowed to cure for 12 hours at 3°C.

The cured jerky samples were then dried in a dehydrator (Model 778SS LEM^{TM}) at 70°C for 6 h. After drying, the samples were cooled to ambient temperature. Each sample was evaluated for consumer product acceptance and purchase intent. Additionally, each sample was analyzed for pH, moisture content, color (L*, a*, and b* values), lipid stability (TBARS), aerobic plate count, *Escherichia coli* (*E. coli*), *Staphylococcus aureus* (*S. aureus*) and *Campylobacter* spp. at 7 d intervals for 28 d.

2.2 Sensory Evaluation

Each preparation was evaluated for consumer product acceptance and purchase intent. Using a 9-point hedonic scale, sixty-eight consumers evaluated the jerky for acceptability of flavor, texture, taste, saltiness, bitterness and overall liking. Consumers also completed an acceptability and purchase intent questionnaire.

2.3 pH Test

Samples were evaluated for pH with a probe electrode portable meter (Model 2000 VWR Scientific). Calibration of the pH meter was accomplished using pH 7 and pH 4 standardization buffers before use.

2.4 Moisture Content

Moisture content was determined according to the method of the Association of Official Analytical Chemists (AOAC 2000). Each 3 g sample was dried in an air oven (Model 26 Precision Thelco) at 102°C for 24 h.

2.5 Color Test

Color was measured at three different locations on each sample with a Minolta colorimeter (Model CR-10 portable) using an 8 mm aperture, 10° observer angle, D65 illuminant source in terms of L* (white = 100, black = 0), a* (+40 = red, -40 = green), b* (+40 = yellow, -40 = blue).

2.6 TBARS Test

The thiobarbituric acid-reactive substances (TBARS) method (Tarladgis, 1964) was used to measure lipid oxidation. Thiobarbituric acid reacts with the oxidation products of fat in solution to form malonaldehyde, which was measured on a spectrophotometer (Beckman Du-640) at 530 nm. The TBA value was expressed as mg malonaldehyde (MDA)/kg tissue.

2.7 Microbial counts

The microorganisms were determined following the standards of AOAC (2000). For this study, jerky was assayed for four undesirable microorganisms: aerobic plate count (APC), *E.coli*, *S. aureus* and *Campylobacter* spp.

The following protocol was used for APC, *E. coli* and *S. aureus*. Buffered peptone water (BPW) was added as a diluent option for serial dilutions. Following 3M™ Petri film plating instructions, plates were incubated in a horizontal position, clear side up, in stacks of no more than twenty at 37°C for 24-48 h.

The following protocol (Corry et al., 2003) was used for *Campylobacter* spp. BPW was added as a diluent option for serial dilutions. All samples were mixed with a vortexer for 2 min to release the bacteria. Each 0.1 ml of sample was aseptically transferred and spread onto modified charcoal cefoperazone deoxycholate agar. The inoculated plates were then incubated at 42°C for 48 h under a microaerophilic environment (5%O_2, 10%CO_2, and 85%N_2). Data were collected from countable plates (30-300 colonies per plate). The counted colonies were reported as CFU/g.

2.8 Statistical Analysis

The Proc GLM procedures of SAS windows (SAS, 2003) were used to evaluate the significance of differences of the obtained data. The PDIFF option of LSMEANS was employed to determine significance among treatments. All data are presented as means with standard deviation (SD) and a significance level of $P<0.05$ was used for statistical analysis of means from treatments.

3. Results and Discussion

3.1 Demographic Information

Demographic information of the 68 consumers in this study are presented. All consumers were volunteers solicited through advertisements posted in the Agricultural Sciences building on the McNeese State University Campus. The two largest age groups (18-24 and 45-54 years old) accounted for 63.2% of the total. Female participants (63.2%) exceeded males (36.8%).

3.2 Rank Response and Bitterness Evaluation

To confirm that consumers can distinguish relative saltiness and bitterness, discriminative tests by ranking were performed. Consumers were asked to rank the three preparations in order of most to least salty. The 100% NaCl samples received a plurality of 42.6% as the most salty (Table 1). By a slight margin, the 50% KCl + 50% NaCl samples were selected as the least salty ($P<0.05$) at 36.8% (Table 1). The 100% NaCl and 100% KCl samples received 33.8% and 29.4% of the vote respectively (Table 1).

Samples prepared with 100% KCl were rated the most bitter by 39.7% of consumers (Table 1). The 100% NaCl samples were rated the least bitter at 41.1% (Table 1). These results suggest that the prepared samples were distinguishable for saltiness and bitterness by the consumers.

Table 1. Least squares means for rank responses for saltiness and bitterness

Jerky Treatments	Consumers (N = 68)		
	Number/Percentage		
	1st	2nd	3rd
Saltiness	Most salty	Most salty	Most salty
100% KCl	25/36.8[a]	22/32.4[a]	20/29.4[a]
50% KCl + 50% NaCl	14/20.6[b]	30/44.1[b]	25/36.8[a]
100%NaCl	29/42.6[a]	16/23.5[a]	23/33.8[a]
	1st	2nd	3rd
Bitterness	Most bitter	Most bitter	Most bitter
100% KCl	27/39.7[a]	17/25.0[a]	25/36.8[a]
50% KCl + 50% NaCl	19/27.9[a]	34/50.0[b]	15/22.1[b]
100%NaCl	22/32.4[a]	17/25.0[a]	28/41.1[a]

[a,b] LSMeans with different superscripts within a column is significantly different ($P<0.05$).

3.3 Consumer Acceptability

Using the hedonic scale, participants evaluated the jerky for flavor, texture, taste, saltiness, bitterness and overall liking (Table 2). With reference to flavor, texture, taste and saltiness, scores among all three treatments statistically were not significantly different (Table 2). However, bitterness and overall liking scores were different between treatments. Specifically, jerky prepared with 100% KCl increased bitterness and had favorable scores for overall liking (Table 2). These results suggest that jerky prepared with KCl is a viable alternative to jerky prepared exclusively with NaCl.

Table 2. Least squares means for consumer acceptance scores for sensory attributes and overall liking of three salt solutions

Properties	100% KCl	50% KCl + 50% NaCl	100% NaCl
Flavor	6.65[a]	6.38[a]	6.30[a]
Texture	6.65[a]	6.41[a]	6.53[a]
Taste	6.59[a]	6.32[a]	6.42[a]
Saltiness	5.98[a]	6.1[a]	5.92[a]
Bitterness	5.39[a]	3.75[b]	4.79[a]
Overall liking	8.75[a]	6.13[b]	6.28[ac]

[a,b,c] LSMeans with different superscripts within a row is significantly different (P<0.05).

3.4 Acceptability and Purchase Intent

Using the acceptability and purchase intent questionnaire, consumers evaluated the jerky for acceptability, whether or not they would purchase the product and whether or not they would purchase the product if it claimed to contain reduced sodium, which might impact blood pressure (Table 3).

All three jerky treatments received similar scores with respect to acceptability and purchase intent (P>0.05). However, the 100% KCl jerky samples received the highest scores of 86.8% and 70.6% for acceptability and purchase intent respectively (Table 3). Finally, with respect to whether or not the consumers would purchase the product if it claimed to contain reduced sodium, both reduced sodium jerky treatments received similar scores (P>0.05). However, the jerky samples prepared with 50% NaCl + 50% KCl scored the highest at 64.7% (Table 3). Once again, these results suggest that jerky prepared with KCl is a viable alternative to jerky treated exclusively with NaCl.

Table 3. Least squares means for acceptability and purchase intent questionnaire (N = 68)

	100% KCl Number/Percentage	50% KCl + 50% NaCl Number/Percentage	100% NaCl Number/Percentage
Acceptable			
Yes	59/86.8[a]	55/80.9[a]	55/80.9[a]
No	9/13.2[a]	13/19.1[a]	13/19.1[a]
Purchase			
Yes	48/70.6[a]	47/69.1[a]	45/66.2[a]
No	20/29.4[a]	21/30.9[a]	23/33.8[a]
Purchase + health claim[1]			
Yes	40/58.8[a]	44/64.7[a]	N/A
No	28/41.2[a]	24/35.3[a]	N/A

[a]LSMeans with the same superscripts within a row is not significantly different (P>0.05).

[1]Reduced sodium jerky.

3.5 pH

The initial pH values of each individual jerky treatment ranged from 5.54 to 5.79 (Figure 1). Over the 28 day experimental period, changes in pH over each treatment profile exhibited significant differences ($P<0.05$). However, the pH values in each treatment followed a general decrease over the experimental period (Figure 1).

USDA defines intermediate pH, of which jerky is an example, possess a pH between 4.72 and 6.73 (Jose et al., 1994). The pH value has important implications with respect to safety. Specifically, the pH of properly processed jerky will inhibit or delay the spoilage of various dried meat products from mold and microorganism growth (Leistner, 1987). Initial and final pH values for each of the three treatments fell within or below the defined pH range (Figure1). More specifically, the pH value of the 100% KCl treated jerky was 4.48 at day 28 sampling time.

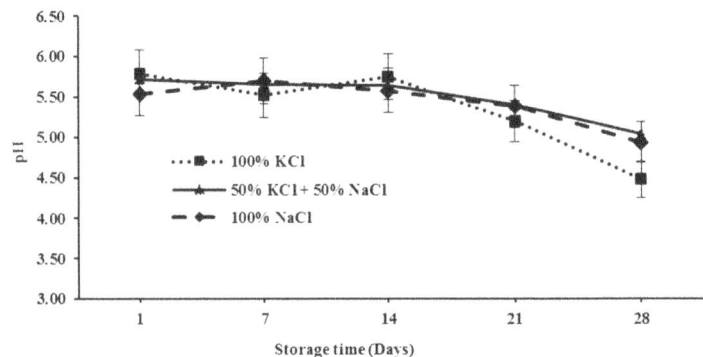

Figure 1. Least squares means for pH values of venison jerky during ambient storage for 28 days. SEM = 0.06

3.6 Moisture Content

Commercial intermediate foods have moisture contents of 20% to 40% (Jose et al., 1994). Moisture content affects sensory properties, stability and safety. Moisture contents of the three salt treatments are shown in Figure 2. The initial water content of each of the three treatments declined during the course of the experiment. Specifically, a respective decline of 17.7%, 14.8%, and 4.6% for 100% KCl, 50% KCl + 50% NaCl and 100% NaCl was detected over 28 days. These values were statistically significant ($P<0.05$).

These results clearly show that jerky treated with KCl experiences materially more moisture loss than that treated with NaCl. Therefore, packaging might become a critical factor in the commercial market for venison jerky.

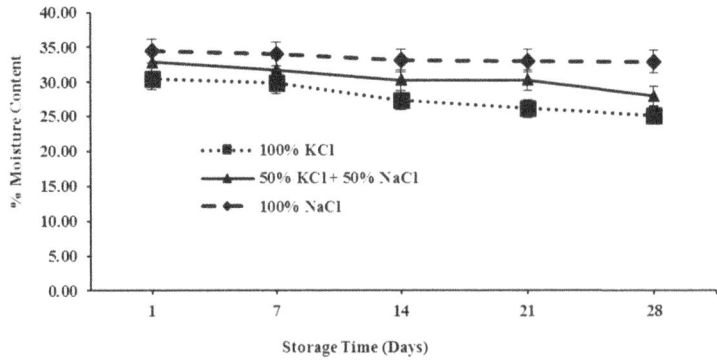

Figure 2. Least squares means for moisture content (%) of venison jerky during ambient storage for 28 days.
SEM = 2.62

3.7 Color Test

Increasing changes in lightness (L*) values represent greater light dispersion and increased lightness and is correlated with changes in meat structure, especially protein destruction (MacDougall, 1982). This is likely due to protein denaturation (Insausti et al., 2001). Results from the present study indicated that no significant differences occurred in the L* values among the three treatments during each week of the storage period (P>0.05) (Table 4). These results suggest that the treatment concentrations of NaCl and KCl used in this study did not significantly affect the meat structure of the venison jerky. These results are similar to those found by Soldatou, Nerantzaki, Kontominas, & Savvaidis (2009).

Redness (a*) values help gauge consumers acceptability of meat product color (Brewer et al., 2002). Typically, the red color of meat transitions to brownish red due to the formation of metmyoglobin. There was a significant difference (P<0.05) observed in a* values between treatments at days 1 and 21. However, no interactions were observed between the treatments and storage periods (P>0.05). The nominal values between the three treatments were small (Table 4). Therefore, color transition among the three different treatments was similar. This suggests that KCl has little impact on initial color or color as it changes through time.

Yellowness is measured in terms of positive b* values. In general, no clear trend in yellowness per treatment was evident during the experimental period. For example, from its initial value, yellowness exhibited alternate declines and increases during the course of experiment (Table 4). However, all three treatments fell within yellowness values between 5.80 and 17.57. Therefore, KCl had no appreciable impact on the yellowness.

Table 4. Least squares means for HunterLab L*, a*, and b* values of venison jerky during ambient storage for 28 days

Parameter	Treatment	Storage time (d)				
		1	7	14	21	28
L*	100% KCl	25.70[a]	30.90[a]	31.40[a]	26.40[a]	28.30[a]
	50% KCl + 50% NaCl	33.20[b]	29.43[a]	32.97[ab]	33.37[b]	30.87[a]
	100% NaCl	32.20[ab]	24.60[a]	24.47[ac]	29.10[ab]	32.90[a]
a*	100% KCl	4.90[a]	6.03[a]	3.33[a]	2.23[a]	2.83[a]
	50% KCl + 50% NaCl	8.85[b]	5.00[a]	3.73[a]	4.87[b]	4.63[a]
	100% NaCl	6.90[a]	4.20[a]	3.17[a]	3.10[ab]	4.67[a]
b*	100% KCl	9.25[a]	14.6[a]	7.07[a]	5.80[a]	7.77[a]
	50% KCl + 50% NaCl	16.50[b]	10.83[a]	17.57[b]	12.67[b]	12.30[ab]
	100% NaCl	14.35[ab]	10.20[a]	7.80[ac]	10.27[ab]	13.37[bc]

[a,b,c]LSMeans with different superscripts within a row is significantly different (P<0.05).

3.7 Lipid stability (TBARS)

Lipid composition constitutes the major determinant for susceptibility to oxidative changes and rancidity development leading to flavor deterioration or what has been termed a *warmed-over* flavor (Sato & Hegarty, 1971). Additionally, oxidation causes off-odors (Lillard, 1987). There was a statistically significant difference (P<0.05) in terms of TBARS values between treatments at days 1, 14, 21, and 28 (Figure 3). However, the data exhibited similar trajectories over the experimental period. That is, TBARS values steadily increased over time which similarly to the study of Melton (1983) and Hamid et al (2010). Additionally, the two treatments containing KCl consistently exhibited TBARS values less the NaCl treatment. This may be due to the presence of NaCl acting as a prooxidant and an increase of lipid oxidation during meat processing (King & Bosch, 1990; Rhee 1999; Rhee & Ziprin, 2001). King and Bosch (1990) found that 2% sodium chloride was more prooxidant compared with 2% potassium chloride in turkey patties. Therefore, the KCl treated jerky represents a viable alternative to jerky treated with NaCl with respect to flavor and odor associated with lipid oxidation.

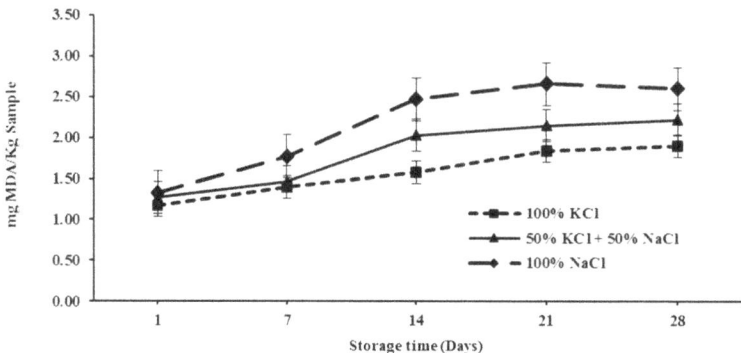

Figure 3. Least squares means for TBARS (thiobarbituric acid-reactive substances) values of venison jerky during ambient storage for 28 days. SEM = 0.90

3.8 Microbial Counts

Jerky is widely regarded as microbial safe. However, gastroenteritis outbreaks have been attributed to consumption of both home-dried and commercially prepared jerky (Nummer et al., 2004). Similar findings by Levine et al. (2001) also reported that food borne pathogens can survive the moderate drying conditions (60-70°C) used by commercial jerky manufacturers. Holley (1985) identified that *S. aureus* can grow in the jerky production process in the initial drying period. However, from our analyses, the presence of *E. coli*, *S. aureus* and *Campylobacter* spp. was not detected over ambient storage for 28 days.

Aerobic plate count is a common method used to indicate the microbiological quality of a food (Stannard, 1997). Aerobic plate counts above 10^7 cfu/cm^2 is commonly used as an indicator of spoilage (Korkeala et al., 1987; Borch et al., 1996). Our study found that aerobic plate counts were detected at levels considered safe for human consumption at 3.21-3.45 log CFU/g for all three jerky treatments over ambient storage for 28 days (Stannard, 1997; Patricia & Azanza, 2005). In general, all three treatments exhibited modest increases in the enumeration of aerobic plate counts over the experimental period (Figure 4). Aerobic plate counts among the three treatments were not statistically significant ($P>0.05$) which is similar to the study of Blesa et al., (2008). These results suggest that KCl provides similar bacterial protection to venison jerky treated with NaCl.

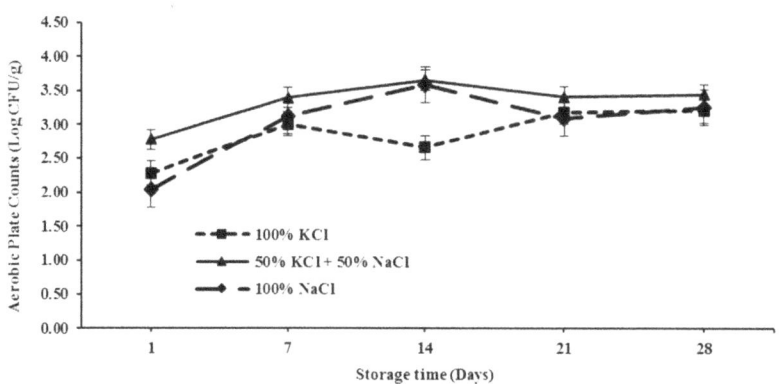

Figure 4. Least squares means for aerobic plate counts of venison jerky during ambient storage for 28 days. SEM = 0.45

4. Conclusions

The results of this study suggest that venison jerky prepared with KCl is a viable alternative to jerky treated exclusively with NaCl. Specifically, participants rated all three treatments similarly with respect to flavor, texture, taste and saltiness. With respect to overall liking, participants rated jerky prepared with 100% KCl as the highest. Additionally, all three treatments received positive participant ratings with respect to acceptability and purchase intent with and without health claims. Therefore, jerky prepared with KCl, is a marketable alternative to traditional jerky.

References

AOAC (Association of Official Analytical Chemists). (2000). Official methods of analysis. In H. William (Ed.). 17th ed. Gaithersburg, MD.: AOAC.

Blesa, E., Aliño, M., Barat, J. M., Grau, R., Toldrá, F., & Pagán, M. J. (2008). Microbiology and physico-chemical changes of dry-cured ham during the post salting stage as affected by partial replacement of NaCl by other salts. *Journal of Meat Science, 78*(1-2), 135-142. http://dx.doi.org/10.1016/j.meatsci.2007.07.008

Borch, E., Kant-Muermans, M. L., & Blixt, Y. (1996). Bacterial spoilage of meat and cured meat products. *International Journal of Food Microbiology, 33*(1), 103-120. http://dx.doi.org/10.1016/0168-1605(96)01135-X

Brewer, S., Jenson, J., Sosnicki, A. A., Field, B., Wilson, E., & McKeith, F. (2002). The effect of pig genetics palatability, color and physical characteristics of fresh loin chops. *Meat Science, 61*, 249-256. http://dx.doi.org/10.1016/S0309-1740(01)00190-5

Bures, D., Barton, L., Kotrba, R., Hakl, J. (2015). Quality attributes and composition of meat from red deer (Cervus elaphus), fallow deer (Dama dama) and Aberdeen Angus and Holstein cattle (Bos taurus). *Journal of the Science of Food and Agriculture, 95*(11), 2299-2306. http://dx.10.1002/jsfa.6950

CDC (Centers for Disease Control and Prevention). (2015). *Stroke in the United States.* Retrieved from http://www.cdc.gov/stroke/facts.htm.

Cordain, L., Watkins, B. A., Florant, G. L., Kelher, M., Rogers, L., Li, Y. (2002). Fatty acid analysis of wild ruminant tissues: evolutionary implications for reducing diet-related chronic disease. *European Journal of Clinical Nutrition*, 56(3), 181–191. http://www.ncbi.nlm.nih.gov/pubmed/11960292

Corry, J. E. L., Atabay, H. I., Forsythe, S. J., & Mansfield, L. P. (2003). Culture media for the isolation of *Campylobacters, Helicobacter* and *Arcobacters.* In J. E. L. Corry, G. D. W. Curtis & R. M. Baird (Eds.). *Handbook of Culture Media for Food Microbiology* (2nd Ed., pp. 271–315). Amsterdam, the Netherlands: Elsevier Science Publication.

Daszkiewicz, T., Janiszewski, P., & Wajda, S. (2009). Quality characteristics of meat from wild red deer (Cervus Elaphus L.) hinds and stags. *Journal of Muscls Foods, 20*(4), 428-448. http://dx.doi.org/10.1111/j.1745-4573.2009.00159.x

Gailani, M. B., & Fung, D. Y. C. (1986). Critical review of water activities and microbiology of drying of meats. *Critical Reviews in Food Science and Nutrition, 25*, 159-183. http://dx.doi.org/10.1080/10408398709527450

Gillette, M. (1985). Flavor effects of sodium chloride. *Food Technology, 39*, 47-52 & 56. http://agris.fao.org/agris-search/search.do?recordID=US8629343

Giordano, G., Guarini, P., Ferrari, P., Biondi-Zoccai, G., Schiavone, B., & Giordano, A. (2010). Beneficial impact on cardiovascular risk profile of Water buffalo meat consumption. *European Journal of Clinical Nutrition, 64*(9), 1000-1006. http://www.ncbi.nlm.nih.gov/pubmed/20588291

Grunert, K. G., Bredahl, L., & Brunsø, K. (2004). Consumer perception of meat quality and implications for product development in the meat sector-A review. *Meat Science, 66*, 259–272. http://dx.doi.org/10.1016/S0309-1740(03)00130-X

Hamid, R. G., Jens, K. S. M, Christina, E. A., & Leif, H. S. (2010). Sodium chloride or heme protein induced lipid oxidation in raw, minced chicken meat and beef. *Czech Journal of Food Science, 28*(5), 364-375. http://curis.ku.dk/ws/files/22477551/182-09_Gheisari_proof-sheet.pdf

Hoffman, L. C., & Wiklund, E. (2006). Game and venison—Meat for the modern consumer. *Meat Science, 74*(1), 197–208. http://dx.doi.org/10.1016/j.meatsci.2006.04.005

Holley, R. (1985). Beef jerky: Fate of *Staphylococcus aureus* in marinated and corned beef during jerky manufacture and 2.5° C storage. *Journal of Food Protection, 48*(2), 107-111. http://www.ingentaconnect.com/content/iafp/jfp/1985/00000048/00000002/art00002

Ingram, M. 1973. The microbiological effects of nitrite. In Krol, B. and Tinbergen, B. J., editors. *Proceedings of the International Symposium on Nitrite in Meat Products* (pp. 63-75). Zeist, The Netherlands.

Insausti, K., Beriain, M. J., Purroy, A., Alberti, P., Gorraiz, C., & Alzueta, M. J. (2001). Shelf life of beef from local Spanish cattle breeds stored under modified atmosphere. *Meat Science, 57*(3), 273-281. http://dx.doi.org/10.1016/S0309-1740(00)00102-9

IOM (Institute of Medicine). (2015). *Dietary Reference 401 Intakes: Water, Potassium, Sodium, Chloride, and Sulfate.* Retrieved from http://www.iom.edu/reports/2004/dietary-reference-intakes-water potassium-sodium-chloride-and-sulfate.aspx.

Jose, F. S., Rafael, G., & Miguel, A. C. (1994). Water activity of Spanish intermediate moisture meat products. *Meat Science, 38*, 341-350. http://dx.doi.org/10.1016/0309-1740(94)90122-8

King, A. J., & Bosch, N. (1990). Effect of NaCI and KCI on rancidity of dark turkey meat heated by microwave. *Journal of Food Science, 55*, 1549-1551. http://dx.doi.org/10.1111/j.1365-2621.1990.tb03565.x

Konieczny, P., Stangierski, J., & Kijowski, J. (2007). Physical and chemical characteristics and acceptability of home style beef jerky. *Meat Science, 76*, 253-257.http://dx.doi.org/10.1016/j.meatsci.2006.11.006

Korkeala, H., Lindroth, S., Ahvenainen, R. & Alanko, T. (1987). Interrelationship between microbial numbers and other parameters in the spoilage of vacuum-packed cooked ring sausages. *International Journal of Food Microbiology, 5*, 311-321. http://dx.doi.org/10.1016/0168-1605(87)90045-6

Leistner, L. (1987). Shelf-stable products and intermediate moisture foods based on meat. In: Rockland LB, Beuchat LR, editors. Water Activity: Theory and Applications to Foods (pp. 295-328). Marcel Dekker, New York: Wiley-Blackwell.

Levine, P., Rose, B., Green, S., Ransom, G., & Hill, W. (2001). Pathogen testing of ready-to-eat meat and poultry products collected at federally inspected establishments in the United States, 1990 to 1999. *Journal of Food Protection, 64*, 1188-1193. http://www.ncbi.nlm.nih.gov/pubmed/11510658

Lillard, D. A. (1987). Oxidative deterioration in meat, poultry, and fish. In A. J. Angelo, M. E. Bailey (Eds.). *Warmed-over flavor of meat* (pp. 41-67). Orlando, Florida: Academic Press.

Lim, H. J., Kim, G. D., Jung, E. Y., Seo, H. W., Joo, S. T., Jim, S. K., & Yang, H. S. (2014). Effect of curing time on the physicochemical and sensory properties of beef jerky replaced salt with soy sauce, red pepper paste and soybean paste. *Asian-Australasian Journal of Animal Sciences, 27*(8), 1174-1180. http://dx.doi.org/10.5713/ajas.2013.13853

MacDougall, D. B. (1982). Changes in the colour and opacity of meat. *Food Chemistry, 9*, 75-78. http://dx.doi.org/10.1016/0308-8146(82)90070-X

Melton, S. L. (1990). Methodology for following lipid oxidation in muscle foods. *Journal of Food Technology, 37*(7), 105-111.

Mickelsen, O., Makdani, D., Gill, J. L., & Frank, R. L. (1977). Sodium and potassium intakes and excretions of normal men consuming sodium chloride or a 1:1 mixture of sodium and potassium chlorides. *American Journal of Clinical Nutrition, 30*, 2033-2040. http://www.ncbi.nlm.nih.gov/pubmed/930873

Nummer, B. A., Harrison, J. A., Harrison, M. A., Kendall, P., Sofos, J. N., Andress, E. L. (2004). Effects of preparation methods on the microbiological safety of home-dried meat jerky. *Journal of Food Protection, 67*, 2337-2341. http://www.ncbi.nlm.nih.gov/pubmed/15508655

Palar, K., & Sturm, R. (2009). Potential societal savings from reduced sodium consumption in the U.S. Adult population. *American Journal of Health Promotion, 24*(1), 49-57. http://www.ncbi.nlm.nih.gov/pubmed/19750962

Patricia, M. A., & Azanza, V. (2005). Aerobic plate counts of Philippine ready-to-eat foods from take-away premises. *Journal of Food Safety*, 25(2), 80-97. http://dx.doi.org/10.1111/j.1745-4565.2005.00554.x

Pszczola, D. E. (2006). Exploring new 'Tastes' in Textures. *Food Technology, 60*(1), 44-55. http://www.ncbi.nlm.nih.gov/books/NBK50965/

Rhee, K. S. (1999). Storage stability of meat products as affected by organic and inorganic additives and functional ingredients. In Xiong YL , Ho CT, Shahidi F, editors. Quality Attributes of Muscle Foods (pp. 95-113). New York: Plenum Publishing Corp.

Rhee, K. S., & Ziprin, Y. A. (2001). Pro-oxidative effects of NaCl in microbial growth-controlled and uncontrolled beef and chicken. *Journal of Meat Science, 57*, 105-112. http://dx.doi.org/10.1016/S0309-1740(00)00083-8

SAS (Statistical Analysis Software). (2003). SAS User's Guide Version 9.1.3. Cary, NC.: SAS Institute Inc.

Sato, K., & Hegarty, G. R. (1971). Warmed-over flavor in cooked meats. *Journal of Food Science, 36*, 1098-1102. http://dx.doi.org/10.1111/j.1365-2621.1971.tb03355.x

Soldatou, N., Nerantzaki, A., Kontominas, M. G., & Savvaidis, I. N. (2009). Physicochemical and microbiological changes of "Souvlaki"—A Greek delicacy lamb meat product: Evaluation of shelf-life using microbial, color and lipid oxidation parameters. *Food Chemistry, 113*, 36-42. http://dx.doi.org/10.1016/j.foodchem.2008.07.006

Stannard, C. (1997). Development and use of microbiological criteria for foods: Guidance for those involved in using and interpreting microbiological criteria for foods. *Food Science and Technology Today, 11*(3), 137-176. http://citeseerx.ist.psu.edu/viewdoc/download?doi=10.1.1.474.2198&rep=rep1&type=pdf

Tarladgis, B. G., Watts, B. M., Younathan, M. T., & Jr. Dugan, L. (1964). Chemistry of the 2-thiobarbituric acid test for determination of malonaldehyde in rancid foods. *Journal of the American Oil Chemists' Society, 37*, 44-48.

Quinton, R. D., Cornforth, D. P., Hendricks, D. G., Brennand, C. P., & Su, Y. K. (1997). Acceptability and composition of some acidified meat and vegetable stick products. *Journal of Food Science, 62*, 1250-1254. http://dx.doi.org/10.1111/j.1365-2621.1997.tb12255.x

USDA-FSIS (U.S. Department of Agriculture–Food Safety and Inspection Service). (2004). Compliance guideline for meat and poultry jerky. USDA-FSIS, Washington, D.C. Retrieved from http://www.fsis.usda.gove/OPPDE/nis/outreach/models/Jerkyguidelines.htm.

US-HHS (US Department of Health and Human Services). (2005). *Dietary guidelines for Americans*. Retrieved from http://www.health.gov/dietaryguidelines/dga2005/document.

WHF (World Heart Federation). (2015). *Stroke*. Retrieved from http://www.worldheartfederation.org/cardiovascular-health/stroke/.

21

Wheat Bran Dietary Fiber: Promising Source of Prebiotics with Antioxidant Potential

Aynur Gunenc[1], Christina Alswiti[1] & Farah Hosseinian[1, 2]

[1]Food Science and Nutrition Division of Chemistry Department, Ottawa, Ontario, Canada

[2]Institute of Biochemistry, Carleton University, Ottawa, Ontario, Canada

Correspondence: Farah Hosseinian, Food Science and Nutrition Division of Chemistry Department, and Institute of Biochemistry, Carleton University, Ottawa, Ontario, Canada. E-mail: farah_hosseinian@carleton.ca

Abstract

The potential of wheat bran (WB) addition as a prebiotic source were demonstrated using yogurt with probiotics (*Lactobacillus acidophilus* and *Bifidobacterium lactis*). Yogurts (with 4% WB) were significantly ($P < 0.05$) different in total bacterial counts (9.1 log CFU/mL), and total titratable acidity % (TTA, 1.4%) compared to controls during 28 days cold storage (4°C). Additionally, WB-total dietary fiber contents and their bound phenolic profiles were investigated as well as the antioxidant activity of WB-water extractable polysaccharides (WEP) was studied. HPLC analysis of alkaline hydrolyzed DF fractions showed that insoluble DF had higher phenolic acids (84.2%) content than soluble DF (15.8%). Also, crude-WEP showed stronger antioxidant activity compared to purified-WEP with an ORAC of 71.88 and 52.48 μmol TE/g, respectively. Here we demonstrate WB has potentials as a source of prebiotics, which may have the potentials for functional foods and nutraceutical applications.

Keywords: Bound phenolics, dietary fiber, DPPH, ORAC, prebiotics, wheat bran

1. Introduction

Increased whole grain consumption is linked to a decreased risk of chronic diseases (Fardet, 2010) such as obesity (Jonnalagadda et al., 2011), type II diabetes (Murtaugh et al., 2003), cardiovascular disease (Mellen, Walsh, & Herrington, 2008) and cancer (Schatzkin et al., 2008). Wheat (*Triticum aestivum)*, is second to rice as the main staple food crop. Most bioactive components (e.g. phenolic acids, alkylresorcinols) and dietary fiber (45%) are present in wheat bran (WB) fractions that represent 14-16% of the grain by weight (Fardet, 2010).

Cereal grain oligosaccharides act as prebiotics (a non-digestible food ingredient/soluble dietary fiber) and increase beneficial bacteria (probiotics) amount in the large bowel, thus improving gut health (Topping, 2007). They are a nutritional substrate for the probiotics in the colon and have the potential to improve host health (Chakraborti, 2011). Meanwhile phenolic compounds (e.g. ferulic and gallic acids) are related to prevention of diseases through some potential mechanisms such as free radical quenching, transition-metal chelation, and stimulation of the antioxidant enzyme system (Aaby, Skrede, & Wrolstad, 2005). Probiotics and prebiotics are commonly used in fermented dairy products (Kanmani et al., 2013) and supplementary research is desired to discover bio-products and their potential use in functional foods or nutraceuticals. Therefore, it was aimed to; 1) investigate the prebiotic effects of WB addition on microbial counts as colony forming units (CFU), and TTA in yogurts with and without probiotic bacteria, 2) measure the antioxidant activity of crude- and purified-WEP by oxygen radical absorbance capacity (ORAC), and 2, 2-diphenyl-1-picryhydrazyl (DPPH) assays along with total phenolic content (TPC), and 3) determine WB-total DF content (SDF, IDF) and analyze the phenolic acids and flavonoid composition of each fiber fraction as well as WB.

2. Materials and Methods

2.1 Materials

Analytical grade solvents including acetone, ethanol, methanol, ethyl ether, HCL, and ethyl acetate were purchased from Caledon Laboratories LTC (Georgetown, ON, Canada). Over 98% pure of fluorescein, mono-and dibasic potassium phosphate, Trolox (6-hydroxy-2,5,7,8-tetramethylchroman-2-carboxylic acid), 2, 2'-azobis

(2-methylpropionamidine) dihydrochloride (AAPH), Folin- Ciocalteu (FC) reagent, α-tocopherol, and 2, 2-diphenyl-1-picryhydrazyl radical (DPPH), NaOH, sodium carbonate phenolphthalein were obtained from Sigma (Oakville, ON, Canada). Protease from *Bacillus licheniform* (saline solution ≥2.4 U/g protein, EC 232-560-9) and α-amylase from *Bacillus licheniformis* (Type XII-A, saline solution ≥500 U/mg protein, EC 232-752-2) were purchased from Sigma-Aldrich (St. Louis, Missouri, USA). The starter cultures, (*Lactobacillus delbrueckii* ssp. *bulgaricus* (B-548; USDA) and *Streptococcus salivarius* ssp. *thermophilus* (14485; ATCC)), probiotics (*Lactobacillus acidophilus* (B-4495, USDA) *and Bifidobacterium lactis* (41405, USDA)), Man Rogosa Sharpe (MRS) broth liquid and MRS agar media were purchased from Oxoid Ltd. (Basingstoke, United Kingdom).

Phenolic acids (gallic, protocatechuic, p-OH-benzoic, chlorogenic, caffeic, vanillic, syringic, p-coumaric, sinapic, ferulic, o-coumaric) and flavanoid standards (pyrogallol, catechin, epicatechin, rutin, quercetin-3-beta-glucoside, epicatechin gallate, myricetin, quercetin, apigenin and kaempherol) were purchased from Sigma-Aldrich (St. Louis, Missouri, USA).

2.2 Prebiotic Activity

2.2.1 Sample Preparation

Wheat bran (WB) was kindly provided by Kraft Canada, freeze dried and kept in sealed plastic bags (-20°C) until further use. Before use, it was ground to 0.5 mm particle size at Agriculture-Canada (Ottawa, ON, Canada) Laboratories using a cyclone sample mill (UDY Corporation, CO, USA).

2.2.2 Milk Preparation

Pasteurized whole milk (3.25%), purchased from a local market, was heated at 85°C for 15 minutes, cooled down to 42°C in a water bath and transferred into 50 mL-sterile test tubes. The starter cultures, probiotics, and 4% WB were added and incubated at 42°C until the yogurt reached ~ pH 5.0 (Santo et al., 2010). All treatments were done in triplicates.

2.2.3 Microbial Cultures

The starter cultures (*Lactobacillus delbrueckii* ssp. *bulgaricus* and *Streptococcus salivarius* ssp. *thermophilus*), probiotic 1 (*Lactobacillus acidophilus*) and probiotic 2 (*Bifidobacterium lactis*) were employed to make all yogurt trials as shown in Table 1. For each strain, sterile aliquots of MRS broth (10 mL) were used to grow microorganisms (incubated at 37°C for 24 h). For stock culture, activated tubes were used after three successive rinses with sterilized distilled water. Then the cultures were diluted with sterilized milk (121°C for 15 min in an autoclave) to obtain a concentration of 6.5 log bacteria cells/mL and then added to the tubes depending on the trials (Table 1) and the initial pH recorded. There were a total of eight yogurt trials including four with WB (4%) and four as controls without WB. Yogurts with WB were compared against corresponding controls. All tubes were incubated at 42°C for fermentation and pH was measured after 4 h and 1 h thereafter. When the pH reached approximately 5, all tubes transferred were stored at 4 °C (Santo, et al., 2010).

Table 1. The experimental design used to evaluate the effect of WB addition on microbial counts in different yogurt trials

Yogurt trials	Sample coding
Y	Y
Y + Pro 1	Y+1
Y + Pro 2	Y+2
Y + Pro 1 & Pro 2	Y+1+2
Y + WB	YB
Y + WB + Pro 1	YB+1
Y + WB + Pro 2	YB+2
Y + WB + Pro 1 & Pro 2	YB+1+2

Y= standard yogurt containing only starter cultures of *Lactobacillus bulgaricus* and *Streptococcus thermophiles*. Pro 1 = probiotic *Lactobacillus acidophilus*. Pro 2 = probiotic *Bifidobacterium lactis*, WB = wheat bran (4%). Equal volumes of each bacterium (6.5 log bacteria cells/mL) were added to the tubes depending on the treatments

2.2.4 Microbial Count

On days 1, 7, 14, 21 and 28, total microbial counts were carried in triplicate for each batch at different dilutions; four serial dilutions of 1 to 10. An aliquot (5 μL) from each dilution was plated on MRS agar dishes using a spread plate method and incubated at 37°C for 24 h (Santo, et al., 2010).

2.2.5 Total Titratable Acidity (TTA) Measurements

Percent TTA for all yogurts were determined on the same days (1, 7, 14, 21 and 28 in triplicate) by titrating each yogurt sample with 0.1 N NaOH using 0.1% (w/v) phenolphthalein as an indicator. Additionally, pH of all yogurts were recorded on each day (Denver Instrument UB-5 pH meter) (Behrad, Yusof, Goh, & Baba, 2009).

2.3 Water Extractable Polysaccharides (WEB) -Antioxidant Activity

2.3.1 WEP-Extraction

Crude-WEP: WB and distilled water were mixed (1:100, w/v), stirred (70°C for 4 h), cooled, centrifuged at 6000 x g for 20 min (Thermo Sorval, Legend XT Series, Fisher Scientific, Nepean, ON, Canada) and the supernatant was retained for further analysis.

Purified-WEP: Starch and proteins/peptides were removed by using α-amylase and protease to the supernatant solution (20 μL/100 mL) and agitated at 37°C for 24 h. Then the supernatant was cooled and centrifuged (6000 x g for 20 min). The supernatant was dialyzed against double distilled water for 48 h and replaced with fresh distilled water every 6 h to separate polysaccharides and other materials with a molecular weight cut-off of 3500 Da (Spectra/Por, CA, USA). The extract was kept at -20°C until further analysis (Escarnot et al., 2011).

2.3.2 Total Phenolic Content (TPC)

Crude- and purified-WEP total phenolic contents were measured by the Folin-Ciocalteu (FC) method (Gao, Wang, Oomah, & Mazza, 2002). Each sample extract of 200 μL was mixed with FC reagent (1.9 mL of 10-fold diluted FC), and then 1.9 mL of sodium carbonate solution (60 g/L) was added to the mixture. All tubes were stored in the dark at room temperature for 2 h, and the absorbance values were recorded at 725 nm (UV-Vis. Spectrophotometer, Cary 50 Bio, Varian Inc., Australia) against a blank of distilled water. Ferulic acid (FA) was employed as standard, and TPC values were calculated as FA equivalents per gram of samples (Gunenc, HadiNezhad, et al., 2013).

2.3.3 DPPH Assay

For DPPH radical scavenging activity of crude- and purified-WEP, Brand-Williams et al. (1995) study was followed. Each sample extract (200 μL) was reacted with DPPH solution (3.8 mL) for 1 h, and then absorbance (A) values of the mixture at 515 nm were read against a blank of pure 95% ethanol. The antioxidant activity was calculated as % discoloration (Gunenc, Tavakoli, et al., 2013; Li et al., 2009).

2.3.4 Oxygen Radical Absorbance Capacity (ORAC) Assay

ORAC values of crude- and purified-WEP were determined by a fluorometric micro plate reader (FLx800™ Multi-Detection Micro plate Reader with Gen5™ software, BioTek Instruments, Ottawa, Canada) described by Gunenc et al., (2013). All analyses were carried out at 37°C with a 20 min incubation and 60 min run-time. A micro plate was prepared containing 20 μL of Trolox standards, sample dilutions, as well as 120 μL of fluorescein (FL) solution. After the incubation of the plate, 60 μL of 153 mM AAPH was added quickly to each well for a final volume of 200 μL. FL micro plate reader was used to read the absorbance during run-time (at an excitation wavelength of 485 nm and an emission wavelength of 525 nm). ORAC values were calculated by the differences of net areas under the FL decay curves between the blank and sample and were expressed as micromole Trolox Equivalents per gram of sample (μmol TE/g) (Gunenc et al., 2015; Ou et al., 2002).

2.4 Total Dietary Fiber (TDF) and Its Phenolics

2.4.1 Extraction of TDF

WB-Total dietary fiber as the sum of soluble dietary fiber (SDF) and insoluble dietary fiber (IDF) was calculated by following the AOAC Official Method (991.43). To remove starch and protein, WB (5 g) was exposed to enzymatic digestions; firstly, heat stable α-amylase (250 μL, boiling water bath for 30 min), then alcalase protease (50 mg/mL, 500 μL, pH 7.5, 60°C for 30 min) and lastly, amyloglucosidase (1500 μL, pH 4.5, 60°C for 30 min). After centrifugation (10, 000 rpm), the residue was washed sequentially with hot water, ethanol (95%), and acetone (95%); filtered and recorded as IDF.

The combined supernatants from the washings was precipitated in ethanol (80%, preheated to 60°C, 4 volumes) overnight and recorded as SDF. The two fractions (IDF and SDF) were placed in a fume hood and dried at

35-40°C overnight to remove organic solvent (Guo & Beta, 2013).

2.4.2 Phenolic Compound Extractions from WB, SDF and IDF

WB, SDF and IDF were alkaline hydrolyzed to release their bound phenolics and followed by liquid-liquid partitioning steps to extract those released phenolic compounds (Gunenc, et al., 2015; Guo & Beta, 2013). More specifically, each sample (WB, SDF, and IDF) were mixed with 2 M NaOH and agitated at room temperature for 4 h. Then, they were acidified (6M HCl) to pH 1.5-2 and liquid-liquid partitioning steps were followed. The dried alkaline extracts were re-dissolved in MeOH and filtered (0.45 μm PTFE) before HPLC analysis (Kim, Tsao, Yang, & Cui, 2006).

2.4.3 HPLC Analysis of Phenolic Compounds from WB, SDF and IDF

The extracts from section 2.4.2 were analyzed via a reverse phase (RP)-HPLC on the Alliance® HPLC system e2695, Separation Module with the 2998 PDA (Waters, Mildford, Massachusetts, USA) with Empower 3 software. All phenolic acid and flavonoids standards as well as the extracts were prepared in methanol. To separate phenolic acids and flavonoids in a single run with a Synergy-Max-RP column, at 35°C, solvent-A (0.01% formic acid:Milli-Q water) and solvent-B (100% acetonitrile) at a flow rate of 1.0 mL/min and a linear gradient program from 90% to 50% solvent A in 35 min was developed. Phenolic acids (280 nm) and flavonoids (320 nm) were identified and quantified using 11 phenolic acids and 10 flavonoid standards (Gunenc, et al., 2015).

2.5 Statistical Analysis

Statistical analysis was performed using Analysis of variance (ANOVA) with Statistical Analysis System (9.2, SAS Institute Inc., Cary, NC). All experiments were accompanied in triplicates. Duncan's Multiple Range test was used when significant ($P < 0.05$) mean comparison was achieved.

3. Results and Discussion

3.1 Prebiotic Activity

3.1.1 Microbial Count (log CFU/mL)

Total microbial counts on days 1, 7, 14, 21 and 28 of fermentations in all yogurt trials were shown in Figure 1. By day 1, microbial counts increased to a range of 8.19-8.31 log CFU/mL in control yogurts, and 8.41-8.65 log CFU/mL in yogurts with WB from an initial bacteria count of 6.5.

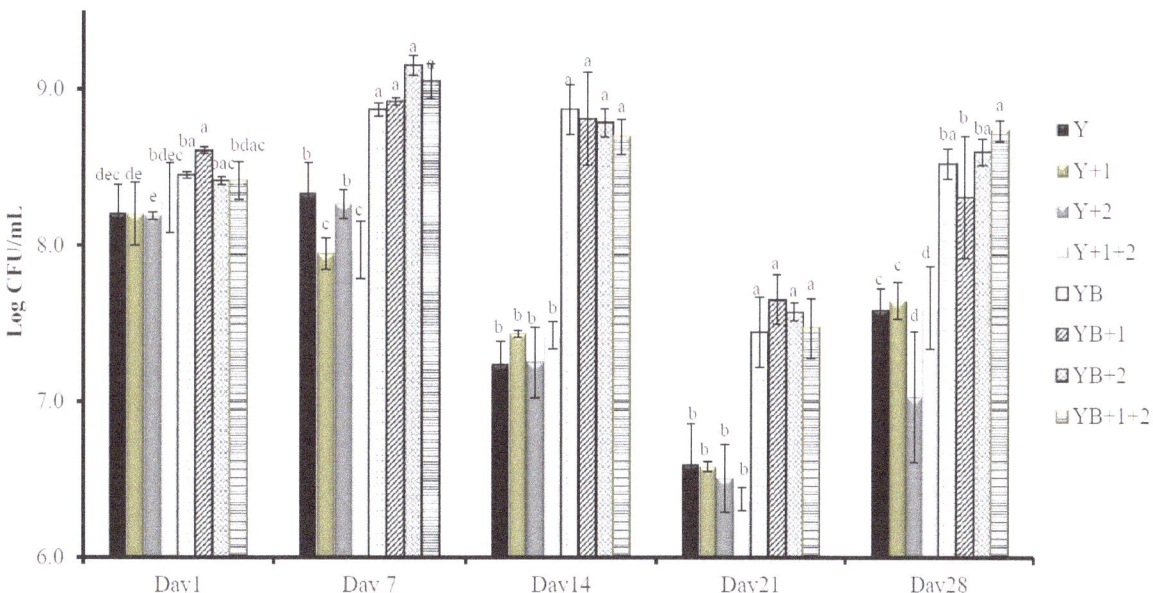

Figure 1. Total microbial count (log CFU /mL) in control yogurts (Y, Y+1, Y+2, Y+1+2) and yogurts with 4% wheat bran (YB, YB+1, YB+2, YB+1+2). Different letters in columns on the same day are significantly different ($P<0.05$)

By day 7, the bacterial growth increased significantly ($P < 0.05$) in yogurts with WB (8.87 and 9.15 log CFU/mL) compared to their corresponding controls (7.94 to 8.34 log CFU/mL). By days 14 and 21, the CFU values in yogurt samples with and without WB continued to be significantly different.

By day 28, yogurt samples lacking WB had significantly ($P < 0.05$) lower total bacteria counts of 7.03-7.65 log CFU/mL compared to 8.31-8.74 log CFU/mL of yogurt samples with WB. It has been suggested that the amount of viable bacteria remaining in yogurt after four weeks of cold storage should be in the range of 6 to 8 log CFU/mL (Vasiljevic & Shah, 2008). The controls remained within the range, but the yogurts containing WB had higher bacteria counts in the range of 8.31 and 8.74 log CFU/mL.

Overall, the yogurts with WB verified significantly higher bacteria counts ($P < 0.05$) during the four weeks of cold storage in comparison to control samples. It can be interpreted that in the presence of WB, there was an increase in microbial viability in sample trials consisting of probiotics. The randomized human study of François et al. (2012) showed that addition of 3g/day WB increased bifidobacteria (1.3 fold) and significantly increased faecal propionic acid levels (Francois et al., 2012). It may be a synergistic, additive or antagonistic effect of WB on probiotics compared to its corresponding controls. These results showed that WB might have a selective effect on increasing probiotics and starters during cold storage from day 1 to day 14, and to a lesser extent on day 21. A decreasing trend in CFU numbers may be a result of nutrient depletion over time or their metabolites (Sengun, Nielsen, Karapinar, & Jakobsen, 2009). For all microbes, certain nutrients such as iron and manganese are necessary to promote viability and growth. At the same time, probiotic bacteria have the ability to bind these elements and reduce accessibility by pathogenic bacteria (Bomba, Nemcová, Mudronová, & Guba, 2002). WB consists of manganese, and iron, with levels of 4-14 and 2.5-19 g/100g (Fardet, 2010).

Additionally, the Food and Drug Administration (FDA) has issued a letter of no objection for the WB extract Brana Vita, giving manufacturers a new prebiotic for a range of food and beverage products (Daniells, 2010). So, yogurts with WB and probiotics might supply micronutrients and oligosaccharides, selectively stimulating these particular bacterial strains (Costa, Queiroz-Monici, Machado-Reis, & Oliveira, 2006).

3.1.2 TTA %

During four weeks storage, TTA of all yogurt trials (Figure 2) with WB additions (YB, YB +1, YB +2, YB +1 +2) showed increasing TTA %, indicating that lactic acid production was increased as a result of the growing number of bacteria. The control yogurts had significantly lower TTA% than corresponding yogurts with WB. On day 28, the yogurt treatment with both probiotics (YB +1 +2) had the highest value from their corresponding control yogurt (Y +1 +2). It might be due to secondary metabolites production (lactate, propionate, or ethanol) and their positive influence on cell viability (Rattanachaikunsopon & Phumkhachorn, 2010). Moreover, volatile aroma compounds like carbonyl compounds might increase cell sustainability without impacting pH levels (Beshkova et al., 2003). The significant drop in pH on day 28 did not seem to intrude cell viability. Also parallel findings has been reported for improved probiotic viability in yogurt samples with low pH levels (4.1) in the presence of a prebiotic source (Agil & Hosseinian, 2012; Santo, et al., 2010).

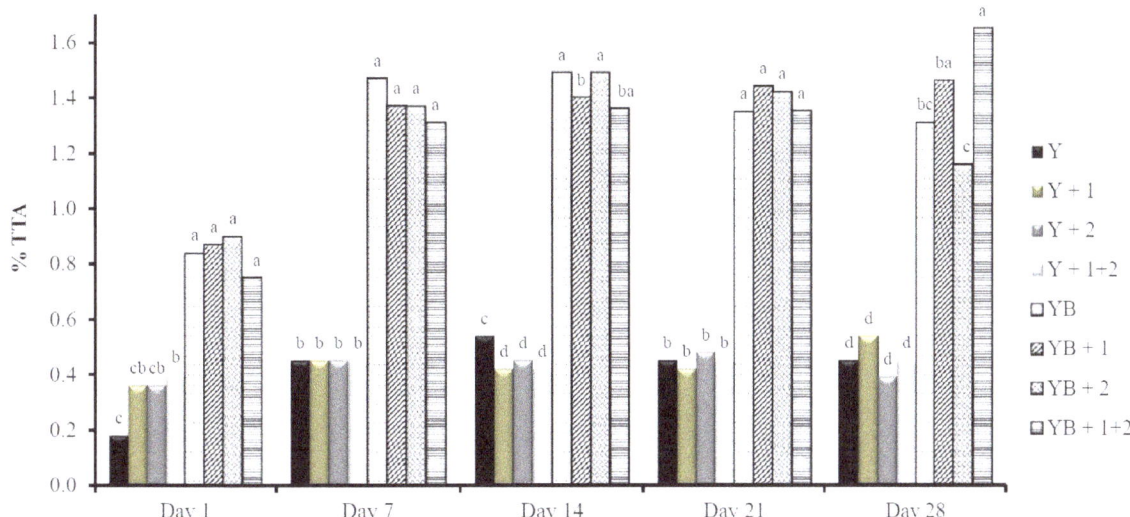

Figure 2. Percent total titratable acidity (%TTA) in control yogurt (Y, Y+1, Y+2, Y+1+2) and yogurt with 4% wheat bran (YB, YB+1, YB+2, YB+1+2). Different letters in columns on the same days are significantly different (*P*<0.05). TTA % values in and. Number 1 and 2 represent probiotic1 and 2. These results, parallel to our findings, suggest that the addition of WB to yogurts caused bacteria to produce more lactic acid, confirming findings obtained from the pH monitoring

3.2 Antioxidant activity of WEPs

3.2.1 ORAC Assay

The ORAC value (Table 2) of crude-WEP (71.88 ± 4.01 µmol TE/g) was higher than purified-WEP (52.48 ± 2.05 µmol TE/g). Moore et al. (2006) reported ORAC values of 45 to 78 µmol TE/g for 20 wheat bran samples (Moore, Liu, Zhou, & Yu, 2006). Also in another study, ORAC values of six wheat ranged from 19.5 to 37.4 µmole TE/g (Okarter, Liu, Sorrells, & Liu, 2010). Our ORAC findings are higher than above mentioned studies. This might be due to different wheat varieties, particle size, and some differences in the experimental procedures. Also, the antioxidant activity could be attributed to the presence of bound phenolic acids in the crude extract (e.g. ferulic and p-coumaric acids) (Hosseinian & Mazza, 2009). Moreover, presence of sugars with acyl groups and/or glycan-polymerization has been reported to have effects on the antioxidant activity of polysaccharides (Rao & Muralikrishna, 2006).

3.2.2 DPPH Assay

The %DPPH of crude- and purified-WEP were 9.65 ± 0.17 and 4.16 ± 0.18, respectively (Table 2). Similar to the study of Hromádková et al (2013), crude-WEP showed high antioxidant activity in both DPPH and ORAC tests (Hromádková et al., 2013). Phenolic components such as phenolic acids have been described to have a significant role in the overall radical scavenging capacity of xylans and xylooligosaccharides from the WB (Veenashri & Muralikrishna, 2011).

Table 2. ORAC, DPPH and TPC values of WEP

Antioxidant assays	Purified WEP	Crude WEP
ORAC (µmole TE/g)	52.48 ± 2.05[b]	71.88 ± 4.01[a]
%DPPH	4.16 ± 0.18[b]	9.65 ± 0.17[a]
TPC (mg FAE/g)	2.96 ± 0.10[b]	6.18 ± 0.84[a]

ORAC = Oxygen radical absorbance capacity as µmole Trolox Equivalent (TE)/g of sample.

DPPH (%) = 2, 2-Dipheny-1-picryhydrazyl radical (DPPH) radical scavenging activity assay

TPC = Total phenolic count was calculated as mg ferulic acid equivalent (FAE)/g of sample

WEP = Water extractable polysaccharides and values are means of triplicates ± standard deviation (SD)

Different letters in rows are significantly different (*P* < 0.05) in Duncan's multiple range tests

3.2.3 TPC

TPC of crude-WEP was over twice that of purified-WEP analyzed by FC method (6.18 ± 0.84 vs 2.96 ± 0.10 mg FAE/g) (Table 2) reflecting its higher antioxidant activity. The purification process probably removed considerable amount of phenolic compounds such as phenolic acids, flavonoids, phenolic acid diacyl glycerols, phenolic aldehydes and ferulates (Fardet, 2010). The most ample phenolic acid is FA followed by di-FA, sinapic acid, p-coumaric acid, caffeic acid and benzoic acid derivatives (Adom & Liu, 2002). About 95% of phenolic compounds in cereal grains are linked to cell wall polysaccharides by ester bonds and classified as dietary fiber-phenolic compounds (Hatfield, Ralph, & Grabber, 1999). Consequently, the addition of WB in yogurt might affect the gastrointestinal tract by acting like potential prebiotics, improving probiotic viability and functioning as antioxidants, especially after colonic fermentation (Fardet, 2010).

3.3 TDF and its Phenolics

3.3.1 TDF

TDF content (53%) was counted as a sum of SDF (6%) and IDF (47%). The content and proportion of both DF fractions are diverse among different types of cereals. This might be due to differences in seed morphology among most cereal grains (including wheat, maize and barley). Most of them have higher IDF than SDF, with WB dietary fiber around 44.5% (Fardet, 2010). Also, the DF content of WB (containing both outer layers and germ) and WB (only outer layers) were found 52.2% and 59.3%, respectively (Frolich & Asp, 1981). Our results are in close range with the above studies and literature (Stevenson, Phillips, O'Sullivan, & Walton, 2012).

3.3.2 Phenolic Compounds from WB, SDF and IDF

Table 3 shows the bound phenolic acid and flavonoid contents (mg/g of sample) of WB, IDF, and SDF. All three samples (WB, SDF, and IDF) were hydrolyzed prior to liquid-liquid extraction of bound phenolic acids. WB phenolic acid content (2.64 mg/g) was higher compared to that of total dietary fiber (IDF+SDF; 1.46 mg/g). This was an expected result since each step of fractional DF extraction resulted in some phenolic acid loss (Guo & Beta, 2013). IDF had higher total phenolic acid content compared to SDF fraction.

Table 3. HPLC profiles of bound-phenolic compounds in the fractions of both soluble/insoluble dietary fibers and WB

Phenolics	SDF*	IDF	WB
Gallic	0.10[a]	0.15[b]	0.24[c]
Proto-catechuic	nd	0.06[cb]	nd
p-OH-benzoic	0.03[c]	0.08[cb]	0.17[d]
Chlorogenic	nd	nd	0.89[b]
Vanillic	0.01[cd]	nd	0.04[e]
Syringic	0.01[cd]	0.03[d]	0.05[e]
p-coumaric	0.02[cd]	0.07[cb]	nd
Sinapic	nd	nd	0.11[de]
Ferulic	0.06[b]	0.84[a]	1.13[a]
Total phenolic acids (PA)	**0.23 ± 0.02**	**1.23 ± 0.01**	**2.64 ± 0.25**
Catechin	0.01[cd]	0.07[cb]	0.10[de]
Rutin	0.03[c]	0.07[cb]	0.13[de]
Quercetin-3-beta glucoside	0.03[c]	0.08[cb]	nd
Epicatechin gallate	nd	nd	0.12[de]
Total Flavonoid (FC)	**0.07 ± 0.01**	**0.22 ± 0.09**	**0.35 ± 0.12**
Total phenolics (PA+FC)	**0.30 ± 0.02**	**1.45 ± 0.05**	**2.99 ± 0.17**

*Values are means of triplicates, and different letters in columns are significantly different ($P < 0.05$) in Duncan's multiple range tests. SDF = Soluble dietary fiber fraction, IDF = Insoluble dietary fiber fractions, WB = wheat bran, nd = not detected.

Most total content of phenolic compounds (PA + FC) in both IDF and SDF fractions were found in insoluble fractions as in our previous study (Gunenc, et al., 2015). For example, 84.24% PA, 75.86% FC and 82.86% of total phenolic compounds were found in IDF fractions.

Nine phenolic acids including gallic, protocatechuic, p-OH-benzoic, chlorogenic, vanillic, syringic, p-coumaric, sinapic and ferulic acids were determined in HPLC analyses of WB, IDF, and SDF. Ferulic acid was the predominant phenolic acid, and mainly found as bound form in IDF (93.3%). Also, gallic acid was predominant in SDF. Our findings are in parallel with the study of Guo and Beta (2013).

With HPLC, Four flavonoid peaks were identified and respectively assigned as catechin, rutin, quercetin-3-beta glucoside, and epicatechin gallate (Table 3). IDF had more flavonoids (0.22 mg/g) than SDF (0.07 mg/g). Those biologically active components have not received much consideration as the phytochemicals in fruits and vegetables although the increased consumption of whole grain products has been linked to a diminished risk of chronic diseases (Liu, 2007). Our flavonoid content findings were in the same range reported by Singh, Sharma and Sarkar (2012) and close range with the study of Feng and McDonald (1989) in which the mean flavonoid content of four wheat classes were characterized and reported to be 0.29 mg/g, and our corresponding flavonoid content of total DF and WB were 0.29 and 0.35 mg/g respectively (Feng & Mc Donald, 1989; Singh, Sharma, & Sarkar, 2012).

4. Conclusions

This study showed that WB enhanced bacterial survival and growth in yogurt over 28 days cold storage period at 4°C. The overall increase in TTA% values in yogurts containing WB during the storage period suggest that WB could be consumed by probiotics. Furthermore, crude-WEP showed stronger antioxidants activity than purified-WEP as well as IDF has higher phenolic content compared to SDF. Consequently, DF can act as a carrier of phenolic compounds. It can be concluded that WB might be used as a potential source of prebiotics with higher antioxidant activity for functional foods and nutraceutical applications. Further investigations are needed for nutritive values (mineral solubility) and sensory evaluation of yogurt samples with WB addition.

Acknowledgments

This study was supported by the Ontario Ministry of Agriculture, Food and Rural Affairs (OMAFRA).

References

Aaby, K., Skrede, G., & Wrolstad, R. E. (2005). Phenolic composition and antioxidant activities in flesh and achenes of strawberries (fragaria ananassa). *Journal of Agricultural and Food Chemistry, 53,* 4032-4040. http://dx.doi.org/10.1021/jf048001o

Adom, K. K., & Liu, R. H. (2002). Antioxidant activity of grains. *Journal of Agricultural and Food Chemistry, 50*(21), 6182-6187. http://dx.doi.org/10.1021/jf0205099

Agil, R., & Hosseinian, F. (2012). Dual Functionality of Triticle as a Novel Dietary Source of Prebiotics with Antioxidant Activity in Fermented Dairy Products. *Plant Foods for Human Nutrition, 67,* 88-93. http://dx.doi.org/10.1007/s11130-012-0276-2

Behrad, S., Yusof, M. Y., Goh, K. L., & Baba, A. S. (2009). Manipulation of probiotics fermentation of yogurt by cinnamon and licorice: Effects on yogurt formation and inhibition of helicobacter pylori growth in vitro. *World Academy of Science, Engineering and Technology, 60,* 590-594. http://dx.doi.org/10.1.1.193.4755

Beshkova, D. M., Simova, E. D., Frengova, G. I., Simov, Z. I., & Dimitrov, Z. P. (2003). Production of volatile aroma compounds by kefir starter cultures. *International Dairy Journal, 13,* 529-535. http://dx.doi.org/10.1016/S0958-6946(03)00058-X

Bomba, A., Nemcová, R., Mudronová, D., & Guba, P. (2002). The possibilities of potentiating the efficacy of probiotics. *Trends in Food Science and Technology, 13,* 121-126. http://dx.doi.org/10.1016/S0924-2244(02)00129-2

Chakraborti, C. K. (2011). The status of synbiotics in colorectal cancer. *Life Sciences and Medicine Research, 20,* 1-15.

Costa, G. E. A., Queiroz-Monici, K. S., Machado-Reis, S. M. P., & Oliveira, A. C. (2006). Chemical composition, dietary fiber and resistant starch contents of raw and cooked pea, common bean, chickpea and lentil legumes. *Food Chemistry, 94,* 327-330. http://dx.doi.org/10.1016/j.foodchem.2004.11.020

Daniells, S. (2010). Prebiotic wheat bran extract gets FDA GRAS no objection. Nutraingredients-USA.com. Retrieved from http://www.nutraingredients-usa.com/Suppliers2/Prebiotic-wheat-bran-extract-gets-FDA-GRAS-no-objection."

Escarnot, E., Aguedo, M., Agneessens, R., Wathelet, B., & Paquot, M. (2011). Extraction and characterization of water-extractable and water-unextractable arabinoxylans from spelt bran: Study of the hydrolysis conditions for monosaccharides analysis. *Journal of Cereal Science, 53*(1), 45-52. http://dx.doi.org/10.1016/j.jcs.2010.09.002

Fardet, A. (2010). New hypotheses for the health-protective mechanisms of whole-grain cereals: what is beyond fiber? *Nutrition research reviews, 23*(1), 65-134. http://dx.doi.org/ 10.1017/S0954422410000041

Feng, Y., & Mc Donald, C. E. (1989). Comparison of flavonoids in bran of four classes of wheat. *Cereal Chemistry, 66*(6), 516-518. http://dx.doi.org/10.1007/s13197-011-0276-5

Francois, I. E. J. A., Lescroart, O., Veraverbeke, W. S., Marzorati, M., Possemiers, S., et al. (2012). Effects of a wheat bran extract containing arabinoxylan oligosaccharides on gastrointestinal health parameters in healthy adult human volunteers: a double-blind, randomised, placebo-controlled, cross-over trial. *British Journal of Nutrition, 108*(12), 2229-2242 2214 p. http://dx.doi.org/10.1017/S0007114512000372

Frolich, A., & Asp, N. G. (1981). Dietary fiber content in cereals in Norway. *Cereal Chemistry, 58*(6), 524-527. http://dx.doi.org/10.1371/journal.pone.0039361

Gao, L., Wang, S., Oomah, B. D., & Mazza, G. (2002). Wheat quality elucidation. In P. Ng, Wrigley, C.W. (Ed.), *Wheat Quality: Antioxidant Activity of Wheat Millstreams* (pp. 219-233). St.Paul, MN: American Association of Cereal Chemists International.

Gunenc, A., HadiNezhad, M., Farah, I., Hashem, A., & Hosseinian, F. (2015). Impact of supercritical CO_2 and traditional solvent extraction systems on the extractability of alkylresorcinols, phenolic profile and their antioxidant activity in wheat bran. *J. Funct. Foods, 12*, 109-119. http://dx.doi.org/10.1016/j.jff.2014.10.024

Gunenc, A., HadiNezhad, M., Tamburic-Illincic, L., Mayer, P. M., & Hosseinian, F. (2013). Effects of region and cultivar on alkylresorcinols content and composition in wheat bran and their antioxidant activity. *Journal of Cereal Science, 57*, 405-410. http://dx.doi.org/10.1016/j.jcs.2013.01.003

Gunenc, A., Tavakoli, H., Seetharaman, K., Mayer, P. M., Fairbanks, D., et al. (2013). Stability and antioxidant activity of alkylresorcinols in breads enriched with hard and soft wheat brans. *Food Research International, 51*(2), 571-578. http://dx.doi.org/10.1016/j.foodres.2013.01.033

Guo, W., & Beta, T. (2013). Phenolic acid composition and antioxidant potential of insoluble and soluble dietary fiberextracts derived from select whole-grain cereals. *Food Research International, 51*, 518-525. http://dx.doi.org/10.1016/j.foodres.2013.01.008

Hatfield, R. D., Ralph, J., & Grabber, J. H. (1999). Cell wall cross-linking by ferulates and diferulates in grasses. *Journal of the Science of Food and Agriculture, 79*, 403-407. http://dx.doi.org/ 10.1002/(SICI)1097-0010(19990301)79:3<403::AID-JSFA263>3.0.CO;2-0

Hosseinian, F. S., & Mazza, G. (2009). Triticale bran and straw: Potential new sources of phenolic acids, proanthocyanidins, and lignans. *Functional Foods, 1*, 57-64. http://dx.doi.org/10.1007/s11130-012-0276-2

Hromádková, Z., Paulsen, B. S., Polovka, M., Košťálová, Z., & Ebringerová, A. (2013). Structural features of two heteroxylan polysaccharide fractions from wheat bran with anti-complementary and antioxidant activities. *Carbohydrate Polymers, 93*(1), 22-30. http://dx.doi.org/ 10.1016/j.carbpol.2012.05.021

Jonnalagadda, S. S., Harnack, L., Liu, R. H., McKeown, N., Seal, C., et al. (2011). Putting the whole grain puzzle together: Health benefits associated with whole grains—Summary of American Society for Nutrition 2010 Satellite Symposium. *American Society for Nutrition, 141*(5), 10115-10225. http://dx.doi.org/10.3945/jn.110.132944

Kanmani, P., Satish Kumar, R., Yuvaraj, N., Paari, K. A., Pattukumar, V., et al. (2013). Probiotics and its functionally valuable products—A Review. *Critical Reviews in Food Science and Nutrition, 53*(6), 641-658. http://dx.doi.org/10.1080/10408398.2011.553752

Kim, K.-H., Tsao, R., Yang, R., & Cui, S. W. (2006). Phenolic acid profiles and antioxidant activities of wheat bran extracts and the effects of hydrolysis conditions. *Food Chemistry, 95*, 466-473. http://dx.doi.org/10.1016/j.foodchem.2005.01.032

Li, W., Hosseinian, F. S., Tsopmo, A., Friel, J. K., & Beta, T. (2009). Evaluation of antioxidant capacity and aroma quality of breast milk. *Nutrition, 25*(1), 105-114. http://dx.doi.org/10.1016/j.nut.2008.07.017

Liu, R. H. (2007). Whole grain phytochemicals and health. *Journal of Cereal Science, 46*, 207-219. http://dx.doi.org/10.1002/9780470277607.ch15

Mellen, P. B., Walsh, T. F., & Herrington, D. M. (2008). Whole grain intake and cardiovascular disease: A meta-analysis. *Nutrition, Metabolism and Cardiovascular Diseases, 18*, 283-290. doi: 10.1136/bmj.i2716

Moore, J., Liu, J.-G., Zhou, K., & Yu, L. L. (2006). Effects of genotype and environment on the antioxidant properties of hard winter wheat bran. *Journal of Agricultural and Food Chemistry, 54*(15), 5313-5322. http://dx.doi.org/10.1021/jf0603811

Murtaugh, M. A., Jacobs, D. R., Jacob, B., Steffen, L. M., & Marquart, L. (2003). Epidemiological support for the protection of whole grains against diabetes. *Proceedings of the Nutrition Society, 62*, 143-149. http://dx.doi.org/10.1079/PNS2002223

Okarter, N., Liu, C.-S., Sorrells, M. E., & Liu, R. H. (2010). Phytochemical content and antioxidant of six diverse varieties of whole wheat. *Food Chemistry, 119*, 249-257. http://dx.doi.org/10.1016/j.foodchem.2009.06.021

Ou, B., Huang, D., Hampsch-Woodill, M., Flanagan, J. A., & Deemer, E. K. (2002). Analysis of antioxidant activities of common vegetables employing oxygen radical absorbance capacity (ORAC) and ferric reducing antioxidant power (FRAP) assays:a comparative study. *Journal of Agricultural and Food Chemistry, 50*, 3122-3128. http://dx.doi.org/10.1021/jf0116606

Rao, R. S. P., & Muralikrishna, G. (2006). Water soluble feruloyl arabinoxylans from rice and ragi; changes upon malting and their consequence on antioxidant activity. *Phytochemistry, 67*, 91-99. http://dx.doi.org/10.1016/j.phytochem.2005.09.036

Rattanachaikunsopon, P., & Phumkhachorn, P. (2010). Lactic acid bacteria: Their antimicrobial compounds and their uses in food production. *Annals of Biological Research, 1*, 218-228. http://dx.doi.org/10.1007/s12393-012-9051-2

Santo, A. P., Silva, R. C., Soares, F., Anjos, D., Gioielli, L. A., et al. (2010). Acai pulp addition improves fatty acid profile and probiotic viability in yogurt. *International Dairy Journal, 20*, 415-422. http://dx.doi.org/10.1016/j.idairyj.2010.01.002

Schatzkin, A., Park, Y., Leitzmann, M. F., Hollenbeck, A. R., & Cross, A. J. (2008). Prospective study of dietary fiber, whole grain foods, and small intestinal cancer. *Gastroenterology, 135*, 1163-1167. http://dx.doi.org/10.1053/j.gastro.2008.07.015

Sengun, I. Y., Nielsen, D. S., Karapinar, M., & Jakobsen, M. (2009). Identification of lactic acid bacteria isolated from tarhana, a traditional Turkish fermented foods. *International Journal of Microbiology, 135*, 105-111. http://dx.doi.org/10.1016/j.ijfoodmicro.2009.07.033

Singh, B., Sharma, H. K., & Sarkar, B. C. (2012). Optimization of extraction of antioxidants from wheat bran (Triticum spp.) using response surface methodology. *Journal of Food Science and Technology, 49*(3), 294-308. http://dx.doi.org/10.1007/s13197-011-0276-5

Stevenson, L., Phillips, F., O'Sullivan, K., & Walton, J. (2012). Wheat bran: its composition and benefits to health, a European perspective. *International Journal of Food Sciences and Nutrition, 63*(8), 1001-1013. http://dx.doi.org/10.3109/09637486.2012.687366

Topping, D. (2007). Cereal complex carbohydrates and their contribution to human health. *Journal of Cereal Science, 46*(3), 220-229. http://dx.doi.org/10.1016/j.jcs.2007.06.004

Vasiljevic, T., & Shah, N. P. (2008). Probiotics-from metchnikoff to bioactives. *International Dairy Journal, 18*, 714-728. http://dx.doi.org/10.1016/j.idairyj.2008.03.004

Veenashri, B. R., & Muralikrishna, G. (2011). In vitro anti-oxidant activity of xylo-oligosaccharides derived from cereal and millet brans – A comparative study. *Food Chemistry, 126*(3), 1475-1481. http://dx.doi.org/10.1007/s13197-010-0226-7

Permissions

The contributors of this book come from diverse backgrounds, making this book a truly international effort. This book will bring forth new frontiers with its revolutionizing research information and detailed analysis of the nascent developments around the world.

We would like to thank all the contributing authors for lending their expertise to make the book truly unique. They have played a crucial role in the development of this book. Without their invaluable contributions this book wouldn't have been possible. They have made vital efforts to compile up to date information on the varied aspects of this subject to make this book a valuable addition to the collection of many professionals and students.

This book was conceptualized with the vision of imparting up-to-date information and advanced data in this field. To ensure the same, a matchless editorial board was set up. Every individual on the board went through rigorous rounds of assessment to prove their worth. After which they invested a large part of their time researching and compiling the most relevant data for our readers.

The editorial board has been involved in producing this book since its inception. They have spent rigorous hours researching and exploring the diverse topics which have resulted in the successful publishing of this book. They have passed on their knowledge of decades through this book. To expedite this challenging task, the publisher supported the team at every step. A small team of assistant editors was also appointed to further simplify the editing procedure and attain best results for the readers.

Apart from the editorial board, the designing team has also invested a significant amount of their time in understanding the subject and creating the most relevant covers. They scrutinized every image to scout for the most suitable representation of the subject and create an appropriate cover for the book.

The publishing team has been an ardent support to the editorial, designing and production team. Their endless efforts to recruit the best for this project, has resulted in the accomplishment of this book. They are a veteran in the field of academics and their pool of knowledge is as vast as their experience in printing. Their expertise and guidance has proved useful at every step. Their uncompromising quality standards have made this book an exceptional effort. Their encouragement from time to time has been an inspiration for everyone.

The publisher and the editorial board hope that this book will prove to be a valuable piece of knowledge for researchers, students, practitioners and scholars across the globe.

List of Contributors

Ernst-August Nuppenau
Institut für Agrarpolitik und Marktforschung, Justus-Liebig-University, Germany

Rabie Khattab
Food Science Department, Faculty of Agriculture (Saba Basha), Alexandria University, Alexandria, Egypt
Department of Process Engineering & Applied Science, Dalhousie University, B3H 4R2 Halifax, NS, Canada

Amyl Ghanem and Marianne Su-Ling Brooks
Department of Process Engineering & Applied Science, Dalhousie University, B3H 4R2 Halifax, NS, Canada

Galbraith, J. K.
Alberta Agriculture and Rural Development, 5712-48 Avenue T4V 0K1, Camrose, Alberta Canada

Aalhus, J. L., Juárez, M., Dugan, M. E. R. and Larsen, I. L.
Agriculture and Agri-Food Canada, 6000 C&E Trail, T4L 1W1, Lacombe, Alberta, Canada

Aldai, N.
Food Science and Technology, Faculty of Pharmacy, Universidad del País Vasco/Euskal Herriko Unibertsitatea, 01006 Vitoria-Gasteiz, Spain

Goonewardene, L. A. and Okine, E. K.
Department of Agricultural, Food and Nutritional Science, University of Alberta, Edmonton, Alberta, Canada, T6G 2P5, Canada

Dong Han
Auburn University, Auburn, AL, USA

Inyee Han and Paul Dawson
Department of Food, Nutrition andPackaging Sciences, Clemson University, Clemson, SC, USA

Jomana Khawandanah and Ihab Tewfik
Department of Life Sciences, Faculty of Science and Technology, University of Westminster, London, UK

J. M. Tirado-Gallegos, D. R. Sepulveda-Ahumada and P. B. Zamudio-Flores
Fisiología y Tecnología de alimentos de la Zona Templada, Centro de Investigación en Alimentación y Desarrollo, A.C.-Unidad Cuauhtémoc. Cuauhtémoc, Chihuahua, México

M. L. Rodríguez-Marin
Conacyt Research Fellow at: Universidad Autónoma del Estado de Hidalgo, Ciudad Universitaria, Centro de Investigaciones Químicas, Carretera Pachuca-Tulancingo Km 4.5, 42183 Mineral de la Reforma, Hidalgo, México

F. Hernández-Centeno
Departamento de Ciencia y Tecnología de Alimentos, Universidad Autónoma Agraria Antonio Narro, Buenavista, Saltillo, Coahuila, México

V. Espinosa-Solis
Universidad Autónoma de San Luis Potosí. Coordinación Académica Región Huasteca Sur de la UASLP, km 5, carretera Tamazunchale-San Martin, 79960, Tamazunchale, S.L.P. México

R. Salgado-Delgado
Departamento de Posgrado en Ingeniería Química y Bioquímica, Instituto Tecnológico de Zacatepec, Zacatepec, Morelos, México

Rafidah A. Ramli, Azila Azmi, Noorsa R. Johari and Syuhirdy M. Noor
Faculty of Hotel & Tourism Management, Universiti Teknologi MARA Pulau Pinang, Malaysia

Saheeda Mujaffar and Alex Lee Loy
Food Science and Technology Unit, Department of Chemical Engineering,The University of the West Indies, St. Augustine, Trinidad and Tobago, West Indies

Tomoko Osera
Hygiene and Preventive Medicine, Graduate School of Life Science, Kobe Women's University, Japan
Takakuradai Kindergarten attached to Kobe Women's University, Japan

Setsuko Tsutie
Clinical Nutrition Management, Graduate School of Life Science, Kobe Women's University, Japan

Misako Kobayashi
Takakuradai Kindergarten attached to Kobe Women's University, Japan

Nobutaka Kurihara
Hygiene and Preventive Medicine, Graduate School of Life Science, Kobe Women's University, Japan

Ojinnaka, M. C.
Department of Food Science & Technology, Michael Okpara University of Agriculture, Umudike, Abia State, Nigeria

Emeh, T. C. and Okorie, S. U.
Department of Food Science & Technology, Imo State University, Owerri, Imo State, Nigeria

Keri Lipscomb, James Rieck and Paul Dawson
Food, Nutrition and Packaging Sciences, Clemson University, United States

Poloko Stephen Kheoane, Clemence Tarirai and Carmen Leonard
Department of Pharmaceutical Sciences, Tshwane University of Technology, Pretoria, South Africa

Tendekayi Henry Gadaga
Department of Environmental Health Science, University of Swaziland, Mbabane, Swaziland

Richard Nyanzi
Department of Biotechnology, Tshwane University of Technology, Pretoria, South Africa

Angeline W Maina, John M Wagacha, Francis B Mwaura
School of Biological Sciences,Universityof Nairobi, Kenya

James W Muthomi
Department of Plant Science and Crop Protection, University of Nairobi, Kenya

Charles P Woloshuk
Department of Botany and Plant Pathology, Purdue University, United States

Amegovu K. Andrew
Department of Food Science & Technology, College of Applied and Industrial Sciences, University of Juba. P. O. BOX 83, Juba, South Sudan

Nadia Manzo, Fabiana Pizzolongo and Raffaele Romano
Dept. of Agricultural Sciences, University of Naples Federico II, Portici (NA), Italy

Loredana Biondi, Donatella Nava and Federico Capuano
Istituto Zooprofilattico Sperimentale Del Mezzogiorno, Portici (NA), Italy

Alberto Fiore
School of Science, Engineering & Technology Division of Food & Drink, Abertay University, Dundee, UK

George. E. Inglett, Diejun Chen and Sean. X. Liu
Functional Food Research Unit, National Center for Agricultural Utilization Research, USDA-ARS. 1815 N University Street, Peoria, IL 61604, USA

Keagile Bati
Department of Biological Sciences, University of Botswana, Botswana

Oarabile Mogobe and Wellington R. L. Masamba
Okavango Research Institute, University of Botswana, Botswana

Justina Serwaah Owusu
Department of Dietetics and Nutrition, Robert Stempel College of Public Health and Social Work, Florida International University, Miami, FL, USA
Department of Nutrition and Food Science, University of Ghana, Accra, Ghana

Esi Komeley Colecraft
Department of Nutrition and Food Science, University of Ghana, Accra, Ghana

Richmond Aryeetey
School of Public Health, University of Ghana, Accra, Ghana

Joan A. Vaccaro and Fatma G. Huffman
Department of Dietetics and Nutrition, Robert Stempel College of Public Health and Social Work, Florida International University, Miami, FL, USA

Alemneh Kabata
Hawassa University, School of Nursing and Midwifery, Ethiopia

Carol Henry, Susan Whiting and Nigatu Regassa
College of Pharmacy and Nutrition, University of Saskatchewan, Saskatoon, S7N 5A2. Canada

Debebe Moges and Afework Kebebu
Hawassa University, School of Nutrition, Food Science and Technology,Ethiopia

Robert Tyler
Department of Food and BioProduct, University of Saskatchewan, 9 Campus Drive, Saskatoon, S7N 5A5. Canada

Wannee Tangkham and Frederick LeMieux
Department of Agricultural Sciences, McNeese State University, Lake Charles, LA 70609, USA

Aynur Gunenc and Christina Alswiti
Food Science and Nutrition Division of Chemistry Department, Ottawa, Ontario, Canada

Farah Hosseinian
Food Science and Nutrition Division of Chemistry Department, Ottawa, Ontario, Canada
Institute of Biochemistry, Carleton University, Ottawa, Ontario, Canada

Index

www.ingramcontent.com/pod-product-compliance
Lightning Source LLC
Chambersburg PA
CBHW080412190526
45161CB00003B/214